Ecological Studies

Analysis and Synthesis

Edited by

W. D. Billings, Durham (USA) F. Golley, Athens (USA)
O. L. Lange, Würzburg (FRG) J. S. Olson, Oak Ridge (USA)
H. Remmert, Marburg (FRG)

Volume 50

Zeev Reiss · Lukas Hottinger

The Gulf of Aqaba

Ecological Micropaleontology

With 207 Figures

Springer-Verlag
Berlin Heidelberg New York Tokyo 1984

Professor Dr. Zeev Reiss
Department of Geology
The Hebrew University of Jerusalem
Jerusalem/Israel

Professor Dr. Lukas Hottinger
Geologisch-paläontologisches Institut
der Universität Basel
Bernoullistraße 32
CH-4056 Basel/Switzerland

ISBN 3-540-13486-7 Springer-Verlag Berlin Heidelberg New York Tokyo
ISBN 0-387-13486-7 Springer-Verlag New York Heidelberg Berlin Tokyo

Library of Congress Cataloging in Publication Data. Reiss, Zeev. The Gulf of Aqaba (Elat). (Ecological studies; v. 50). 1. Micropaleontology – Aqaba, Gulf of. 2. Paleoecology – Aqaba, Gulf of. I. Hottinger, Lukas. II. Title. III. Series. QE719.R45. 1984. 560'.916533. 84-13860.

Typesetting, printing and bookbinding: Brühlsche Universitätsdruckerei, Giessen
2131/3130-543210

Contents

Acknowledgments

The research on the Gulf of Aqaba carried out between 1970 and 1983 and on which the present summary is based has been supported by many scientific institutions and funding agencies in Israel and abroad: Israel National Academy of Sciences and Humanities; Israel–U.S.A. Binational Science Foundation; Israel Ministry of Energy and Infrastructure; Israel Oceanographic and Limnological Research Ltd.; The Hebrew University's Authority for Research and Development; The Hebrew University's Departments of Geology, Oceanography, Zoology and Botany; Lady Davis Foundation (grants to The Hebrew University); Levy Eshkol Fund for Oceanography; Gudalsky (Hebrew Univ.) Fund for Exchange of Scientists; USA National Science Foundation (OCE); U.S. Office of Naval Research; The Geological Society of America; Danish Ministry of Education (Tribute to Danes Fund); Rask-Ørsted Foundation, Denmark; Geological Institute of the University of Copenhagen; Swiss National Science Foundation; Geological-Paleontological Institute of the University of Basel; ZWO, Holland; Geological Institute of the University of Utrecht; Centre National de Recherches Scientifiques, France; Graduierten-Förderung, Federal Republic of Germany; Deutsche Forschungsgemeinschaft; GEO-Magazine.

The research programs would not have been possible without the facilities of the Hebrew University's H. Steinitz Marine Biological Laboratory in Elat and without the dedicated assistance of its staff.

The previously published illustrations are reproduced here with the kind permission of the various publishers. Among the illustrations included here, ninety-eight were originally drafted by Mr. P. Grossmann (Jerusalem). Mrs. E. Porath (Jerusalem) typed the final manuscript.

The preparation of the manuscript for printing was made possible through a generous grant by the Israel Ministry of Energy and Infrastructure, Directorate for Earth Science Research.

The following publishers and scientific institutions owning copyrights kindly permitted the reproduction of parts of or entire figures and tables as listed below. For bibliographic references see corresponding index.

Akademie-Verlag (Berlin-East): Tables D.4, D.5
American Association Advancement of Science (Washington. D.C.): Figs. C.3–C.6
American Meterological Society (Boston, Mass.): Fig. D.16
Biologische Anstalt Helgoland (Hamburg): Fig. F.38

Catena-Verlag (Cremlingen, FRG): Fig. D.2
Cushman Foundation, U.S. National Museum (Washington, D.C.): Figs. G.41, G.46, G.47
Edition Muséum Natl. Hist. Nat. (Paris): Figs. G.16, G.20, G.22, G.39
Elsevier Scient. Publ. Co. (Amsterdam): Figs. C.1, C.3–C.6, D.45, D.53, E.1–E.3; Tables E.2, E.4, H.9, H.10, H.16
Empresa Nacional Adaro (Madrid): Figs. E.4, H.7, H.13–H.15; Table H.2
Ferrara University (Ferrara, Italy): Fig. G.53
Gauthier-Villars (Montrouge, France): Figs. D.49, D.52
Geobios, Editions Université Claude Bernard (Lyon): Figs. F.11, F.25, F.28
Geological Society of Danmark (Kopenhagen): Figs. G.43, G.44
Geological Survery of Israel (Tel Aviv): C.2, C.7
IRL Press (Oxford): Figs. D.44, D.45, D.49, D.51
Macmillan Journals Ltd (London): Figs. G.48, G.49
Micropaleontology Press, Amer. Museum Nat. Hist. (New York): Table G.1
Paleontological Dept. Barcelona University (Barcelona): Fig. G.9
Palynological Research, Univ. of OFS (Bloemfontein, South Africa): Fig. H.11
Paul Parey Scient. Publ. (Berlin, New York): Figs. D.39, D.54, D.55, H.8, H.12
Pergamon Press Ltd (Oxford): Figs. D.20–D.26, D.35–D.38, D.40–D.48, G.51; Tables D.1, D.2, D.6, E.2
Quaternary Research, University of Washington (Seattle): Figs. H.2–H.5
Schweiz. Paläont. Abh. (Basel): Figs. G.38, G.40
Schweiz. Paläont. Ges. (Basel): Figs. D.5, G.7
Springer Verlag (Heidelberg, Berlin, New York): Figs. D.27–D.34, D.45, G.34; Tables D.3, D.7, D.8
Utrecht Micropaleontology Bulletin, Dept. Stratigr. Paleontol. Utrecht State University (Utrecht): Figs. F.15c, F.19c, F.26, F.31, G.8, G.10, G.11, G.19, G.21, G.23, G.30, G.31, G.42, G.45, G.52, G.54, H.17, H.18
Weizmann Science Press of Israel (Jerusalem): Figs. D.10–D.15, D.17–D.19, E.14, E.16, E.17

A. Introduction

The reconstruction of ancient marine environments is an important task of the Earth Sciences, of great interest to geology, pure and applied, to oceanography and climatology, as well as to such fields as ecology and evolution. In principle, such reconstructions are based on the actualistic approach of "the present is the key to the past." Since the deciphering of past environmental changes, either natural or man-made, has considerable bearing on planning, the past may provide a key to the future.

Paleoenvironmental interpretations in the marine realm are based to a great extent on assemblages of fossils recovered from ancient sediments in outcrops and from land-based or submarine drillings.

Observable lateral sequences of assemblages in contemporaneous strata, the known or assumed functional or adaptive significance of preserved skeletal structures, as well as their evolution through time and space, provide by themselves indications of ancient environmental settings. In some cases even present-day ecological considerations may be derived from analysis of fossils.

On the other hand, wherever possible, attempts are being made to interpret ancient environments by comparison of fossil associations with living populations and Recent sediment assemblages in terms of the – largely empirical – correlation of the latter with water mass or sea-bottom characters, regarded as major ecological factors. This applies mainly to Cenozoic rocks, since difficulties arise with regard to earlier deposits for which no analogues exist at present, and which contain fossils of organisms which are either extinct or whose phylogenetic relationships are doubtful.

During the last years, independent lines of evidence for paleoenvironmental interpretation based on fossils have been successfully used, especially stable oxygen and carbon isotope biogeochemistry of calcium carbonate shells in terms of water temperature, sea-level changes, evaporation/precipitation ratios and salinity, as well as effects of water-quality-dependent, selective dissolution of carbonate or silicate skeletal material in the water column or on the sea bottom. In this manner, physicochemical factors which can no longer be measured directly are inferred from paleontological and geochemical evidence.

The major part of the paleontological evidence in paleoenvironmental interpretations, especially of Cenozoic marine sediments, is provided by the so-called microfossils, amongst which the most important are the carbonate shell-bearing benthic and planktic Foraminiferida, the planktic Coccolithophorida, the siliceous shell-bearing planktic Radiolaria and planktic or benthic Diatomeae, as well as the Pteropoda with aragonitic shells.

The great abundance in remains of most of these groups in very small volumes of sediment or rock, their wide geographical distribution, rapid phylomorphogenetic evolution over long geological time intervals, high specific diversity in time and space, as well as the apparent close correlation of their distribution patterns with ecologically well-defined environments make these groups ideal paleoecological tools.

On the other hand, although the above-mentioned groups, especially the Foraminifera and Coccolithophorida are probably among those with the best-documented and most complete record, very little is known of their biology and hence of their ecological limitations, while nearly nothing is known of their functional test morphology. Furthermore, most of the ecological and paleoecological research on these groups carried out in the past pertain to marine environments with significant bathymetric and seasonal gradients of ecological factors, where rates of deposition are low and where calcium carbonate dissolution occurs at depth.

Numerous deep-sea cores were studied from marginal, partly silled basins with special conditions, like the Mediterranean or the Red Sea. However, since little relevant information (such as, e.g., population dynamics of plankton as compared with hydrological characteristics) was available from these areas, paleoceanographic interpretations of biogenic constituents in core sequences were based mainly on extrapolation of knowledge derived from the open ocean.

Most of the recent compositional and distributional patterns of the different groups obtained until the early 1970's in various places of the world were interpreted mainly as a function of temperature, salinity or – in the case of the benthic Foraminifera – of characters of the substrate, as well as, to a much lesser extent, of such factors as pH, nutrients, food requirements and light. All distributional studies of benthic Foraminifera indicated for many taxa a definite depth pattern, generally attributed to temperature, salinity and water energy gradients, as well as to changes in the nature of the substrate. However, in many cases, in marine environments with greatly differing physicochemical characteristics, bathymetric zonation boundaries occur at or around certain depths which cannot be directly correlated in a simple manner with the position of temperature, salinity or energy boundaries. Moreover, while planktic organisms conform generally well to climatic zones, the composition of certain assemblages could often not be explained simply and directly in terms of temperature or salinity gradients.

Laboratory experiments on the relevant organisms have provided important clues, but were not sufficiently accompanied by field studies of the same environment from which the laboratory cultures originated. Conversely, field observations were compared with laboratory results on specimens from quite different environments.

Although the giant size and complex structure of tropical, shallow-water benthic Foraminifera, whose remains are major carbonate producers, have intrigued micropaleontologists for many years, biologists almost completely neglected the subject until the end of the 1960's. Until the early 1970's symbiosis of Foraminifera with algae was considered with scepticism or even rejected. Progress in work on stable isotopes of oxygen and carbon in foraminiferal shells during the latter 1960's and early 1970's raised some serious questions about the nature of the materials which can be used for stable isotope-derived paleotemperatures. Since the shells of some Foraminifera turned out to be in isotopic inequilibrium with the surrounding sea water and since hermatypic corals were shown to be relatively depleted in ^{18}O and ^{13}C, a depletion attributed to symbiont (light)-dependent activity, the problem of symbiosis in Foraminifera became of great paleoenvironmental importance.

An adaptive significance of symbiosis in foraminiferal shell architecture and its evolution was suggested by some micropaleontologists, but until fairly recently very little work has been done on this subject.

A.1 Background and History of Research

From earlier, albeit scanty hydrographic measurements in the Gulf of Aqaba (Elat) (see Oren 1962), it was known that no significant changes of temperature with depth occur over the year in the warm and permanently highly saline Gulf. Work by Said (1949, 1950a, b), Reiss (1959), Reiss et al. (1961), and Por and Lerner-Seggev (1966), indicated that benthic foraminiferal faunas in the narrow and deep Gulf are abundant, highly diversified, and depth-distributed. This, despite the lack of significant changes in temperature and salinity and notwithstanding the very steep submarine slopes. A variety of substrates on which Foraminifera live were recorded in the Gulf. These facts, as well as the easy accessibility of the various environments in the Gulf of Aqaba and the existence in Elat of the well-equipped H. Steinitz Marine Biological Laboratory of the Hebrew University of Jerusalem, make the Gulf of particular interest to studies of microfaunas, especially, Foraminifera, in tropical regions. Stimulated by a reconnaissance survey with W. W. Hay (Univ. of Miami) during his visit to Israel in 1968, Z. Reiss initiated in 1969 a research program on the Foraminiferida in the Gulf of Aqaba. The project was soon joined by other colleagues from Israel and from abroad, especially the groups of L. Hottinger (Geol.-Pal. Instit., Univ. Basel), H.-J. Hansen (Geol. Instit., Univ. Copenhagen), C. W. Drooger (Geol. Instit., Univ. of Utrecht), J. J. Lee (City Coll., Univ. New York). Over the years the program developed into a more ramified Micropaleontology-Ecology-Paleoecology Program, dealing – in addition to benthic and planktic Foraminiferida – with Coccolithophorida and Pteropoda, as well as with some corals and molluscs.

Fairly soon it became apparent that despite earlier important work in the Gulf (see Emery 1963, Friedman 1968, Mergner and Schuhmacher 1974, Morcos 1970 and references therein) very little is known of pertinent environmental characteristics, such as physicochemical parameters, current regime, nutrient and plankton distribution, primary production, biomass, deeper bottom topography or subbottom structure, etc.

In 1974 an interdisciplinary and comprehensive **D**ata **C**ollecting **P**rogram in the Gulf of **E**lat-Aqaba (**DCPE**) was initiated and carried out until 1977 under the direction of Z. Reiss. It was based on the H. Steinitz MBL and on the participation of a number of scientists and graduate students from the Hebrew University (Jerusalem), the Weizmann Institute of Science (Rehovoth), and the Israel Oceanographic and Limnological Research Corp., Ltd. (Haifa). This program provided a large body of oceanographic information of great importance to ecological studies in the Gulf. Thus, both programs mentioned were closely interrelated.

Until the late 1970's only one deep-sea core was raised in the Gulf of Aqaba by the R.V. VEMA (Rosenberg-Herman 1965). It provided little information, however, since it was taken in the deepest part of the Gulf and penetrated turbidites.

In 1977, the R.V. ATLANTIS II, of Woods Hole Oceanographic Institution, with D.A. Ross as chief scientist, raised seven deep-sea cores in the Gulf of Aqaba and several more outside the Straits of Tiran. These cores were studied by Z. Reiss and his co-workers at the Dept. Geology of the Hebrew University in the light of the oceanographic knowledge obtained previously in the Gulf. The cores provided information of considerable interest not only to the deciphering of the Late Quaternary paleoceanographic history of the Gulf of Aqaba and northernmost Red Sea, but also to paleoceanographic interpretations of other tropical-subtropical seas (Project **CATE** – **C**ores **A**tlantis **T**wo **E**lat).

Finally, in situ field experiments and laboratory studies on benthic and planktic Foraminifera from the Gulf, their biology and shell isotope geochemistry were carried out at the Hebrew University's MBL, Elat and Dept. of Geology, Jerusalem, and have resulted in a better understanding of this important paleoenvironmental method.

A.2 List of Participants and Their Contributions

About 60 scientists have actively participated since 1969 in the above-mentioned research programs, i.e.[1]:

The Hebrew University of Jerusalem

O. Adar	Euphausiacea.
A. Almogi-Labin	Foraminifera and Pteropoda in plankton and sediments; pteropods in cores.
J. Erez	Statistical analysis of foraminiferal sediment assemblages. In situ and laboratory experiments on photosynthesis of symbionts, calcification and stable isotope fractionation in benthic and planktic Foraminifera and corals. Life cycle in Foraminifera, experiments on ecological tolerance of planktic Foraminifera.
H. Frenkel	Distribution of sediment foraminiferal assemblages.
Z. Garfunkel	Geology and bathymetry.
B. Golandsky	Tintinnida.

1 Institutional affiliation at the time of active participation

E. Halicz	Foraminiferal sediment- and core assemblages. Mangrove Foraminifera. Taxonomy of benthic Foraminifera.
H. Harpaz	Water chemistry.
D. Kahan	Photo- and thermotaxis in benthic Foraminifera.
J. Klinker	(F.R.G. Grad. Förderung) – Shipboard chief scientist. Circulation pattern, nutrients and biomass. Siphonofores.
C. Kropach	Shipboard scientist.
B. Lazar	Carbonate sedimentology.
I. Levanon-Spanier	Nutrients, primary production.
B. Luz	Stable isotopes in Foraminifera and Pteropoda. Paleoceanography.
N. Noy	Living epiphytic Foraminifera and their stable isotopes.
E. Padan	Primary production.
M. S. Pardo-Por	Copepoda.
L. Perelis-Grossovicz	Distribution of foraminiferal sediment assemblages, taxonomy of benthic Foraminifera.
M. Rahat	Photo- and thermotaxis in benthic Foraminifera.
Z. Reiss	Direction DCPE Program, coordination Micropaleontology-Ecology-Paleoecology Program, Direction Project CATE, Circulation pattern. Calcareous plankton. Paleoceanography. Taxonomy and ultrastructure of Foraminifera.
Y. Shapiro	Chaetognatha.
A. Shemesh	Carbonate system and stable isotopes in water.
M. Shilo	Symbionts in Foraminifera.
A. Winter	Coccolithophorida in plankton, sediments and cores.
M. Wolf	Stable isotopes in cores.
A. Zmiri	Photo- and thermotaxis in benthic Foraminifera.
T. Zohary	Population dynamics of benthic Foraminifera.
M. Zweig-Strykovski	Distribution of foraminiferal sediment assemblages. Taxonomy of benthic Foraminifera.

Israel Oceanographic and Limnological Corp., Lt., Haifa

Z. Ben-Avraham	Geophysics, bathymetry.
A. Hecht	Physical oceanography.
B. Kimor	Microphytoplankton. Tintinnida.

Geological Survey of Israel, Jerusalem

G. Almagor	Geology.
A. Ehrlich	Benthic diatoms.
J. Hall	Geophysics and bathymetry.

The Weizmann Institute of Science, Rehovoth

D. Anati	Physical oceanography.
G. Assaf	Physical oceanography.
N. Paldor	Physical oceanography.

Geologisch-paläontologisches Institut, Universität Basel, Switzerland

R. Buchmann	Nitrate and phosphate measurements in situ.
D. Dreher	Ultrastructure of protoplasm and symbionts in nummulitids.
L. Hottinger	Distribution of substrates. Taxonomy, functional morphology, and distribution of larger Foraminifera. Early diagenesis as ecological factor.
S. Leutenegger	Distribution, ultrastructure of symbionts, protoplasm and cell envelopes of larger Foraminifera.
G. Levinson	Taxonomy, structure and distribution of sessile perforated Foraminifera.
R. Reber	Direction of diving activities, underwater photography.
G. Scheidegger	Distribution of benthic Foraminifera, reproduction in selected species. Ahermatypic corals.
P. Steiner	Inventory of Bryozoans.
P. Wettstein	Functional morphology of corals.
H. Fricke	(Max Planck Institute, Seewiesen, F.R.G.) associated with the group for the exploration of the deeper parts of the shelf by means of the research submersible GEO.

Geological Institute, University of Copenhagen, Denmark

B. Buchhardt	Stable isotopes in benthic Foraminifera.
P. Dalberg	Stable isotopes in benthic Foraminifera.
H.-J. Hansen	Distribution of benthic Foraminifera in sediments, TEM of symbionts, calcification and stable isotopes in benthic and planktic Foraminifera. Wall structures and gas diffusion in Foraminifera. Taxonomy of foraminifera (SEM).
N. Mikkelsen	Coccolithophorida in plankton.
A. Rosenkrands-Larsen	Biometry and ecology of benthic Foraminifera.

Geological Institute, University of Utrecht, Holland

C. W. Drooger	Biometry of benthic Foraminifera.
W. Y. Y. Fermont	Biometry and stable isotopes in benthic Foraminifera.
J. E. Meulenkamp	Biometry of benthic Foraminifera.
E. Thomas	Biometry of benthic Foraminifera.
G. Y. van der Zwan	Stable isotopes in benthic Foraminifera.

Woods Hole Oceanographic Institution, Woods Hole, Mass., USA

D. A. Ross — Chief scientist, ATLANTIS II cruise.

City College, City Univ., New York, USA

J. J. Lee — Taxonomy and biology of symbionts in benthic Foraminifera.

University of Hawaii, Honolulu, USA

P. Hallock-Muller — Symbiosis and shell adaptation in benthic Foraminifera.

A.3 Purpose of the Book

A large number of publications summarizing the results obtained during the last 12 years have been published and are still in press. These contributions are specialized and they are scattered over many scientific journals in various countries. For these reasons, an integrated summary of the results obtained hitherto appears both useful and timely. The present book is an attempt in this respect.

It is in the spirit of this attempt, which incorporates only the work published by the above-mentioned group of scientists (and which by no means represents an exhaustive treatment of the various subjects), that mostly the group's own publications are cited. For a full bibliography (which would be disproportionately large for this book) the reader should consult the references cited in the individual publications mentioned here. Usually only publications of authors outside our group are cited both localized in the text and at the end of chapters or subchapters while publications of participants in the research projects mentioned above are cited in most cases at the end of relevant chapters only.

The substance of the book is subdivided into five major parts, i.e., the oceanographic background relevant to the ecological-micropaleontological aspects; composition and distribution of calcareous plankton and the environmental factors affecting them; description of the benthic environments and discussion of the mechanisms regulating their genesis and distribution; distribution of benthic Foraminifera and their ecological relationships, including symbiosis; and a paleoceanographic reconstruction of the Gulf of Aqaba during the past 150,000 years. A brief summary of the major parts is given in the following synopsis.

References: Emery (1963), Friedman (1968), Mergner and Schuhmacher (1974), Morcos (1970), Oren (1962), Por and Lerner-Seggev (1966), Reiss (1959), Reiss et al. (1961), Said (1949, 1950 a, b).

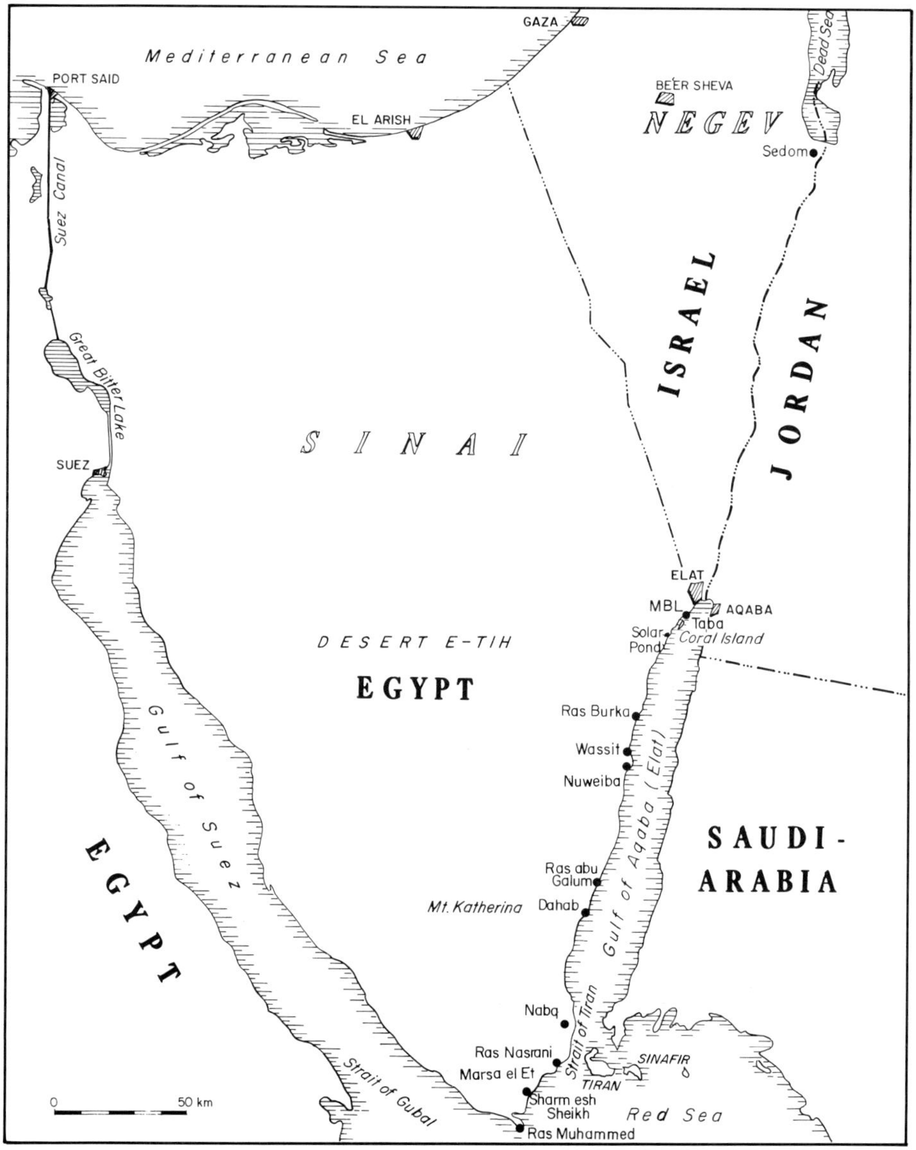

Fig. B.1. General location map of the Gulf of Aqaba

B. Synopsis

The **geological structure** and the **bathymetry** of the Gulf of Aqaba are determined by the Cenozoic break-up of the once continuous Arabian-African platform and left-lateral strike-slip movements along the Dead Sea Rift. The Gulf of Aqaba represents an early stage of formation of a plate boundary (between Arabia and Sinai) by a transform fault. Faulting produced a rugged topography along the margins of the Gulf, as well as narrow rhomb shaped grabens forming the deep basins. The history of the Gulf of Aqaba as part of the Dead Sea transform within the Gulf of Aden–Red Sea–Dead Sea–Jordan Rift system has considerable bearing on the bathymetry, distribution of physiographic elements, circulation pattern, as well as of land-derived material, hence on depth-related substrates and benthic organisms. Nature, distribution and high rates of accumulation of the mainly biogenic sediments in the deeper parts of the Gulf are of great importance to considerations of the significance of planktic sediment assemblages and their paleoceanographic interpretation.

The arid and hot **climate** and **high evaporation** rates, as well as the presence of the 252-m-deep sill of Tiran, separating the Gulf of Aqaba from the Red Sea, leads (in addition to a prevailing wind-driven, NE–SW surface circulation) to a thermohaline and, because of the Coriolis effect, counter-clockwise ("inverse estuarine") **circulation pattern.** This pattern fills the Gulf of Aqaba with upper Red Sea water and causes a hypersaline water column with high temperatures to depths of 1,830 m and with an insignificant thermocline, halocline, or pycnocline. A vertical vortex of deep waters is caused probably both by wind stress and by transient inflow of highly saline Red Sea waters at sill depth. Residence of upper waters is about 1 year, of the deep water 3–8 yr. The weak bathymetric and seasonal gradients in open waters of the Gulf of Aqaba of temperature (extreme: 20.5°–27.3 °C), salinity (40.3‰–41.6‰) and density (27.0–29.5 σt) are of special interest to ecological studies. (Higher salinities and lower temperatures occur in lagoons, sabkhas, mangroves and heliothermal ponds at the margins of the Gulf.) In the absence of rivers entering the Gulf, the circulation pattern also determines the distribution of the Red Sea-water-derived, low levels of **nutrients**, phosphate, nitrate-nitrite, and silicate. **Primary production** and chlorophyll a are consequently low and are attributable – in addition to diatoms, dinoflagellates, and cyanobacteria – to a large extent to coccolithophorids and picoplankton. Circulation pattern, salinity, and low consumption lead to a highly **oxygenated water column**, with a pH up to 8.4. Low biomass, negligible apport of abiogenic, suspended matter from land, high insolation, and weak cloudiness restricted to the short winter period, lead to clear and blue waters with very deep light penetration. The Gulf of Aqaba is a moderately **oligotrophic** sea with oceanic gyre center characteristics, in contrast to the southern Red Sea and the Gulf of Aden, which in their characters resemble productive oceanic gyre margins. These characteristics and differences are also reflected in the plankton.

The paleoenvironmentally significant (and preservable in sediments) **plankton** in the Gulf of Aqaba is calcareous, i.e., Foraminifera, Coccolithophorida and Pteropoda. Despite massive winter blooms of diatoms, the latter are not preserved in sediments, because of silica dissolution in the warm waters. Radiolaria are extremely rare.

In spite of the great depth of the Gulf, populations of **planktic Foraminiferida** are composed of epipelagic species. Nearly all of them are spinose, symbiont-bearing and characteristic of tropical-subtropical to warm-temperate seas. Deeper water, nonspinose, "pitted" or "smooth", in part keeled, species appear in the northern Red Sea proper and become abundant in its southern part and in the Gulf of Aden. Correlation with ecological gradients along this stretch, as well as comparison with faunas from other seas, indicate that this pattern is largely determined by the nutritional setting and probably by negative phototropy of deeper plankton. Because of deep light penetration (base of euphotic zone at 170 m), influencing symbiosis with dinophyceans, the epipelagic Foraminifera live in the Gulf at depths which are greater than in other seas.

Seasonal peaks in abundance of planktic Foraminifera seem to coincide mainly with abundance of the – in the Gulf nearly exclusively herbivorous and omnivorous (mainly epipelagic!) – copepods, as well as of diatoms and dinoflagellates. Seasonal dominance of three major species of Foraminifera is correlatable with seasonality in temperature and primary production in the waters. Differences in hydrographic details, between the (wider) southern basin and the northern (narrow) segment of the Gulf, as well as indications of upwelling at its northern tip (hydrography, nutrients, carbonate system) are also reflected by abundance patterns of Foraminifera.

Assemblages in **bottom sediments** agree in composition with the seasonally and bathymetrically integrated plankton data, with such modifications only as determined by (also laboratory-examined) life cycles of individual species (lunar or semi-lunar). This, as well as the abundance of aragonitic pteropod shells, indicates that no dissolution of the highly dissolution-prone epipelagic foraminiferal shells occurs at depth in the warm, highly oxygenated and $CaCO_3$-oversaturated waters of the Gulf. The plankton/benthos ratio of Foraminifera in sediments is depth-dependent, apparently a function of minimum depth requirements for reproduction and diel migration and of increasing abundance of planktic specimens, as well as of the general decrease of benthic species and specimens with depth.

Symbionts of planktic Foraminifera (dinophytes similar to those in hermatypic corals) seem to contribute considerably to primary production of the waters. In situ and laboratory experiments have shown that light, photosynthesis of symbionts, and shell calcification are interrelated. However, experiments with photosynthesis inhibitors (DCMU) indicate that light by itself enhances calcification. Planktonic Foraminifera have been shown to have in general very high rates of calcification, in comparison with benthic ones. Experiments have also demonstrated minimum temperatures for reproduction in individual species and upper salinity limits (about 50‰) for planktic Foraminifera.

Studies on planktic Foraminifera from the Gulf of Aqaba raise some problems still under investigation. Thus, the shells of living planktic Foraminifera caught in tows in the Gulf seem to be out of oxygen isotopic equilibrium by about 0.5‰. On the other hand, specimens of the same populations, grown in the Elat laboratory, deposit their shells in oxygen isotopic equilibrium with the sea water, in accordance with the paleotemperature curve originally derived from molluscs.

Coccolithophorid floras in the Gulf of Aqaba are highly diverse, with strong dominance of *Emiliania huxleyi*. In the Gulf of Aden *Gephyrocapsa oceanica* dominates. Comparison of relative abundances of these species with ecological parameters along the stretch Gulf of Aden – Red Sea – Gulf of Aqaba, as well as stable isotopes and paleontological information derived from deep-sea cores in the Gulf and the Red Sea (see below), indicate that the major factors affecting this pattern are nutrient content and probably the ability of *E. huxleyi* to use particulate matter as nutrient source in oligotrophic environments. High salinities affect *G. oceanica* less than (the more euryhaline in lower reaches) *E. huxleyi*.

Standing crop of coccolithophorids in the Gulf of Aqaba decreases with depth, although it is still fairly high at about 400 m (light penetration!). It is fairly low throughout the year, correlates negatively with insolation (photoinhibition?) and reaches a maximum in early winter. Summer standing crop is similar to that in transitional waters, winter standing crop to that in productive equatorial waters. Species diversity seems to be temperature-dependent. Comparison with primary production, chlorophyll a and microphytoplankton data indicates that coccolithophorids are major contributors to primary production in the Gulf of Aqaba. It is also noteworthy that coccolithophorids serve as important food organisms for copepods and for certain species of pteropods.

Pteropod faunas are low diversity ones, composed of tropical-subtropical, mostly epipelagic to shallow mesopelagic, exclusively herbivorous species. High concentrations of pteropods occur mainly in the upper 200 m. The average abundance corresponds to that in oceanic gyre centers, while maximum abundances are similar to those in intermediately productive regions. Seasonality of abundance is connected mainly with primary production, chlorophyll a and phytoplankton blooms. Seasonal dominance of individual species is attributed either to summer oligotrophy, or to the nature of winter phytoplankton, i.e., microphytoplankton or coccolithophorids, serving as food for the different species. Thus, fertility of the waters is a major factor affecting composition of pteropod faunas.

Relative abundance of pteropod species in bottom sediments reflects the seasonally and bathymetrically integrated specimen counts in the plankton. Discrepancies between plankton and sediment assemblages were found only in turbidites, where mechanical concentration of elongated species occurs.

Since the life cycle of pteropods is longer by an order of magnitude than that of Foraminifera, the latter reflect in sediments more clearly the conditions in different basins of the Gulf at different seasons; pteropod assemblages are more homogeneously distributed.

The studies in the Gulf of Aqaba suggest that **fertility** of the sea and **light penetration** are major ecological factors and are more important than hitherto realized in determining of composition of assemblages of calcareous plankton.

Benthic Foraminifera which are extremely important tools for paleoenvironmental interpretations, were intensively studied in the Gulf of Aqaba by various methods: in situ observations and experiments, qualitative and quantitative examination of assemblages from plants, soft and hard bottoms, as well as from box cores and grab samples, biometry and populations dynamics of selected groups, laboratory experiments concerning different biological aspects, including symbiosis and stable isotope fractionation in the shell.

Since in addition to the properties of the water the **characters of the substrate** are of paramount importance to the distribution of benthic Foraminifera, a considerable effort was made to investigate by deep SCUBA diving the varied mosaic of hard and soft bottom types characterizing the neritic realm of the Gulf of Aqaba. This, in order to better understand the complexities of the interplay between biotic and abiotic processes determining nature and distribution of the various substrates, such as, e.g., bottom topography, the sedimentological significance of perennial vegetation or coral growth, the local rates of sedimentation, especially as determined by productivity of shell-bearing organisms and their destruction, early diagenesis and burrowing activity in soft subsurface sediments and their influence on nutrient recycling. **Light penetration** plays a major role in determining the distribution of substrates, at least to depths of about 130 m. Vegetal cover on soft bottoms, mainly *Halophila*, reaches 60–70 m depth, symbiont-bearing stony corals, and coralline algae on hard bottoms penetrate to 130 m, while below the zones of main light impact encrusting bryozoans form bioherms on hard substrate. On exposed parts of the shelf edge and of the continental slope, bryozoan mounds are overgrown with symbiont-free corals, which construct up to 2-m-high bioherms.

Larger Foraminifera with complex shell structure react in a highly specific manner to the different benthic environments and, therefore, the composition of assemblages and the distribution pattern of particular species reflect simultaneously bottom types and the light gradient. As a response to the nutrient-poor, shallow-water environment in the Gulf of Aqaba, the major part of benthic foraminifera adopt a K-mode life strategy designed to keep up the carrying capacities over considerable periods of time. The requirements of this strategy are met by benthic Foraminifera with certain adaptations, repeated over and over again throughout their evolutionary history. Thus, the asexual phase in reproduction is emphasized, resulting in voluminous megalospheric schizozoites (with large parts of the parental cytoplasm), kept and protected in brood chambers until calcification of the earliest chambers is completed. Free-swimming small and unprotected gametic zygotes would need rapid synthesis of large quantities of biomass in each individual in order to reach minimum size for benthic settlement. In order to yield a sufficient number of schizozoites, parent individuals have to reach considerable size, achieved by a (for protists) unusually long vegetative life period with numerous growth stages covering often a whole year. Biosynthesis of protoplasm and storage products are supported by endosymbionts which provide a primary source of energy and partially recycle nutrients. It seems that reproduction in the Gulf of Aqaba is coordinated with the yearly cycle, in early spring, when the highest amounts of nutrients necessary for the high rates of biosynthesis in the earliest phases of growth of the offspring are available.

Chamber construction mode of the foraminiferal shell in K-strategists is adapted to growth rates which are kept as constant as possible throughout ontogeny. Alternating "orbitoid" arrangement of chamberlets in annular and "chess-board" ("gypsinid") arrangement in concentric growth are the most widespread adaptations, also beneficial in competition for space of sessile, encrusting or arborescent species.

Large-sized foraminiferal K-strategists with their long vegetative cycle must adapt more closely to substrate properties in their niche than small-sized, and short-lived forms. Some of these adaptations are reflected in shell morphology and especially in shell shape as related to the disposition of functional shell openings for the extrusion of pseudopodial ectoplasm. Hence shell morphology is related to motility and, through the latter, to substrate. The presence of canal systems in bottom-dwelling Foraminifera and the lack of such systems in epiphytic species supports this view.

Most larger Foraminifera possess unicellular plants as symbionts, whose ultrastructure in situ indicates specific, exclusive, and permanent host–symbiont relationships. All symbionts within the foraminiferal hosts are "naked," lacking any cell envelope. On the other hand, diatoms isolated from their host construct frustules in culture. Larger peneroplids are symbiotic with dinoflagellates possessing a pair of short flagella which enable them to move in the lacunal system of the host protoplasm and thereby to regulate light requirements for photosynthesis. Other porcellaneous Foraminifera host species-specific chlorophyceans, diatoms or rhodophytes, distinguishable by their ultrastructure. The perforate nummulitids and amphisteginids contain different types of diatoms, housed in vacuoles in the host protoplasm and located in the internal pore pits of the shell. Although plugged by organic material, the pores enhance direct gas exchange with the ambient environment. In elphidiids, isolated chloroplasts occur and produce starch grains in the form of isolated bodies which might be used as storage products by the host.

The specific **light requirements** for differently pigmented algal symbionts may be responsible for the specific depth distribution of foraminiferal host species. Diatom-bearing species occur to greater depths than dinoflagellate-bearing ones. The lower limits of host distribution seem to reflect more definitely the minimum light requirements of the symbionts than the upper ones, where additional factors play an important role, such as, e.g., autonomous movement of symbionts in the shell (in soritids) or retraction of symbiont-bearing protoplasm into the interior of the shell (in alveolinids) as a means of regulating irradiation. Moreover, phototaxis of the host may regulate light requirement of symbionts by crawling into or out of the shade of plants or boulders. Dependent upon light penetration, upper limits of depth distribution of individual species reflecting photoinhibition through excessive light intensity are different from place to place; thus, the upper limit of distribution of *Heterostegina depressa* in the Gulf of Aqaba is about 20 m, while in Hawaii or in the Maldive Islands it may reach the surface.

Foraminiferal shells exhibit systematic changes in **morphology related to depth.** The most widely observed trend is a general flattening of the shell in such unrelated groups as soritids, amphisteginids, or nummulitids. In a succession of closely related species, the flatter ones occur at greater depths. Other trends are the increase with depth in mean diameter of the megalospheric proloculus and changes in protoconch parameters in certain species, the decrease with depth in degree of involution, as well as diminishing with depth of thickness of individual lamellae in several unrelated forms. There is no simple causal explanation of the general depth-related flattening of the test, since different processes are involved in producing it. Flattening of the test in perforate lamellar Foraminifera results from diminishing thickness of individual lamellae on the one hand and from changes in chamber shape during ontogeny on the other. While changes in chamber shape during ontogeny reflect processes of growth, lamellar thickness may depend upon rates of calcification, which are correlatable (in situ experiments) directly with light and water energy levels and indirectly with rates of photosynthesis on one hand and with stable oxygen isotope ratios on the other.

Stable oxygen isotope ratios in shells of porcelaneous Foraminifera indicate carbonate deposition near equilibrium with the ambient sea water, whereas disequilibrium values ("vital effects") are observed in shells of perforate, lamellar species. This finding may point to a fundamental difference in the biological process of calcification in imperforate and perforate Foraminifera, as reflected by the different ultrastructure of the mineralized shell. These differences correlate well with quite distinct construction modes of chambers and shells and support, therefore, the basic hierarchy of presently accepted systematics of the Foraminiferida.

In certain perforate-lamellar, symbiont-bearing, benthic Foraminifera the specific $\delta^{18}O$ disequilibrium is additionally modified according to depth. This depletion is interpreted as related to photosynthetic activity of symbionts according to light quality and to incorporation of isotopically light-metabolic CO_2. Indeed, **carbon isotope** ratios in these shells are also in disequilibrium and indicate depth-related trends. On the other hand, $\delta^{13}C$ values in the shells of the same individual species at the same water depth vary – at least in epiphytes – with sampling site as well as with the location of the living specimens (leaves, stems, or near roots). Overall metabolic activity of the host–symbiont system influences the organism's internal pool of metabolic CO_2 and its share in the skeleton. However, departure from isotopic equilibrium in the shell may also reflect, at least in part, feeding habits and rates.

In situ radio tracer experiments with ^{45}Ca and ^{14}C indicate correlation between **calcification rates** of the foraminiferal host and photosynthetic rates of the symbionts. This supports the relationship between lamellar thickness and depth of habitat. On the other hand, experiments in the light on planktic foraminifera, using DCMU as photosynthetic inhibitor, have shown that calcification continues even after photosynthesis has been blocked and thus is related at least in part to light intensity by itself.

Some larger Foraminifera (planorbulinids) and the sessile, encrusting and arborescent perforate forms, as well as the majority of smaller Foraminifera, lack symbionts. Nevertheless in some of these symbiont-free groups the species distribution is also clearly depth-related. This is of particular interest considering the relatively homogenous water column, in terms of temperature, salinity, and oxygen, of the Gulf of Aqaba. Within the photic zone, the specific light requirements of food organisms may explain such a zonation.

Multivariate analysis of biogenic constituents in sediments sampled along a traverse to 40 m depth in a comparatively exposed area (off Ras Burka) resulted in a threefold depth zonation of foraminiferal shells identified at generic level only. This tends to show that thanatocoenoses in the sediment largely reflect the environment of life and the biocoenoses and that sediment transport across the slope is negligible in the Gulf of Aqaba.

Qualitative and quantitative analysis of **foraminiferal sediment assemblages** along traverses over the whole shelf and upper slope toward the Elat Deep shows a distinct depth zonation. Selected benthic species, particularly larger Foraminifera, replace each other in dominating the benthic thanatocoenoses in parallel, depth-related sequences, with such modifications as introduced by influence of soft and hard substrate, or of water energy levels. Generally, porcelaneous larger and smaller Foraminifera dominate the assemblages between 0 and 50 m on soft substrates, while nummulitids, together with amphisteginids, are the dominant element between 50 and 100 m depth on soft and hard bottoms. Below 100 m depth, the frequency of larger Foraminifera diminishes rapidly in favor of smaller benthic and planktonic forms. In the immediate neighborhood of hard bottom, sediment samples from 130–200 m depth have yielded in some places high percentages of fragmented, sessile arborescent foraminiferal species. Free-living forms on these deep hard bottoms are hardly known, since sampling of such faunas at these depths is impossible with the available gear.

Coinciding range limits of selected benthic species in the Gulf of Aqaba are at about 70, 130, and 220 m depth. They may serve as boundaries for **depth zonations.** The 70-m boundary corresponds roughly to the lower limit of permanent plant cover or of coral carpets serving as substrates for epiphytic or hard-bottom communities respectively. To some extent, the 70-m boundary may be over-emphasized for methodological reasons, since 70 m is the depth limit of SCUBA diving with compressed air and, therefore, of direct observation and sampling under visual control. The 130-m boundary corresponds roughly to the lower limit of symbiont-bearing corals and Foraminifera, at light levels of about 0.5% of surface light intensity. The depth of this latter limit depends, of course, upon the transparency of the water, which is exceptionally great in the Gulf of Aqaba. The 220-m boundary in the Gulf is located on the upper part of the continental slope and is defined by the upper range limit of a few, rare genera. Below 220 m and down to more than 800 m, the foraminiferal faunas are fairly constant in composition, in agreement with the weak gradients of temperature, salinity, or oxygen in the deep waters.

The **succession** of neritic foraminiferal **communities** in the Gulf of Aqaba is similar to those observed and sampled by the same methods on the Maldive Islands in the Indian Ocean, and this despite the open ocean water conditions on the latter. On the other hand, on the Maldives there is a greater diversity of important species, while some species replace others in the succession observed in the Gulf of Aqaba. Although the limits of distribution of selected index species are at slightly different depths as compared to the Gulf of Aqaba, the general depth succession of dominant forms on the Maldives is the same between 0–70 m, i.e. discoidal larger peneroplids are followed by alveolinids first and by nummulitids deeper down. At 60–70 m cyclical nummulitids appear. Diversity on the Maldives is somewhat greater and certain species occurring in the Gulf of Aqaba are replaced by others.

Wall structure and basic shell shape of the foraminiferal depth-dependent dominants in the Gulf of Aqaba are the same for each major depth zone not only in the Indian Ocean, but apparently also in all tropical-subtropical seas. This indicates that the combination of shell shape and nature of the wall has an ecological meaning which is the legitimation for the extension of comparisons of depth successions to fossil communities. As a prerequisite the latter must have a comparable diversity, repeatedly developed in Late Paleocene-Middle Eocene, Oligocene-Early Miocene and in Late Miocene times. In stratigraphic sequences reflecting tendencies of shallowing or deepending during short time intervals or in formations permitting a reconstruction of the (mainly neritic) paleogeography, the basic succession in larger foraminifera is repeated in the same order with respect to the depth gradient: discoidal-porcelaneous; fusiform porcelaneous; thick nummulitid; flat nummulitid. During early Tertiary times, an additional unit of extremely shallow, low-energy environments is characterized by larger agglutinated foraminifera, but has no extant analog.

The significance of the faunistic differences between the Gulf of Aqaba and the open Indian Ocean is not understood at present. The foraminiferal species in the Gulf are not endemic. The limit of what may be called an "Arabian" foraminiferal province seems to run through the open Indian Ocean in a N–S direction, meeting the continent somewhere between Karachi and Bombay. In this area very low oxygen content of the water column rises to neritic depths and might constitute a barrier, sealing off the neritic faunas of the Persian Gulf, the Red Sea and the East African continental margin from exchange with the open Indopacific province. The "Arabian" foraminiferal faunal province seems to be characterized not only by particular species, but also by lower diversity in general and by a higher variability of the individual species representing alone and exclusively their genera. The limits of this foraminiferal province, particularly in Indian waters, seem to coincide with distribution limits of a number of bivalves and gastropods. A problem deserving further study is the fact that *Planorbulinella* living at present in the Gulf of Aqaba has biometric characters – used in biostratigraphy – of Miocene affinities, while *Heterocyclina* living in the Gulf is nearest to Oligocene relatives.

The information obtained from the above-mentioned studies of the present-day Gulf of Aqaba and northernmost Red Sea has greatly facilitated the paleoceanographic interpretation of eight deep-sea cores raised in 1977 in the Gulf and in the northern Red Sea by the WHOI R. V. Atlantis II.

The shape of the **stable oxygen isotope curve** obtained from planktic Foraminifera and Pteropoda in the cores as compared to the globally established, generalized curve made it possible to recognize a chronostratigraphic framework of isotope stages 1 to 6. Thereby it has been possible to date – even beyond the radio carbon range – several points in the sequence, especially 11 ky (1,000 yr), 18 ky, 75 ky, and 127 ky BP. The presence–absence patterns and especially the numerous and rapid shifts in relative abundance of microfossil species occurring in the cores made it possible to recognize a **biostratigraphic subdivision** by Foraminifera (11 intervals), by Pteropoda (15 intervals) and by Coccolithoporida (13 intervals). The abundance peaks of the different taxa reflect a sequence of environmental events which can be related to the oxygen isotope curve and correlated throughout the basin. The microfossil assemblages are composed of planktic, nektoplanktic and benthic animal taxa, herbivorous or omnivorous and in part symbiont-bearing, as well as of planktic primary producers. Since the different groups are not influenced directly and equally by the same ecological factors their responses are, at least in part, diachronous. In this manner a **multiple stratigraphic zonation** was obtained.

Rates of sedimentation range between 4 und 10 cm per 1,000 yr for the pelagic, silty-sandy calcareous muds recovered by most of the cores raised from water depths of less than 900 m, while much higher rates were determined in turbiditic sequences. The relatively high rate of sedimentation and the multiple stratigraphic framework provide a high resolution in time, of 3,000–7,000 yr.

The difference in $\delta^{18}O$ in foraminiferal and pteropod shells between the last Glacial maximum, about 18 ky BP, and the present is about 1.5‰ in the open ocean, and in excess of 5‰ in the Red Sea and the Gulf of Aqaba. This difference expresses the Glacial-Postglacial change in sea water composition and thus does not provide a direct clue to changes in temperature and/or salinity. However, the presence throughout the core sequences from the Gulf of Aqaba and northernmost Red Sea of shallow-water, benthic tropical-subtropical, symbiont-bearing Foraminifera, whose present-day minimum winter isotherms are about 17 °C, points to the fact that throughout Glacial times **minimum temperature** of the upper waters was never lower than 17 °C, hence 4 °C below the present.

The glacial-postglacial difference in $\delta^{18}O$ values (in well-preserved specimens!) accounts, therefore, only partly for a drop in temperature in the Gulf of Aqaba and reflects mainly a maximum **Glacial salinity** higher by more than 10‰ than the present one of about 41‰.

The total absence of planktic Foraminifera during Glacial maxima in the cores cannot be explained by dissolution of their low-Mg shells, since in the same strata aragonitic pteropods occur in abundance. This absence is probably due to extremely high salinities, a conclusion supported by laboratory experiments, indicating approximately 50‰ as the upper salinity tolerance limit for planktic Foraminifera.

Most of the **planktic Foraminifera, Pteropoda** and **Coccolithophorida** living at present in the Gulf of Aqaba occur throughout the cores. An exception is *Globigerinoides sacculifer,* absent in Glacial intervals, probably because of water temperatures lower than 17 °C at depth of reproduction. Three species of Foraminifera, seven of Pteropoda, and four species of Coccolithophorida occurring in Glacial, Interglacial, and early Postglacial intervals have not been found living in the Gulf. Most of these species are known at present from areas of higher biological productivity than the Gulf of Aqaba. Apparently one pteropod species has been prevented from living in the Red Sea since the last Glacial period because of the establishment of the oxygen minimum in the NW Indian Ocean.

Benthic foraminiferal diversity is considerably lower during Glacial periods than in Interglacial ones.

Sequences in time of abundance peaks of individual species or higher taxonomic groups provide important clues to changes in the environment. Comparison of relative abundances of planktic taxa in present-day oceans and especially along the stretch Gulf of Aden–Red Sea–Gulf of Aqaba with the sequences through time in the cores, indicates that during Glacial times the upper waters in the Gulf were not only more saline, but also much more fertile. This conclusion is corroborated by abundance in Glacial intervals of the Gulf of Aqaba cores of benthic Foraminifera indicative of low-oxygen conditions near the bottom, as well as by the $\delta^{13}C$ of their shells. **Fertility of the photic zone** in the Gulf of Aqaba increased during Glacial times mainly because of higher fertility in the Gulf of Aden, as well as because of reduced water exchange over the sill at lowered sea level, resulting in longer residence times of the water and in higher input/output ratios of nutrients. In the northernmost Red Sea core an extraordinary abundance of planktic diatoms in strata of the last glacial maximum is indicative of high fertility and upwelling. **Stratification** of the water column in the Gulf seems to have been stronger during glacial times, as indicated by the difference in $\delta^{18}O$ between planktic and benthic Foraminifera, and as deduced from the calculated difference in salinity within the upper waters by parametrisation of sea straits. The present-day "blue desert" seems to have developed about 4,000 yr ago.

Although **unconformities** have been recognized in the cores, they are neither synchronous, nor restricted to Glacial maxima. It seems, therefore, that during the last 150,000 yr the Gulf of Aqaba was continuously connected with the Red Sea and the latter with the Gulf of Aden, through the 134-m-deep sill at Bab-el-Mandeb. Tectonic movements may also have contributed to the continuing connection.

While knowledge of the present-day Gulf of Aqaba has facilitated the reconstruction of its environmental history throughout the Late Quaternary, microfossil assemblages from deep-sea cores in the Gulf have furnished clues for a better understanding of the ecology of various living planktic and smaller benthic species. Thus, the evidence from the cores concerning the very highly saline Gulf of Aqaba during Glacial maxima – conditions for which no present-day analog exists – makes it possible to derive upper salinity tolerance limits for a number of species of Foraminifera, Pteropoda and Coccolithophorida living at present in the world oceans.

In general, geological history, climate and strait dynamics create in the Gulf of Aqaba an environmental setting which represents a unique **"natural laboratory"** for the study of the influence on micropaleontologically important organisms of such factors as oligotrophy, light, symbiosis, water energy or character of substrate. In many respects (including calcareous plankton) the Gulf of Aqaba bears similarities with oceanic gyre centers and may thus serve as a model for this kind of ancient environment.

The biological insight and the ecological understanding derived from the investigation of the Gulf of Aqaba lead to new interpretations and views concerning the meaning of certain microfossils.

C. The Gulf of Aqaba – a Rift-Shaped Depression

The Gulf of Aqaba is the northern extension of the Red Sea und is part of the about 6,000-km-long **Syrian–African rift system** which extends from Mozambique to Turkey. The Gulf occupies the southern segment of the approximately 1,000-km-long, so-called Dead Sea rift. This rift is a plate boundary of transform type, connecting the Red Sea, where sea-floor spreading occurs, with the Zagros–Taurus zone of continental collision. The Gulf of Aqaba is one of the two rift systems (the Gulf of California is the other) where a mid-ocean ridge system passes into a transform system and runs into a continent (Fig. C.1).

The Dead Sea transform (or rift) was formed – like the Red Sea and the Gulf of Suez – in Cenozoic times by the breakup of the once continuous Arabo-African continent. As indicated by matching Cambrian to Pleistocene rock formations on opposite sides of the rift, a left-lateral strike-slip movement displaced the Arabian plate northward by 105 km in relation to the Sinai plate. The major rift faults are arranged in an en-echelon pattern, producing rhomb-shaped, pull-apart morphotectonic depressions with local crustal separation ("leaky transform"), and a structural relief exceeding in places 5 km. The 177-km-long, 14–25-km-wide and 1,830-m-deep Gulf of Aqaba is the most spectacular among these depressions. From radiometric dates on basaltic rocks on land an age of 20–30 my BP is derived for the initial continental breakup.

The motion along the Dead Sea strike slip accommodates most of the Red Sea spreading (the opening along the Gulf of Suez being small) and the age of the earliest magnetic anomalies in the Red Sea is about 20 my BP. From this it is inferred that about 65 km of the Dead Sea slip occurred before the end of the Miocene, the last 40 km having been completed – following a standstill – since Early Pliocene times, at a rate of about 1 cm yr^{-1}.

Pre-rift formations bordering the Gulf of Aqaba are Precambrian basement rocks (plutonic and metamorphic) with few relicts of unmetamorphozed supracrustal late Precambrian to earliest Cambrian, as well as sedimentary formations of a stable platform (Fig. C.2). The latter, 1–1.3 km thick, are composed of mainly clastic Early Paleozoic and predominantly carbonatic Cretaceous-Eocene rocks. Marine beds and evaporites of Miocene age, deposited during the first slip phase, are widespread mainly along the eastern side of the Gulf, while small patches of sediments occur along the northern part of the Gulf, suggesting an early "protogulf".

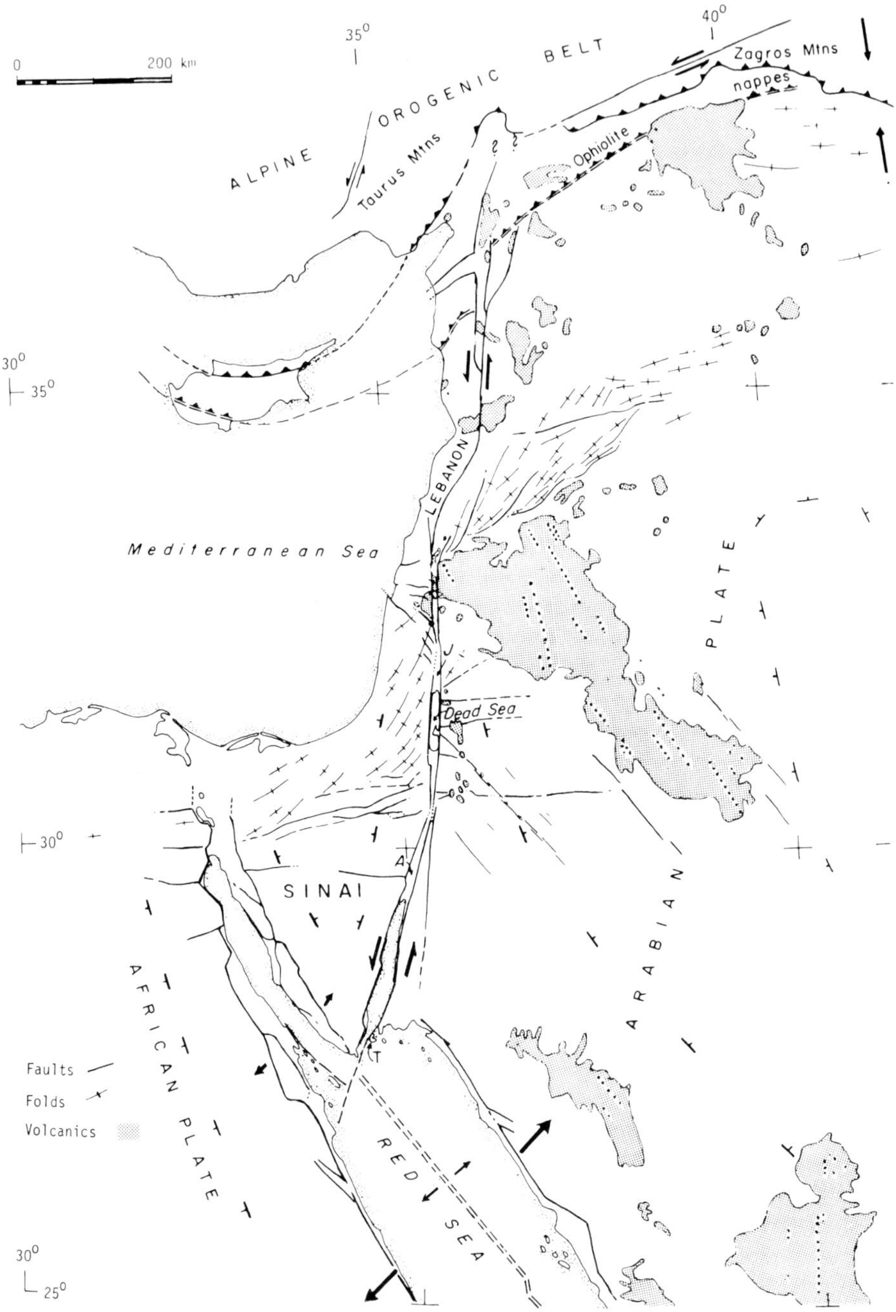

Fig. C.1. General geological setting of the Dead Sea transform. *J* Jordan Valley, *A* Arava Valley, *T* Island of Tiran, *G* Gulf of Suez, *GA* Gulf of Aqaba (Garfunkel 1981)

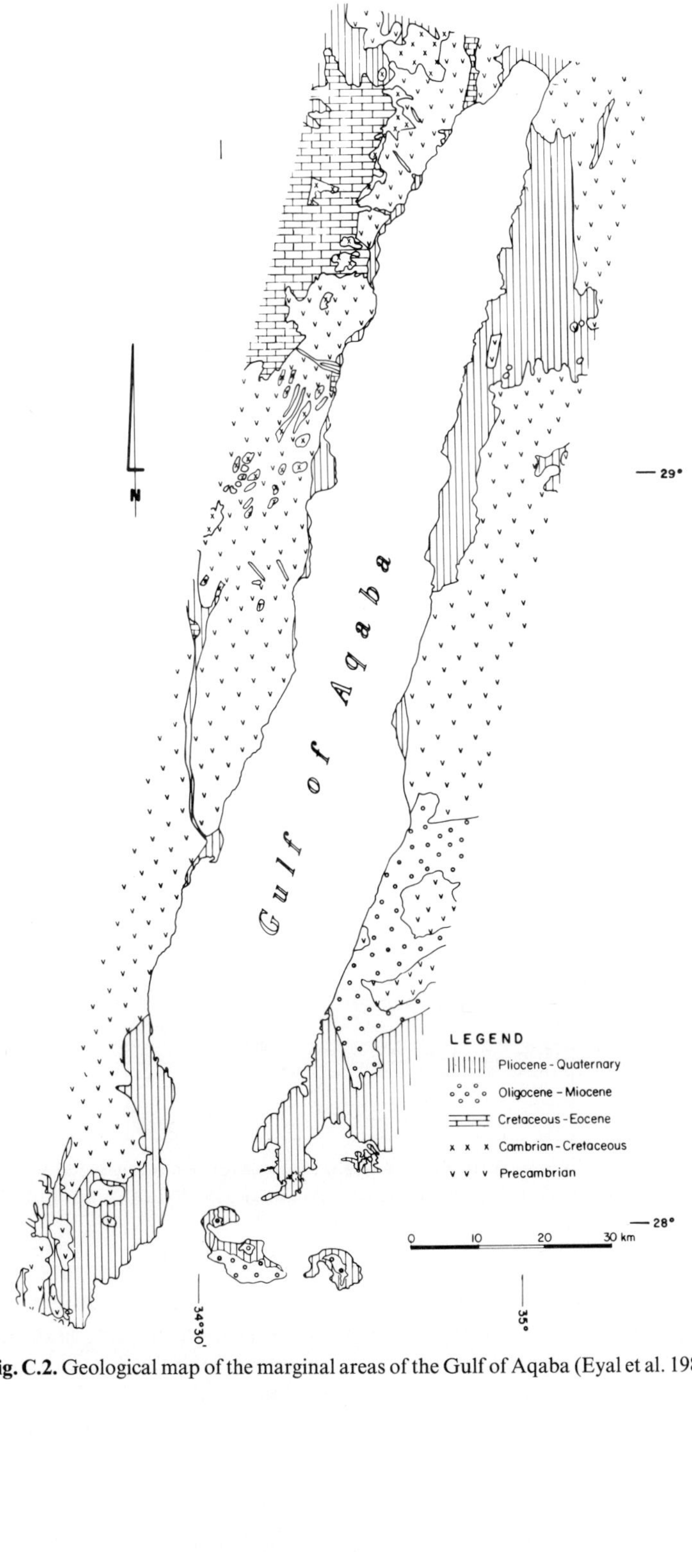

Fig. C.2. Geological map of the marginal areas of the Gulf of Aqaba (Eyal et al. 1980)

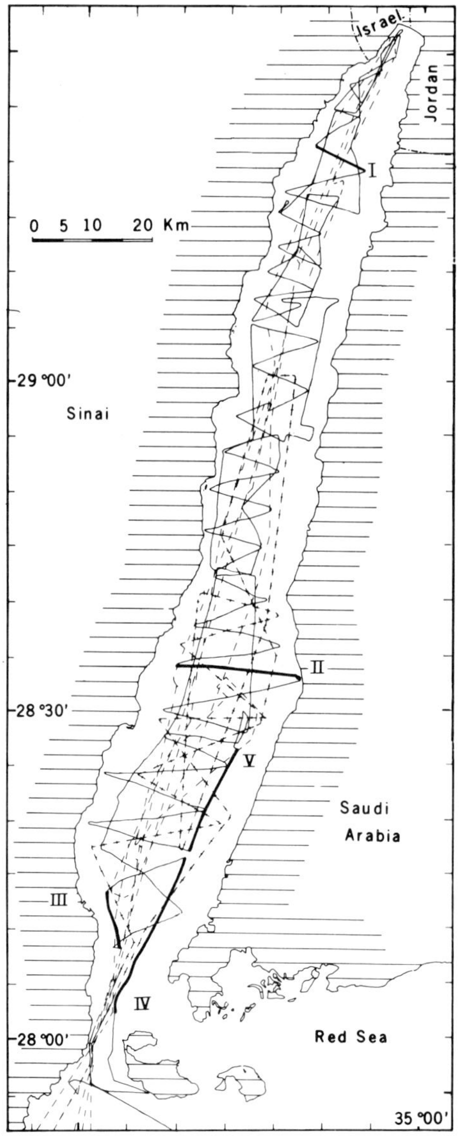

Fig. C.3. Geophysical track map of the Gulf of Aqaba. Surveys carried out by the R. V. Vema (1958), Aragonese (1961), Ramona (1976), *(solid lines)* and Atlantis II (1977) (Ben-Avraham et al. 1979 b)

Formation of marginal faults was followed in the Late Miocene by erosion and truncation of a structural relief probably exceeding 1 km in the Sinai and Elat region. Sediments of the younger slip phase (mainly alluvium, lake and playa beds) along the Gulf are thin, occur on both sides of the Gulf, and overlap marginal normal faults at some distance from the shore, thus indicating inactivity of these faults. In places there is evidence of some continuing activity and displacement of land surface. Small rhomb-shaped grabens between marginal faults, sub-

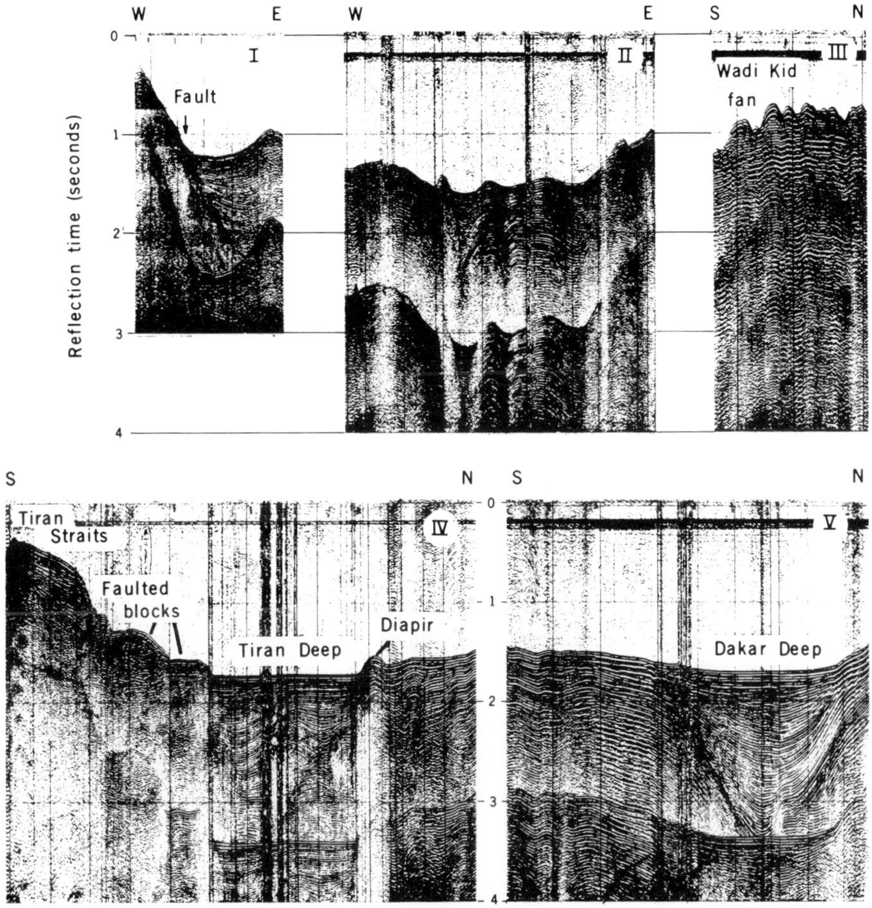

Fig. C.4. Five continuous air-gun seismic profiles (4,900 cm^3) in the Gulf of Aqaba (see Fig. C.3 for locations). Profile *I* traverses the Elat Deep and shows the well-stratified ponded sediments of the deep separated by a fault from the poorly and irregularly stratified sequence on the marginal block. Profile *II* runs through the sill between the central and southern basins. The basinal sediments are folded; the fold forming the prominent hill may be of diapiric origin. Note the contrast between the basinal and marginal sediments. Profile *III* is through the fan off Wadi Kid. Note the strong irregular reflectors and the incised submarine canyons. Profiles *IV* and *V* cross the southern basin from the Tiran Straits of the Dakar Deep. A diapiric structure is shown north of the Tiran Deep, whereas the deep itself is separated from the straits by several structural steps. There is a gradual transition from undeformed to deformed sediments along the southern margin of the Dakar Deep (Ben-Avraham et al. 1979 b)

parallel to the trend of the Gulf, contain slices of Paleozoic-Eocene sedimentary cover, which was down-dropped, and accommodate a left slip of a few kilometers.

The lands bordering the Gulf were uplifted, at the latitude of Elat the uplift reaching about 4 km. The uplifted areas were extensively eroded, exposing the Precambrian basement. At present mountains on the Sinai side of the Gulf rise abruptly, delimited by coastal faults, to heights of more than 1.5 km within 10 km from the shore and to more than 2.5 km further inland.

All primary land forms produced by tectonic movements are modified and degraded by mass wasting, erosion and sedimentation. Large alluvial fans were built in front of the fault scarps, especially on the Sinai side, continuing into submarine cones in the Gulf. Sinuosity of the coast line is due in part to fault scarps, formation of embayments or "marsas" ("sharms") due to drawning of 30–40 m deep canyons (Sharm es-Sheikh, Marsa el At), and to young, tectonically triggered fault-scarp slumping, producing baylets.

On the Sinai side of the Gulf three major **elevated terraces** of fossil coral reefs are present between +30 m and +15 m m.s.l. (mean sea level) and were dated between 306 and 81 ky BP. The earliest date of a fourth terrace at +2 m to −1 m m.s.l. is 6.5 ky BP and its seaward side includes the living reef (the northernmost in the world).

Detailed geophysical surveys have clarified structure and sedimentary fill of the Gulf (Figs. C.3 and C.4). On the floor of the Gulf three major rhomb-shaped, deep grabens, arranged in en-echelon and trending N 20°–25° E, are filled by more than 3 km of sediments. They are flanked by marginal blocks, delimited by normal faults which control also much of the coasts. The sills separating the major deep basins from each other, the undulations of the sea floor within these basins, as well as the 800 m-wide and 80–252 m-deep straits of Tiran separating the Gulf of Aqaba from the Red Sea proper, are compressional features like pressure ridges, half domes and complex folds produced by the collision of the two sides of the strike-slip faults. The northward and landward continuation of the Gulf is the Arrava Valley which extends to the Dead Sea (Figs. C.5 and C.6).

Two major parts of the Gulf of Aqaba can be discerned, a northern, narrow segment (14–17 km width) and a southern, broader one (23–25 km wide) with the greatest depth in the Gulf, 1,830 m. The Gulf shallows gradually toward the north, while in the south the sill of Tiran rises fairly abruptly (Figs. C.6 and C.7).

The margins of the Gulf are very steep. Its eastern side descends rapidly toward the deep basins (25°–30° slope), while on the western side the descent is broken by sloping terraces at varying depth (average slope 16°, in shallow parts as little as 7°). These differences and asymmetry are due to position and distance from shore of the faults and fault blocks, and to build-up of large alluvial submarine cones (in continuation of alluvial fans) to depths of more than 700 m on the western side.

Submarine terraces at 8, 10, 20, 40, 75–90, 120–140, and 220–250 m depth below sea level determine in part the slope relief particularly in areas of low sedimentation rates. In the northern Gulf of Aqaba (Elat Nature Reserve) the recent sediment cover of the −8 m terrace is carried away by currents revealing a slightly sloping hardground consisting of coral rubble and sand cemented and filled by carbonate. Similar hardgrounds at −8 and −10 m are found south of Dahab and in the Ras Muhammed area. Living coral patch reefs (e.g., the "Moses Rock" in the Elat Nature Reserve) grow on this solid platform, while others cover the −40 m terrace to a height of 10–15 m. In places (e.g., El Kura Bay, Dahab) the −40 m terrace is covered by a carpet of living coral colonies, reaching a thickness of 10 m. Thus, the bathymetry often does not reflect the original terrace surfaces and is difficult to interpret. The −40 and −75–90 m terraces occur on both sides of the Gulf of Aqaba, indicating absence of differential tectonic movements after the formation of these terraces. The 120–140 m terrace may correspond to the last maximum glacial sea level (18,000 yr BP). Near Elat, this terrace is built of a vertical, 3–5 m-thick cliff with caves and overhangs. Below the cliff a sandy slope is present, below which there occur reef knolls down to a depth of approximately 150 m.

V-shaped canyons, 150 m deep and in places 1.5 km wide, dissect the submarine cones and the pre-existing fringing reefs.

Because of the steep slopes, coral reefs of fringing and patch type occur in narrow bands along portions of the coast. They are absent in front of actively transporting wadi mouths, but are present at the edges of alluvial cones. Some coral reefs overlie pre-existing alluvial fans and are cut by furrows across the reef flat (fluvial relief?). The furrows are blocked seaward in places by coral growth forming pits and pools several meters deep on the reef flat. Except on the western side of the southern segment of the Gulf there is no true coastal plain (Fig. C.5). The moderate and gradual shallowing in the northernmost segment is due to mass transport from surrounding land. There are no well-defined continental shelves, although on the western side of the Gulf the more strongly sloping, deeper part of the descent can be separated from a shallower (less than 180 m deep), terraced and more moderately steep one. For reasons of convenience the shallower part is referred to here as "shelf". Its width varies between 0.5 and 2 km and its edge is situated between 100 and 200 m.

Three **major basins**, corresponding to the three en-echelon grabens, occupy the interior of the Gulf. They are separated by shallow sills and correspond to three parts of the Gulf which are also distinct structurally and sedimentologically. The floors of the basins are undulating and five conspicuous deeps, separated by intrabasinal sills, are present within them.

The northern basin has a simple bathymetry and structure and is dominated by the 50-km-long, 3–8-km-wide and 900-m-deep Elat Deep (Figs. C.4 and C.5). In the Elat Deep, sedimentation is fast enough to maintain a flat bottom. However, the deeper reflectors dip to the east and south, the dips increasing downward as a result of continuing tilting.

The central basin is 35 km long, 6 km wide and includes the deepest points in the Gulf, viz. the 1,830 m deep, 6–8-km-long and 2-km-wide Aragonese Deep and the 1,550-m-deep, 4–5-km-long and 1-km-wide Arnona Deep. The central basin is varied in bottom relief and in structures (see Figs. C.4 and C.5). Most of the basinal sediments were warped and now form an elongated anticlinal fold which extends between the Aragonese and Arnona Deeps and is expressed in the bottom topography. Sedimentation here occurs at a slower rate than the deformation, in contrast with the northern basin. The transition between the central and southern basins is not abrupt.

The southern basin, about 55 km long and 6 km wide, includes the Dakar and Tiran Deeps, with depths of 1,400 m and 1,300 m respectively (Figs. C.4 and C.5). This basin is the widest and also structurally the most complex. The deep basin occupies the eastern side of the Gulf. West of it is a wide terrace at water depths of 800–1,000 m on which the large fans of Wadi Dahab and Wadi Kid were built. Different parts of the southern basin experienced variable degrees of subsidence. The sedimentary sequence filling the flat-bottomed Dakar Deep thins dramati-

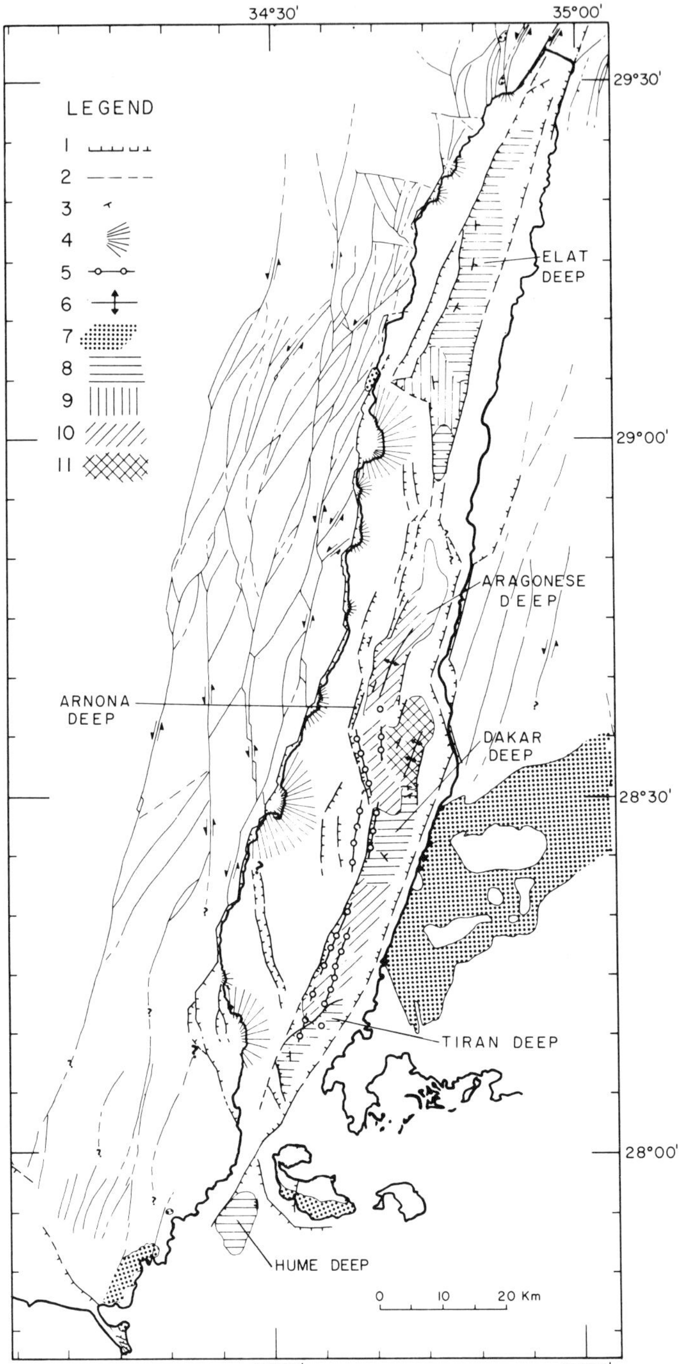

Fig. C.5. Structural elements of the Gulf of Aqaba. *1* Main active faults, downthrown side shown; *2* main inactive faults of rift margins; *3* dip of basinal sediments; *4* alluvial fans and submarine cones; *5* diapiric folds; *6* main anticlines; *7* outcrops of Neogene (mainly Miocene) sediments; *8* basinal areas with a flat sea floor; *9* disturbed southern part of northern basin; *10* folded and warped basinal areas; *11* uplifted basinal area at sill between central and southern basins (Ben-Avraham et al. 1979 b)

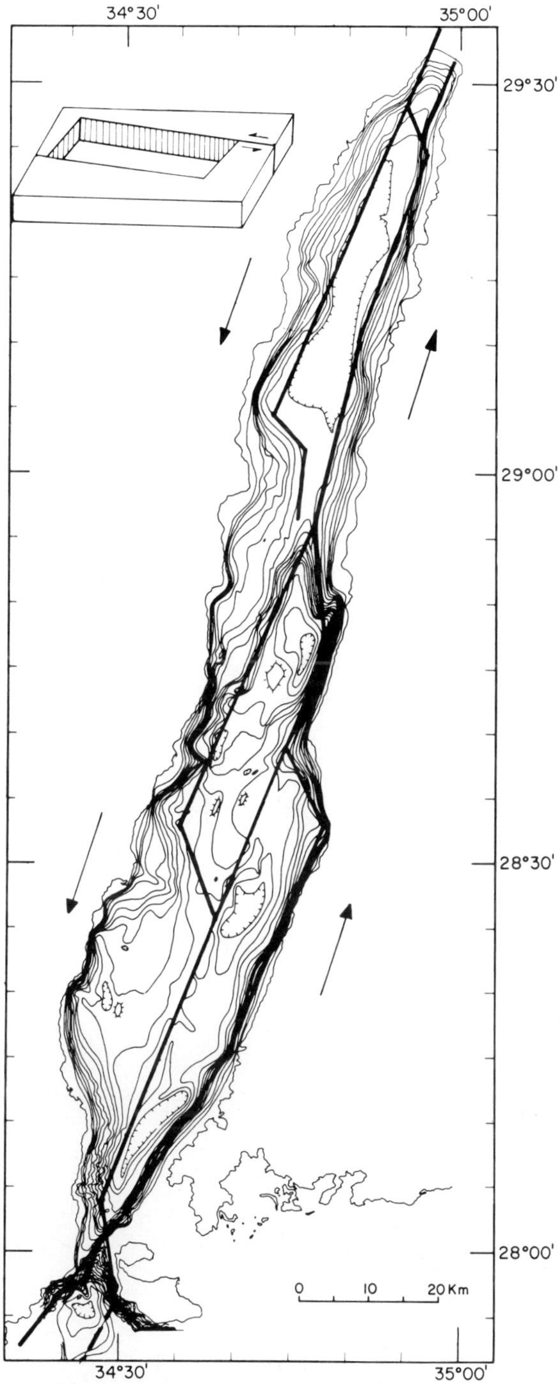

Fig. C.6. Generalized model for structure of Gulf of Aqaba: the three basins are interpreted as three en-echelon rhomb-shaped grabens produced by strike slip (see *inset*) (Ben-Avraham et al. 1979 b)

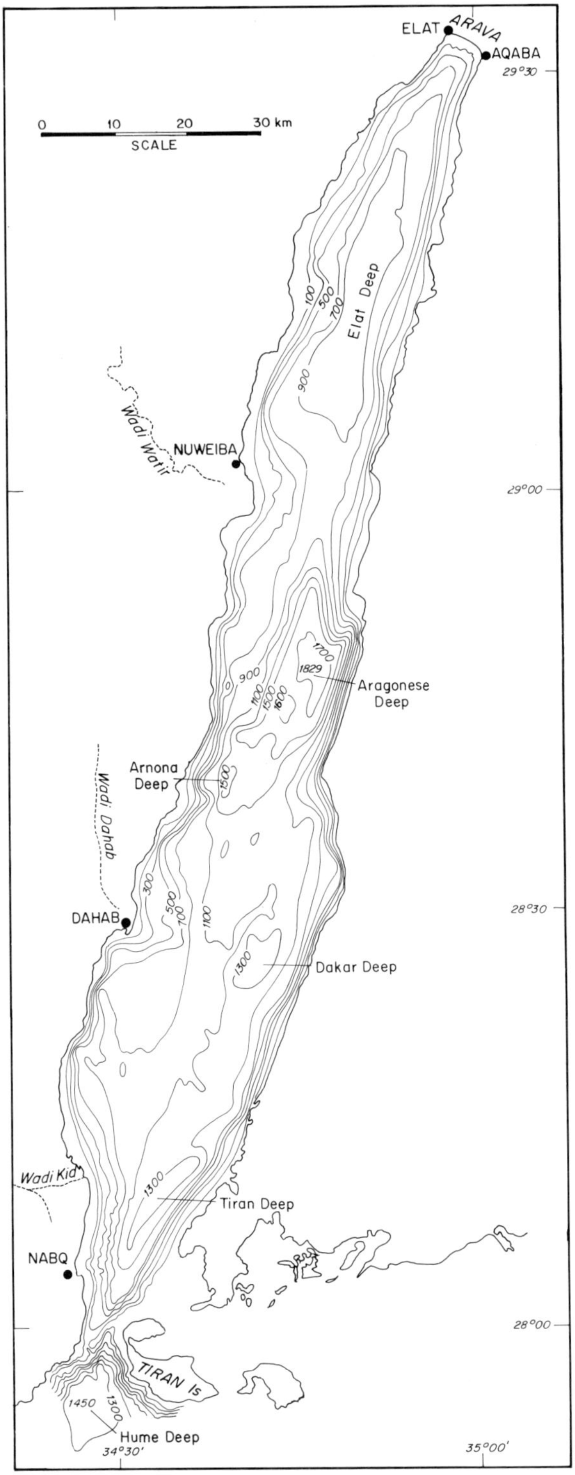

Fig. C.7. Bathymetric chart of the Gulf of Aqaba. (Modified from Hall and Ben-Avraham 1978, in Eyal et al. 1980)

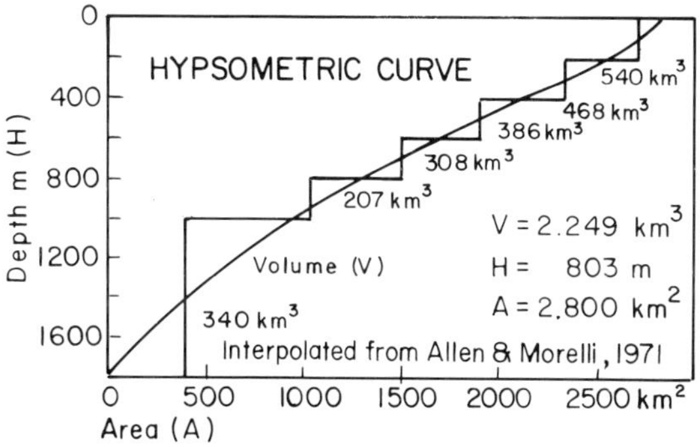

Fig. C.8. Hypsometric curve of the Gulf of Aqaba

cally toward structural highs on the north and south, but is not bounded by faults, while the Tiran Deep to the south is bounded by faults on all sides.

South of the Straits of Tiran the 1,450-m-Hume Deep links the Gulf of Aqaba with the axial depression of the Red Sea. The northern side of this Deep slopes at more than 40°. Generally, the complex bottom relief of the basins is due to intrabasinal faults, grabens, folds, arching of sediments, canyons, and, in places, diapiric structures, possibly belonging to Miocene evaporites (Fig. C.4).

The fact that the basins are shorter than the total slip which took place along the Dead Sea rift is another indication that they were formed during the second phase of movement, hence 4–5 my BP. The hypsometric curve of the Gulf of Aqaba is a reflection of the structural history. It shows that more than 20% of the Gulf's volume is comprised in depths greater than 800 m (Fig. C.8).

Abiogenic and biogenic constituents of **bottom sediments** form as series of zones with increasing water depth and distance from the shore. Detrital sediments from land are brought into the Gulf from the slopes on both sides, into deltas and cones through wadis on the western side and from the Arrava Valley in the north. The coarsest material remains on the beaches and progressively finer particles reach greater distance from the shore and greater depth, where they are strongly diluted by biogenic material. Uncemented boulders of basement formations cover steep slopes at depth between 10 and 40 m (e.g., in front of the MBL, Elat or in the bay at the mouth of Wadi Dahab).

Grain size decreases from beach to about 50 m depth in both biogenic and abiogenic particles, including foraminiferal shells which form a major fraction of the sediment (50–80%). Deeper down, to about 100 m, grain size increases again due to the abundance of benthic larger Foraminifera. From about 150 m, grain size decreases due to the dominance of smaller Foraminifera, including planktic ones. Poorest sorting occurs where abiogenic and biogenic constituents are quantitatively about equal.

Calcareous silts, micas and small quartz grains are carried by currents far into open waters and are added on the sea floor to foraminiferal-pteropod muds, forming a thick, fine-grained cover. Depending upon location with regard to shore, wadi mouths, and currents, the percentages of the abiogenic material are extremely variable. From faunal evidence it seems that across-the-slope gravitational transport is less important than would be expected with such steep slopes. On the other hand, there is transport through canyons, as well as long-shore transport of sediments.

Aeolian transport of sand is due mainly to the chimney effect of N–NE winds blowing for days along the Arava Valley. On the larger alluvial fans (Wadi Dahab, Wadi Kid) which protrude from the main coastline into the sea and are thus particularly exposed to northerly winds, considerable amounts of sand are raised by winds and redeposited on the lee-side of the fans, thereby influencing sediments on beaches and near the shore. Localized dunes may be formed by wind-transported Mesozoic sands which – blown into the sea – inhibit the growth of fringing reefs.

Northerly winds are responsible for nearshore currents on the eastern coast transporting sediments southward at the foot of the fringing reefs, while flash-flood sediments carried by wadis into the sea during southerly gales are transported northward by reversal of the long-shore current.

On the reef platform, boulder-sized coral rubble is abundant. In some places, large blocks or boulders of reef-limestone accumulate below the reef front, forming the uppermost part of the reef talus. Deeper parts of the talus are covered with coral rubble, cemented by incrusting organisms. Where surfaces of hard bottom are covered by "coral carpets", at 40–80 m depth, coral rubble and coarse sands accumulate in hollows and along depressions which follow perhaps an ancient drainage pattern. Carbonate sands resulting from degradation of reef builders and from autochthonous skeletal material carpet the bottom near reefs. These sands are less well-sorted than the shallow carbonate beach sands because of the latter's winnowing by wave action.

Between the reefs and below them, down to a depth of about 70 m, meadows of halophile plants form traps for fine-grained, dark sediment. The same is true of mangrove forests bodering the western side of the southern segment of the Gulf. Below 100–150 m depth the sediments are brownish yellow silty muds with abundant planktic foraminiferal shells together with pteropods. In the north, these sediments comprise also aeolian material blown from the Arrava.

Pelagic, brown-yellow to gray-green, partly silty foraminiferal-pteropod "oozes" cover the Gulf's bottom down to depths of about 900 m. Only at greater depths are turbidites found (Fig. C.4). There is very little sorting or winnowing in the pelagic sediments, although sorting by shape of pteropods has been observed in turbidites.

The mineralogical composition of the carbonate fraction is mainly coral and molluscan aragonite with benthic foraminiferal and algal high magnesium calcite in areas of less than about 150 m depth, while that in deeper parts of the Gulf is mainly planktic foraminiferal low magnesium calcite and pteropod aragonite.

As far as is known from seismic surveys and from deep sea cores, the subbottom sediments comprise two major types, i.e., (a) Coarsely stratified sediments underlying the marginal slope areas (Fig. C.4). Their reflectors are undulatory, patchy, irregular and rough, often discontinuous, with local unconformities. Since as a whole this sequence resembles the morphologically defined alluvial cones, they are considered to be built mainly of similar material. Their thickness is 1–1.5 s without reaching basement. Observed dips are in part depositional, and tectonic deformation is also apparent. (b) Pelagic sediments and turbidites filling the basins. This sequence is characterized by smooth and continuous reflectors and its thickness is 2–2.5 s with basement not reached, except at the margins. Changes of dip and thickness indicate continuing syn-sedimentary differential movements. Similar sequences also occur in small depressions in marginal areas. The sharp contact between the two sequences, i.e., marginal slope and deeper basin types, is often located at the foot of conspicuous escarpments, indicating this contact to be fault-conditioned. Continuing fault activity seems to maintain the distinction between basins and marginal areas in spite of continuous sedimentation.

Changes through time in type of sedimentation is indicated in deep-sea cores, e.g., at the edge of Wadi Dahab cone where a drilling penetrated a 127 ky BP sequence of normal pelagic sediments overlying graded, arkosic sands of the penultimate glacial.

Rates of sedimentation in the shallow areas are at present difficult to assess. In the basins, at depth shallower than about 900 m, deep-sea cores indicate rates of about 10 cm/1,000 yr while in the deep basin turbidites rates of sedimentation may be four times as high.

The geological history of the rift-shaped Gulf of Aqaba determines its physiographic elements and is thus a major factor influencing circulation pattern, as well as the distribution of land-derived material and hence depth-related substrates for benthic organisms of importance in ecological micropaleontology. The second major factor affecting the distribution and preservation of organisms, both planktic and benthic, in the Gulf is its climatic setting within an arid area.

References: Ben-Avraham et al. (1979 a, b), Erez and Gill (1977), Erez et al. (1983), Eyal et al. (1980), Fricke (1982 b), Garfunkel (1981), Garfunkel et al. (1981), Gvirtzman and Buchbinder (1978), Gvirtzman et al. (1977), Hall (1975), Hall and Ben-Avraham (1978), Hottinger (1977 a), Reiss (1977), Reiss et al. (1977 a, 1978, 1980).

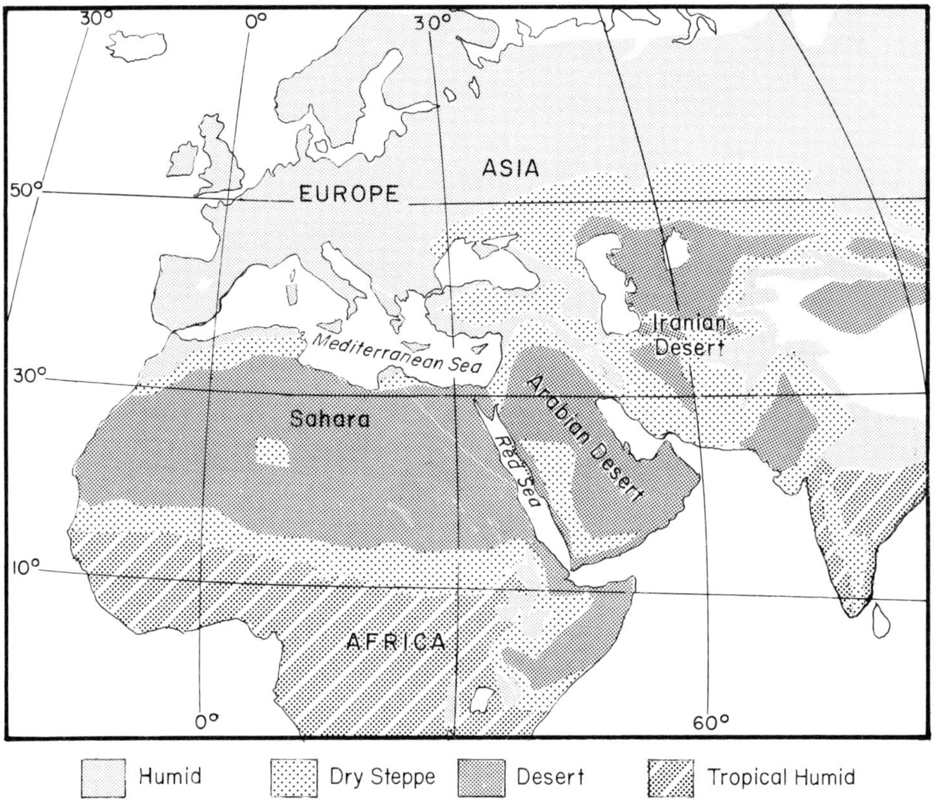

Fig. D.1. African-Arabian deserts surrounding the Red Sea and its gulfs

D. A Desert-Enclosed Sea

D.1 Climate

The Gulf of Aqaba is situated between the Sinai (Et Tih) Desert and the western Arabian (An Nefud) Desert (Fig. D.1). Although at a distance of only about 200 km south of the humid, Mediterranean-influenced region of Israel, the northern tip of the Gulf is already well within the extreme arid zone (Fig. D.2).

Insolation is very strong. Estimated global daily radiation total for the town of Elat ranges between 406 and 668 cal $cm^{-2} d^{-1}$ in April–October and between 313 and 451 cal $cm^{-2} d^{-1}$ in November–March, being highest in June and lowest in December. Cloudiness is very low, only about 3/100 to 1/10 during the long summer (April–early November) and maximum 3/10 during the short winter (late December–March). In the years 1969–1972 sunshine prevailed during 9.2–9.8 h per day on the average.

Summer temperatures on land may reach 45 °C, particularly in the northern part of the Gulf and 47 °C at sharm es Sheikh. Winter temperatures as low as 4 °C were measured in its middle part (Nuweiba; location map Fig. D.3). Mean daily temperatures vary between 15.5 °C (Nuweiba) and 32.3 °C (city of Elat). At sea, south of Elat, at DCPE Station A (Coord. 29°30′ N/34°57′ E) mean daily temperature varies between 14 °C and 29.3 °C (Fig. D.4). About 10 km south of Elat, at Wadi Taba, daily air temperature over the sea close to the shore varies between 17° and 36 °C (Fig. D.5).

The aridity in the area of the Gulf is enhanced by the fact that it is a deep and narrow cleft in the rain shadow of the abruptly rising, 2,000 m-high Sinai massif to the west.

Rainfall in the Elat region is about 22 mm per year, although exceptionally rainy years with as much as 70 mm rainfall may occur (Fig. D.6). The total number of rainy days per year with more than 1 mm of precipitation is 8 on the average, all of them usually concentrated in the winter months, particularly December and February (notable exceptions occur, like of the month of April 1976/77; see Fig. D.7). Generally, the yearly rainfall pattern is rather erratic (Fig. D.6).

Flash floods through major wadis in winter transport terrestrial material into the Gulf, resulting in large deltas and in incision of submarine canyons. Produced by localized heavy rainfalls, most floods activate the wadis only along restricted segments of the coast, thus keeping damage to the coastal road and to the fringing reefs within acceptable limits.

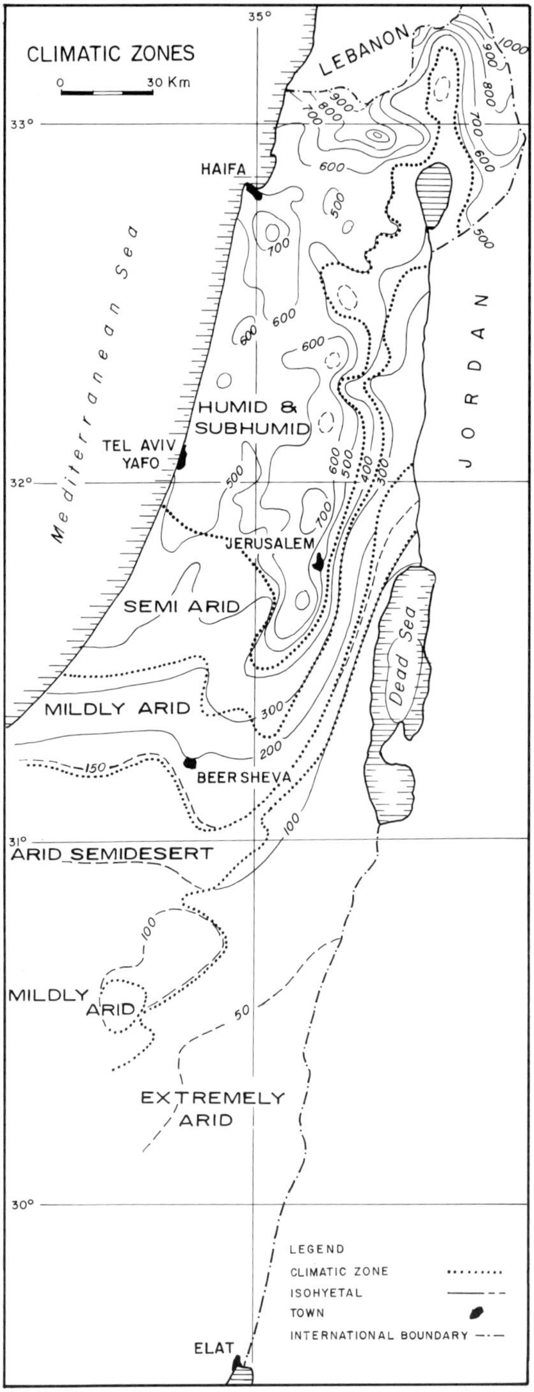

Fig. D.2

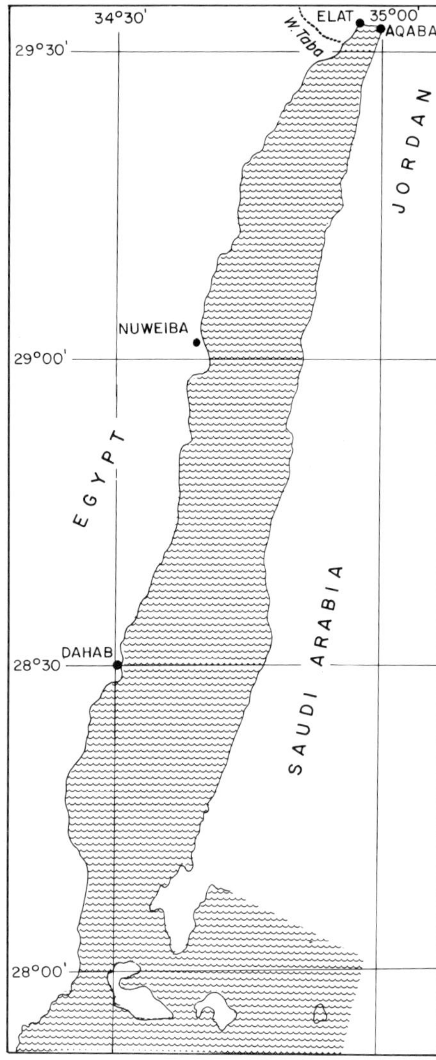

Fig. D.3

Fig. D.2. Climatic zones of Israel (Dan and Yaalon 1982)
Fig. D.3. Sketch map of the Gulf of Aqaba showing location of meteorological stations Elat, Nuweiba and Dahab

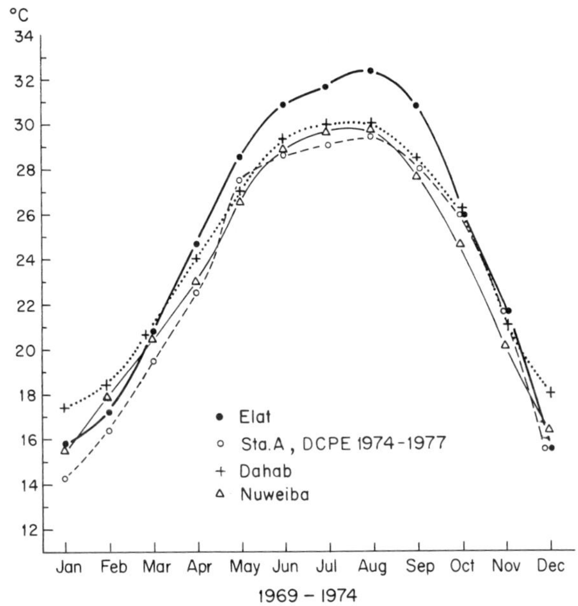

Fig. D.4. Mean daily air temperature at Elat, Nuweiba, Dahab and DCPE-Station A, 1969–1974 (see Fig. D.3)

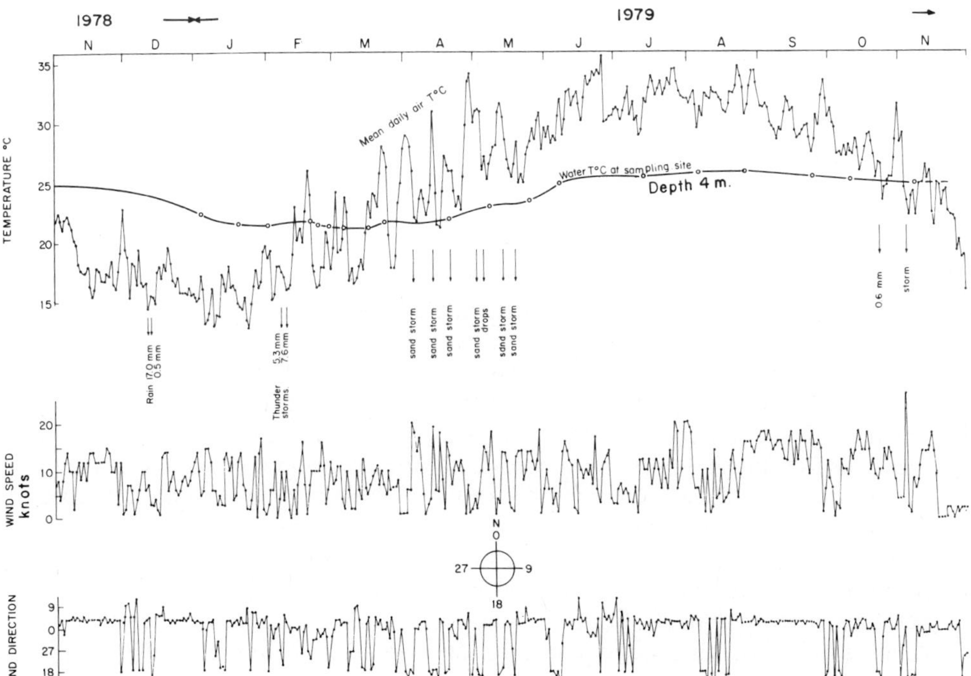

Fig. D.5. Mean daily air temperature, wind speed and direction at Wadi Taba 1978–1979 (Zohary et al. 1980)

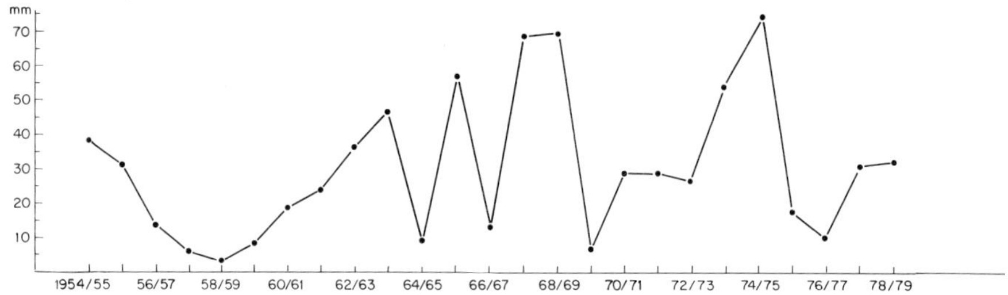

Fig. D.6. Yearly precipitation, city of Elat, 1954–1979. (Data from Israel Meteorological Service)

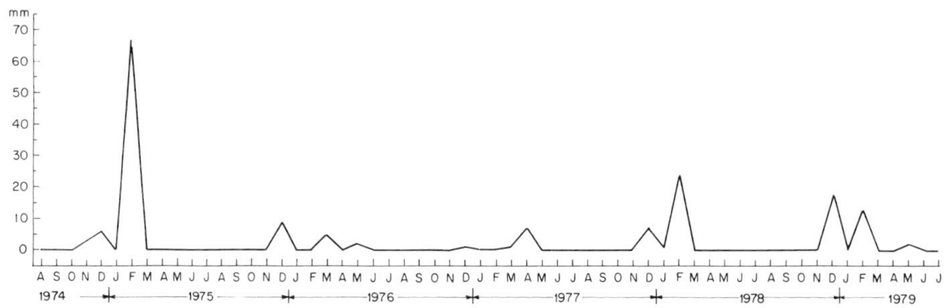

Fig. D.7. Monthly rainfall, city of Elat, 1974–1979. (Data from Israel Meteorological Service)

Annual evaporation rates on land (near Aqaba-Elat) are three times as high as along the Mediterranean shore of Israel. Evaporation over the Gulf is extremely high, about 400 cm y^{-1} (see below).

Relative humidity in shore localities of the Gulf averages 30–55%, the minimum–maximum span being between 5 and 94% (Fig. D.8). The average for the city of Elat is about 35%.

In contrast to the southern and middle Red Sea, where wind direction switches seasonally under the influence of the monsoons, the changes in yearly air pressure over the Gulf of Aqaba and the northern Red Sea are extremely weak (about 12 mbar).

Over most of the Gulf **wind direction** is predominantly from the north (approximately 50% frequency) and from the northeast (approximately 20% frequency). A minor northwestern component, particularly in winter, is present, while winds from the south are subordinate. Average wind speed is about 10–15 knots, maxima attained being 40–45 knots (N–NE) and 28–33 knots (S–SE). Southerly winds in excess of 20 knots were observed mainly in the winter, when, however, gales reaching 45 knots may blow over several days. Changes in wind direction may occur within less than 1 h, producing a rapid rise or drop of air temperature of more than 20 °C and of humidity from one extreme to the other. A detailed record of wind speed and direction over a 1-yr period at Wadi Taba is given in Fig. D.5. At the Straits of Tiran, winds are prevailing northerlies, with significant departures to NW.

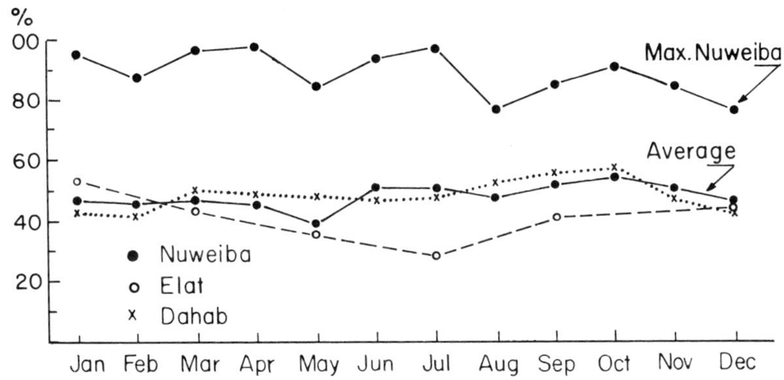

Fig. D.8. Relative humidity at Elat, Nuweiba and Dahab. (Data from Israel Meteorological Service)

Generally, the sea state in the Gulf of Aqaba is calm, particularly in the north, with wave height rarely exceeding 0.5 m. Wave height increases in the afternoon and (because of sea room and fetch) from north to south. Occasionally, during the winter, wave heights of 9–14 m were observed.

The **tides** in the Gulf of Aqaba are due to those in the Red Sea, the direct effect of moon and sun being relatively small. High water is nearly simultaneous over the whole Gulf, but 1 h later at the northern end of the Gulf than in the northern Red Sea. Spring tide range increases from 0.5 m at the entrance of the Gulf to a maximum of about 1.5 m at its head. The average tide range is less than 1 m.

References: Israel Meteorological Service, Reports; Fricke (1982), Friedman (1968), Mergner and Schuhmacher (1974), Reiss (1977), Zohary et al. (1980).

D.2 Hydrography

The peculiarities of the hydrographic pattern, i.e., the small seasonal and bathymetric gradients in temperature (about 6 °C) in the waters, which are warm throughout the year (min. about 21 °C at greatest depth) and uniformly highly saline (about 41‰), the lack of a pronounced oxygen minimum, as well as indications for a counter-clockwise thermohaline circulation in the Gulf of Aqaba have been recognized since the late 19th century through observations made by various expeditions (especially the Pola 1895, Mahabeth 1935, Manihine 1949, Vema 1958, and Snapir 1967) as well as by a few individual investigators (for references see especially Anati, Luksch, Deacon, Mergner and Schuhmacher, Mohamed, Morcos, Oren).

An intensive data-collecting program in the Gulf (DCPE) between 1974 and 1977 from aboard the R. V. Arnona has provided information on physical, chemical, and biological parameters in the whole water column of the Gulf, derived from about 1,000 water samples and about 3,000 plankton hauls obtained at 10

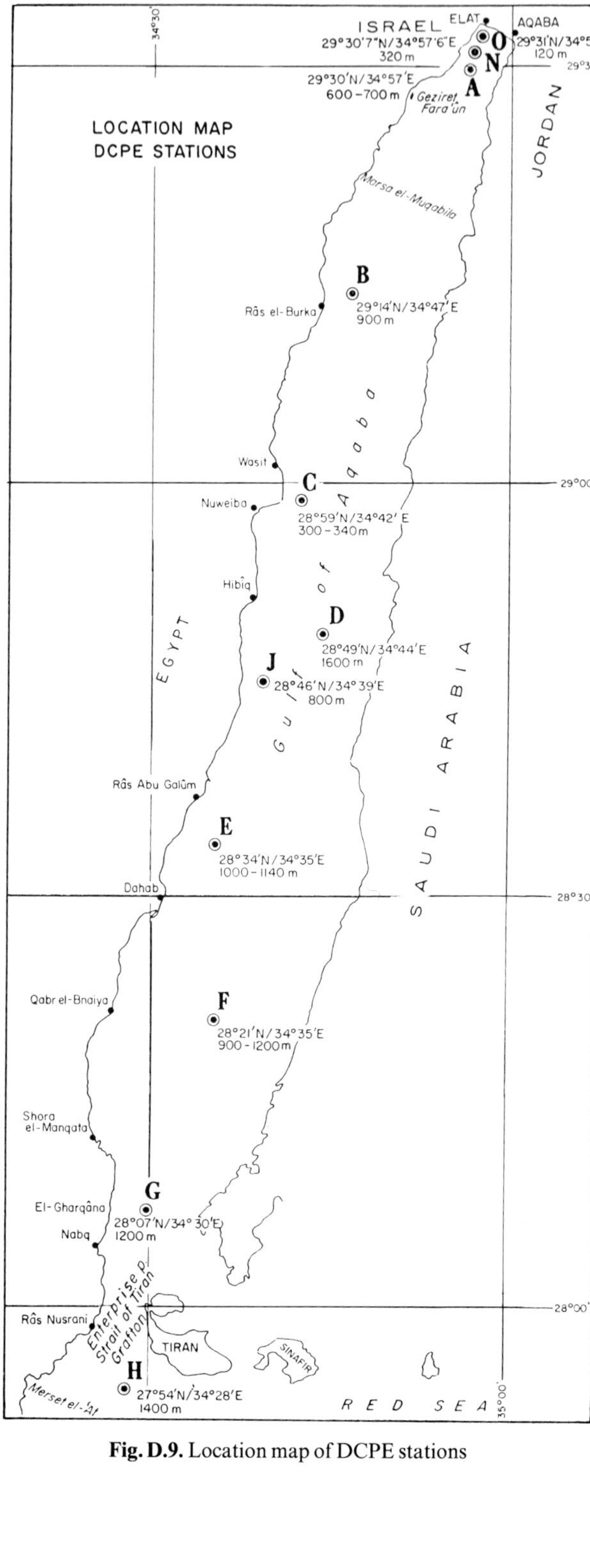

Fig. D.9. Location map of DCPE stations

hydrographic stations along the midline of the Gulf (Fig. D.9). Several stations were occupied during 1978/1979 in studying cross-axial sections in the central sector.

The seasonality in regional climatic conditions is paralleled by that in the water column of the Gulf: a long summer, from April to October, and a short winter, from January to March, separated by a transitional period, November–December (cf. Chap. D.1).

Sea surface temperature within the Gulf varies between 26.7 °C (July) and 20.75 °C (March), although exceptional high extremes of 29.2 °C (May, Station B) and 27.3° (August, Station A) and lows of 20.5 °C (February, Station A) and 20.4 °C (March, Station D) were observed. The average sea surface temperature varies between 21.10° and 23.5 °C. Minimum temperature of deep water is 20.9 °C (April, Station A).

Surface salinity varies between 40.03‰ (November, Station B) and 40.51‰ (April, Station B) with exceptional high salinities up to 41.57‰ observed in August at Station A. The average surface salinity ranges between 40.56 and 40.78‰. Salinities of the deep water are higher than 40.8‰ in summer and, exceptionally, may reach more than 41‰.

The temperature and salinity values mentioned refer to the open waters; much larger variations in both parameters occur in areas of reefs, mangroves, sabhkas, and marginal pools. The temperature, salinity and oxygen characteristics of the Gulf and of the Red Sea immediately outside the Straits of Tiran, based on measurements during 1974/1975, are summarized for summer and winter conditions in longitudinal sections, Figs. D.10–D.13.

During the **summer** season, an upper, temperature-stratified water mass can be distinguished from a more homogeneous deep water mass. Bathymetric difference is about 6 °C and the base of the thermocline is at about 200 m. At any given depth, temperature of the waters in the Red Sea outside the sill is higher by 0.4°–1.0 °C than of those inside the Gulf. Furthermore, at equivalent depths, the waters in the northern sector of the Gulf, corresponding roughly to the narrower segment (which includes the Elat, Aragonese, and Arnona Deeps), are cooler than those in the southern sector, corresponding to the wider segment (which includes the Dakar and Tiran Deeps). A weak halocline is developed in the northern sector between 300 and 600 m depth, with a core of slightly lower salinity being present between 50 and 150 m depth, between Stations B and C. The isohalines ascend toward the head of the Gulf and toward the southern sector, leading to lateral salinity gradients. In the southern sector a high salinity water mass is present at sill depth and can be traced into the Red Sea.

Winter conditions are characterized especially by the extremely weak temperature stratification in the upper water mass, bathymetric differences being only 0.2 °C on the average in the northern sector and 0.7 °C in the southern one. The latter sector is well defined by the fact that the upper water mass is warmer, more stratified, and reaches to greater depth than in the north. Thus lateral temperature gradients from north to south are relatively strong.

Below sill depth (252 m), the waters in the Gulf are cooler than immediately outside the straits by 0.4°–0.7 °C. The northern sector is nearly homohaline, while the southern sector exhibits stratification with steep southward slopes of isoha-

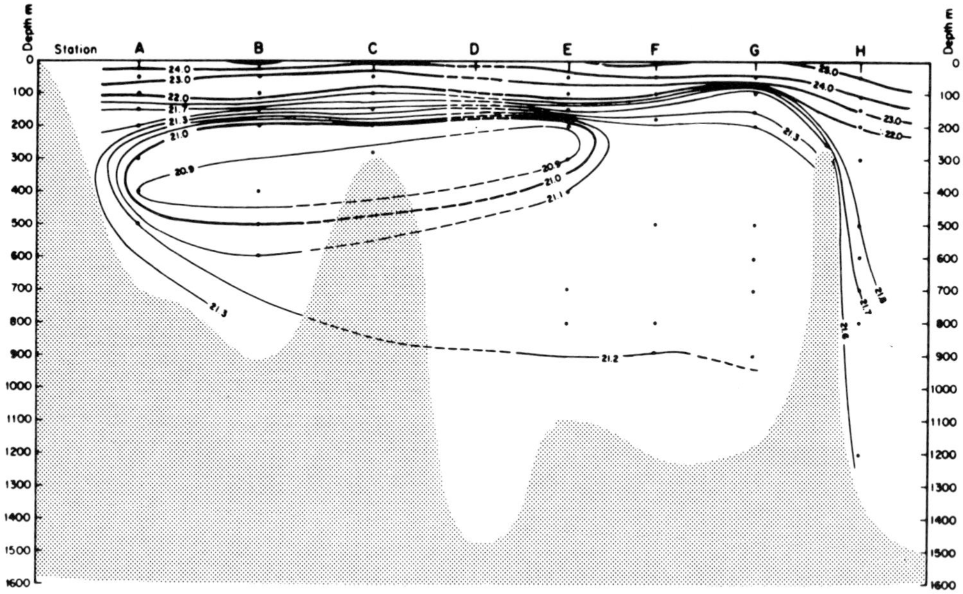

Fig. D.10. Averaged isotherms (°C) April–October 1974. *Shaded area:* simplified bottom profile along station to station track (Klinker et al. 1976)

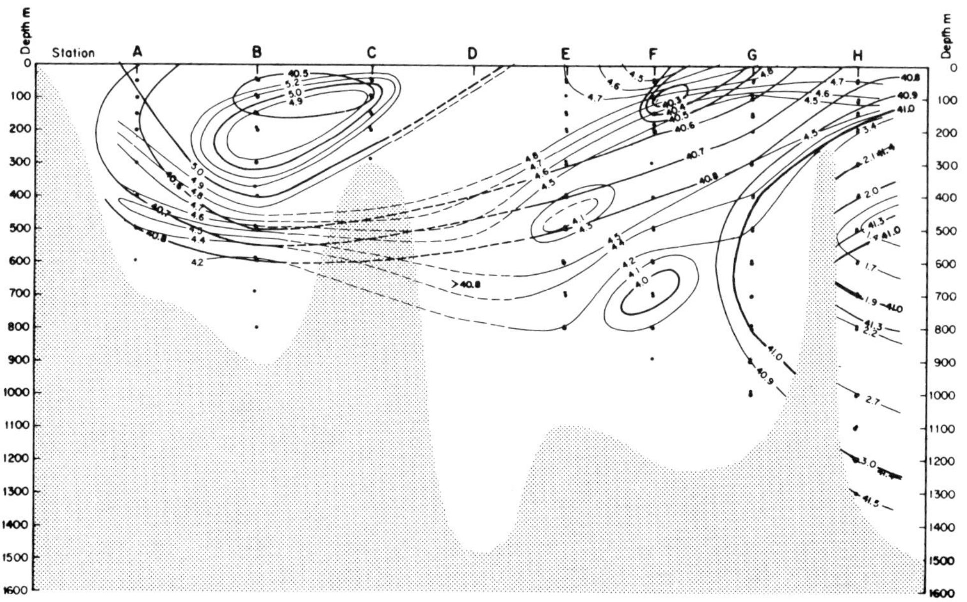

Fig. D.11. Averaged isohalines (‰) and oxygen isopleths (ml l^{-1}) April–October 1974. *Shaded area* see Fig. D.10 (Klinker et al. 1976)

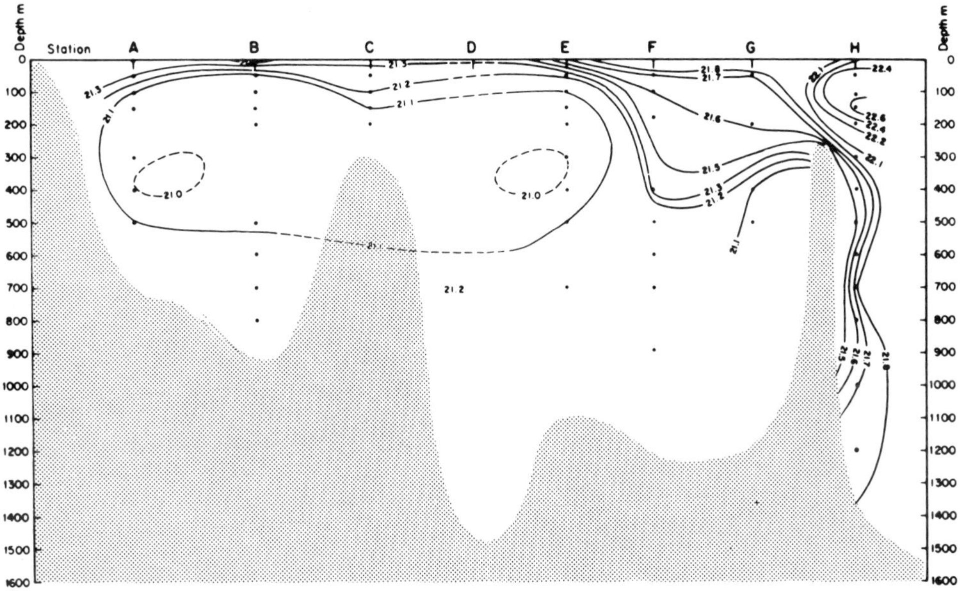

Fig. D.12. Averaged isotherms (°C) January–March 1975. *Shaded area* see Fig. D.10 (Klinker et al. 1976)

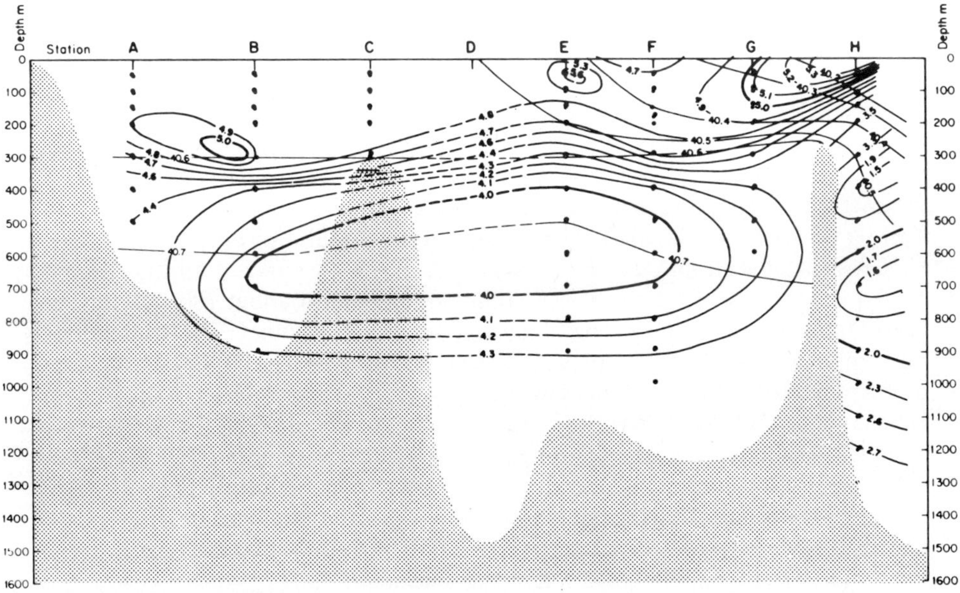

Fig. D.13. Averaged isohalines (‰) and oxygen isopleths (ml l^{-1}) January–March 1975. *Shaded area* see Fig. D.10 (Klinker et al. 1976)

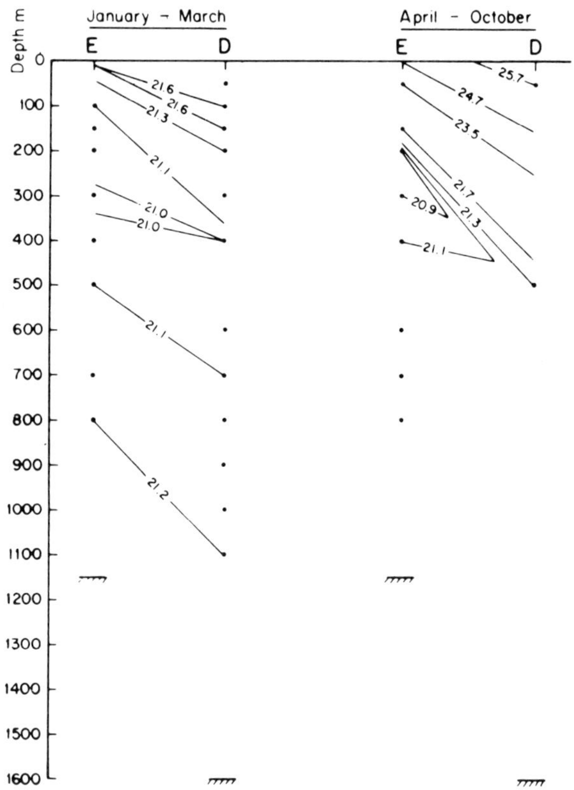

Fig. D.14. Averaged isotherms (°C) April–October 1974 and January–March 1975, DCPE Stations E–D (Klinker et al. 1976)

lines traceable into the Red Sea. Outside the straits, below sill depth, the Red Sea waters are homohaline. The oblique sections, stations E–D (Figs. D.14 and D.15) indicate that during both summer and winter the waters on the eastern (Arabian) side are warmer and less saline than on the western (Sinai) side of the Gulf.

Cross-axial sections near Ras Burka (see Fig. D.9) indicate during summer 1978 a water body of relatively low salinity at a depth of about 100–150 m. This body is still detectable in October, but disappears in November. The origin of this body is not known, but it is noteworthy that between Stations B and C (Fig. D.9) a core of relatively lower salinity has been observed during 1974/1975 (see above). In cross-axial sections at Ras Burka during March 1979, a warm (up to 21.24 °C) and highly saline (up to 41.4‰) tongue is present between about 100 and 150 m depth and is detectable, albeit reduced in intensity, as far south as at Ras Abu Galum. The origin of this tongue, impinging from the Sinai coast, is unknown, but from considerations of volume it is certainly not connected with the hypersaline Solar Lake in the neighborhood of Ras Burka.

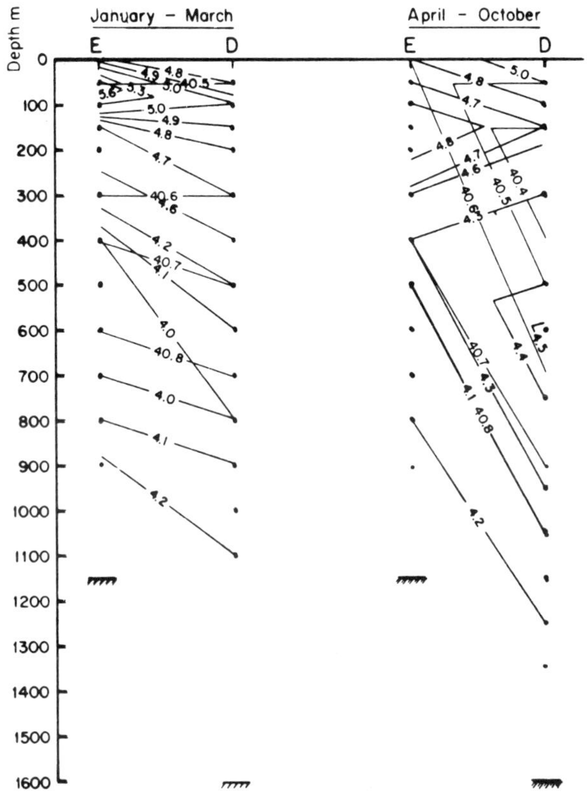

Fig. D.15. Averaged isohalines (‰) and oxygen isopleths (ml l^{-1}) April–October 1974 and January–March 1975, DCPE Stations E–D. *Shaded area* see Fig. D.10 (Klinker et al. 1976)

Oxygen isopleths averaged over the main seasons of 1974/1975 are shown in Figs. D.11, D.13, and D.15. Noteworthy features are the high values of oxygen (up to 6.10 ml l^{-1}) and the lack of an oxygen minimum in the Gulf, the lowest value observed being 3.75 ml l^{-1} at 800 m depth, at Station B, in November (compare with the oxygen minima in the Red Sea immediately outside the sill). The differences between the northern and the southern sectors and between the eastern and the western side are also indicated by oxygen values. Of interest are the high oxygen core in the upper waters in the vicinity of Station B, the low oxygen core in deep waters at Station F (during summer), as well as the configuration of the oxygen isopleths in winter, both in the upper water mass and in the main deep water mass. Oxygen distribution seems to be closely related to primary production and consumption, as well as to salinity in shallower waters.

D.3 Circulation Pattern

From various observations (including diving) it is known that the uppermost water layer is driven by the prevailing northerlies roughly southward through most of the year, and northward during southern storm conditions. The thickness of the wind-driven layer is estimated to be less than 1 m under normal conditions. Wave base during storm conditions is reported in places at about 15 m depth.

Under the arid and hot climatic conditions of the area, evaporation is extremely high and greatly exceeds precipitation and runoff. The evaporative loss is compensated by inflow from the Red Sea through the Straits of Tiran (Enterprise and Grafton Passages) and thereby the main circulation pattern in the Gulf of Aqaba is **thermohaline ("inverse estuarine")**. Calculations based on scanty data obtained by earlier expeditions indicated an evaporation rate of about 200 cm yr^{-1}, i.e., similar to that in the Red Sea. These calculations did not, however, take any heat flux into account.

Meteorological observations during 1972 at near-shore stations (Dahab and Nuweiba) indicated a strong heat flux (280 W m^{-2}) and sea-surface cooling (especially in winter) (Fig. D.16) equivalent to an evaporation rate of 1 cm day^{-1} or 365 cm yr^{-1}, hence twice that in the Red Sea proper.

Using considerations of preservation of mass, the observed heat flux, and the parametrization and dynamics of the Straits of Tiran, where critical interfacial Froude conditions and a state of "overmixing" exist, a still larger evaporation rate, i.e., 420 cm yr^{-1} was calculated for the Gulf of Aqaba, as compared with 180 cm yr^{-1} for the Red Sea. Thus, a considerable volume transport through the Straits of Tiran must take place.

The DCPE data further elucidated the thermohaline circulation pattern in the Gulf of Aqaba, illustrated for the two main seasons 1974/1975 by the accompanying averaged longitudinal density sections (Figs. D.17 and D.18). The water den-

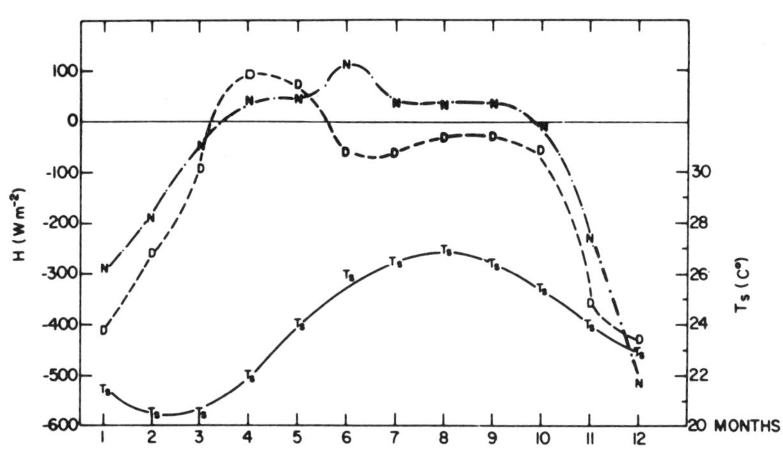

Fig. D.16. Surface temperature (°C) and heat flux (W m^{-2}) at near-shore meteorological stations Nuweiba and Dahab, 1972. T_s surface temperature; *D*, *N* heat flux at Dahab and Nuweiba (Assaf and Kessler 1976)

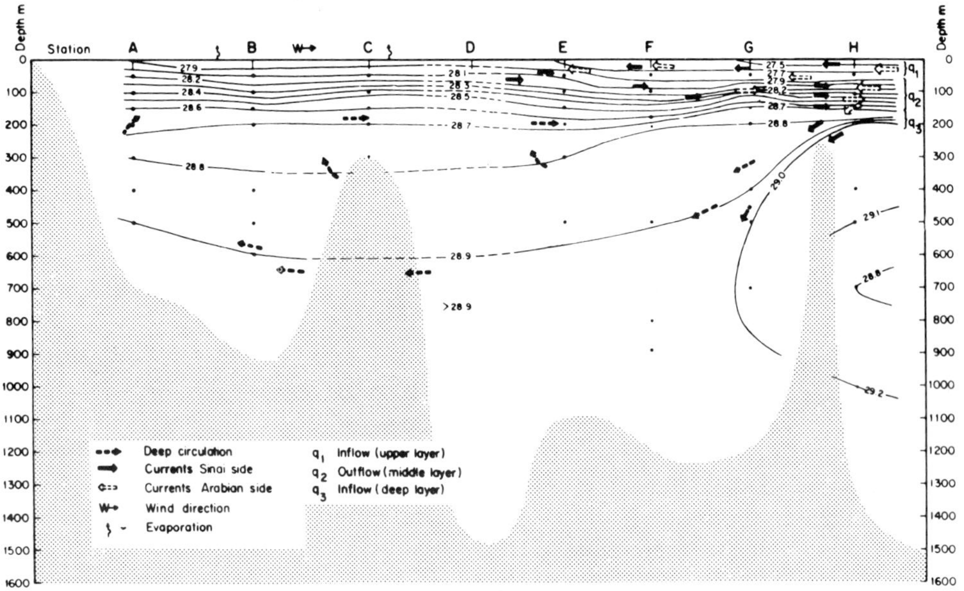

Fig. D.17. Averaged isopycnals (σt) April–October 1974. *Shaded area* see Fig. D.10 (Klinker et al. 1976)

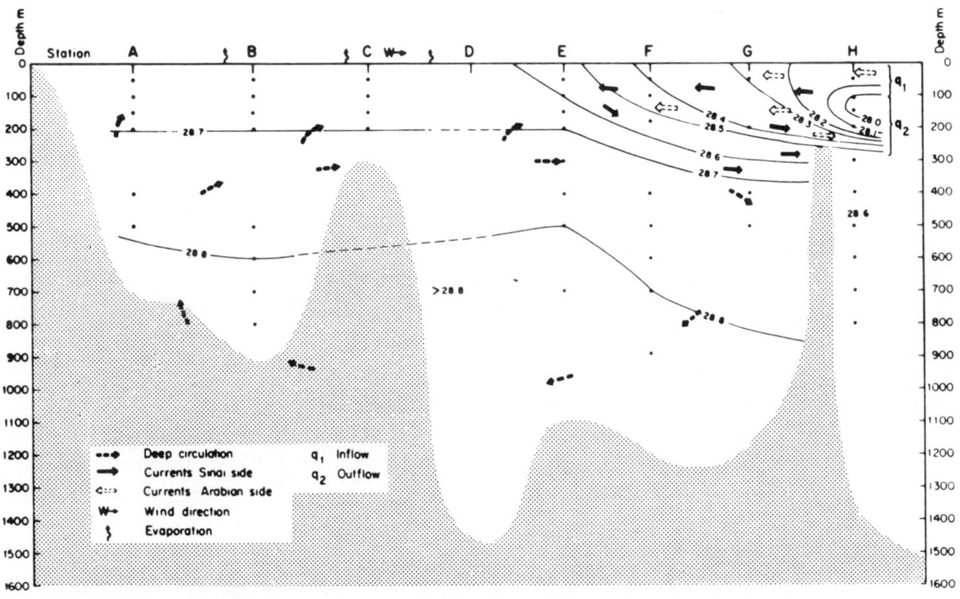

Fig. D.18. Averaged isopycnals (σt) January–March 1975. *Shaded area* see Fig. D.10 (Klinker et al. 1976)

sity ranges between 27.09 σt (surface, Station F, August) to 29.54 σt (500 m depth, Station A, August). During summer the nearly horizontal density stratification of the upper waters in the northern sector is due mainly to temperature. In the southern sector, isopycnals slope gently toward the straits, relatively low density waters being present in the upper 50–75 m. Isopycnal 28.7 descends from south to north, while isopycnals 28.8 and 28.9 descend in the same direction rather strongly in the southern sector, rising slightly toward the head of the Gulf. The deep isopycnal 29.0 σt (due to salinity) can be traced at sill depth from the Red Sea into the Gulf.

During winter, the σt sections follow closely the salinity pattern in the more homothermal waters of the northern sector, and both temperature and salinity patterns in the upper waters in the southern sector and outside the sill. In the southern sector isopycnals rise steeply from the straits towards the north. Isopycnal 26.6 can be traced over the sill into the Red Sea, where it connects with the homopycnal deep water mass. The oblique section E–D (Fig. D.19) indicates that both during summer and winter the waters on the eastern side of the Gulf are less dense than on the western one. The disposition of the isopycnals during the different seasons illustrates the characteristics of the circulation in the Gulf.

During **summer** (April–October) warm, relatively low saline and highly oxygenated upper Red Sea waters enter the Gulf through the Straits of Tiran, flowing northward, against the prevailing winds, because of negative buoyancy flux and in order to compensate for evaporative loss. As these waters proceed northward in the Gulf, they become cooled and more saline due to evaporation and thereby a density slope is formed toward the south due to buoyancy flux and wind stress. A southward return flow is initiated along this slope.

An inflow at sill depth beneath the main outflow of cool and highly saline waters with characteristics of uppermost deep Red Sea waters seem to be a transient (seasonal?) feature. Although isopycnals clearly indicate this inflow, the isotherms (Fig. D.10) suggest outflow. This may be due to the fact that measurements outside the straits (Station H) were made in August, while those inside the Gulf (Station G) were occupied in April and June. On the other hand, the salinity of the deep waters is higher in summer – when stratification persists and no deep water is formed – an increase which can be accounted for only by an inflow of deep Red Sea water.

During **winter** (January–March), warm, relatively low salinity and high oxygen waters from the Red Sea press into the Gulf, against the prevailing winds. The northward flow is cooled and concentrated and rather steep southward density slopes are produced (by evaporation and wind stress), along which a return transport of denser waters takes place beneath the inflow. Since most of the deep water mass in the Gulf has densities which are higher than 28.6 σt (i.e., the density of the water below sill depth outside the Straits of Tiran), outflow of subsill water from the Gulf into the Red Sea is impossible and dense waters sink immediately inside the sill.

The interface between the inflow and the outflow is – dependent upon season and upon distance from the straits – at about 75–150 m depth. As indicated by unpublished surveys during 1977–1979, in the vicinity of the straits, the interface slopes downward toward the Red Sea quite steeply and is situated between 80 and

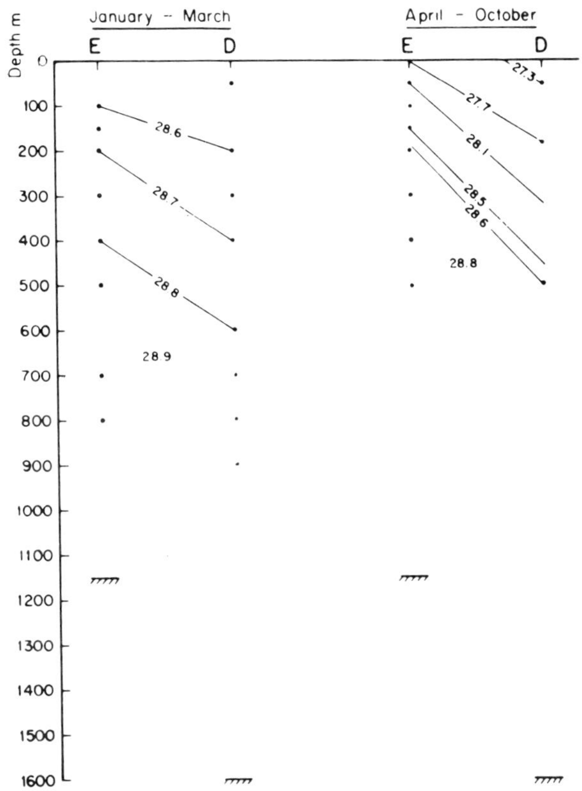

Fig. D.19. Averaged isopycnals (σt) April–October 1974 and January–March 1975 DCPE-Stations E–D (Klinker et al. 1976)

100 m depth. It seems that the shallow Grafton Passage is entirely within the upper water mass, the major two-layered flow being present only in the 252-m-deep Enterprise Passage of the straits.

The distribution of the various parameters indicates that throughout the year the northward inflow from the Red Sea – because of the Coriolis effect – is more important on the Arabian side than on the Sinai side, while for the outflow the opposite is true. Thus, the increase in water density of the outflow occurs actually from a NNE to SSW direction, resulting in a counter-clockwise, "cyclonic" pattern, characteristic of silled basins with evaporative loss.

Temperature, salinity and oxyen patterns seem to indicate a rotational movement in the vertical plane of the deep water. The sense of the rotation is inferred to be clockwise as seen from the west from: (a) theoretical considerations of circulation in a nearly homogeneous basin under permanent wind stress and (b) the increase in salinity in the deep waters of the Gulf in summer, accounted for by inflow and sinking of deep Red Sea water inside the sill.

Distribution of isopleths of density, salinity and oxygen indicates upwelling at the head of the Gulf, an indication supported by other observations (see below) and suggested already by earlier investigators.

Based on the DCPE data obtained during 1974/1975 and on considerations of conservation of mass, estimates of volume transport at the Straits of Tiran yield about 69,000 $m^3 s^{-1}$ in winter, consistent with an estimate of about 71,000 $m^3 s^{-1}$ (at average velocities of 0.63 m s^{-1} in each layer) based on theoretical Froude-Richardson criteria of critical flow.[2] For summer conditions the transport estimates, based on conservation of mass, are twice as large (22,000 $m^3 s^{-1}$) as those derived from critical Froude-Richardson criteria. This may be due to "supercritical" conditions which possibly exist in the lower (150–250 m) layer, consistent with the relative uniformity of this layer, because of turbulence of unstable flow.

DCPE data yield a heat flux through the Straits of Tiran of about 36 W m^{-2} in summer and 207 W m^{-2} in winter, accounting for the high temperature of the Gulf in winter.

Residence time of the waters in the Gulf is calculated from the 1974/1975 data as about 4 months for the upper waters and 3 years for the deep waters, averaging about 1 yr for the whole Gulf. From the data over 3 yr (1974/1977) and seasonal changes, a residence time of about 8 months is calculated for the upper waters (see, however, Chap. D.4). Analysis of stable carbon isotopes in the water column at Station A and considerations of mass balance and primary production (see below) result in a residence time of approximately 8 yr for the central (relatively low oxygen), deep waters of the Gulf.

D.4 Seasonality

The DCPE data have provided for the first time an opportunity to examine the seasonal pattern in the waters of the Gulf on a month-to-month basis. (Some data obtained by the Snapir expedition, 1967, are also included in the following computations.) The vertical distribution of temperature in the northern and southern sectors of the Gulf is given in Fig. D.20, illustrating the weak thermocline in summer, the homothermal stage in winter, and the differences between northern and southern sectors. The monthly variation of temperature and salinity for 1974–1977 is given in Table D.1. Temperature and salinity in the Straits of Tiran themselves are extremely important parameters for considerations of rate of evaporation, water transport and heat flux. Since, however, no detailed published data are available, the variables as given in Table D.1 are estimated from measurements at DCPE Stations G and H, just inside and outside the straits (Fig. D.9).

The annual variation of water temperature and salinity within the upper 300 m in the Gulf is given in Figs. D.21 and D.22. Note that for computing the data as given in Table D.1 and in Figs. D.21 and D.22, the interface between the northbound inflow and the southbound outflow is assumed to be located at about 100 m.

2 A recent evaluation (Murray et al. 1984) of moored current meter data obtained during a 30-day period in February–March 1982 indicates, however, for this latter period a discharge in each of the two layers of approx. 30,000 $m^3 s^{-1}$, the magnitude of the northflowing current being about 30 cm s^{-1} and that of the southflowing one 60 cm s^{-1}

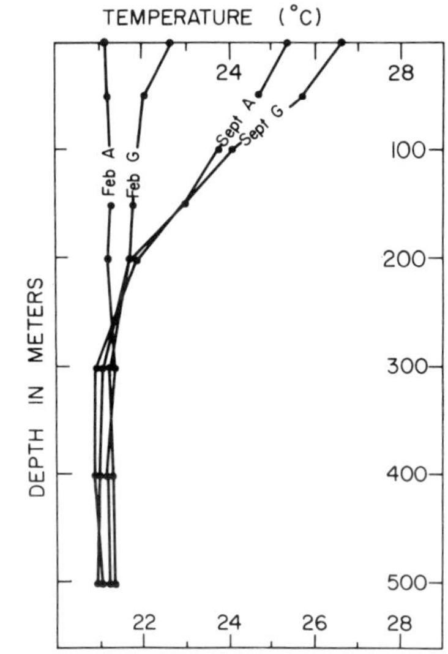

Fig. D.20. Vertical distribution of temperature in winter (February) and in summer (September) 1974–1977 for the north (DCPE-Station A) and for the south (DCPE-Station G) in the Gulf of Aqaba. Although the months of March and August are more typical for winter and summer conditions respectively, the data for February and September are more complete (Paldor and Anati 1979)

Each of the six variables plotted in Fig. D.21 displays its own seasonal pattern; there is a delay in the response of the deeper layers as documented by $\bar{T}$ and especially by T_2^s and T_2. In T_1^s, T_o, and $\bar{T}$ the minimal values occur in March. It is prominent in T_1^s and T_o, less so in $\bar{T}$. T_2^s and T_2, on the other hand, reach their minimal value about 2 months later.

The highest summer temperatures occur in July and August at the surface (T_o) and in July to October in T_1^s, but between August and November in $\bar{T}$, October–November in T_2^s, and November in T_2. The above-mentioned delay is therefore rather conspicuous. It probably results from a combination of an advective effect and the slow downward propagation of heat. The two peaks in T_1^s (July to October) coincide with the most frequent occurrence of southerly winds.

As with the temperature, low winter and high summer values of salinity are observed (Fig. D.22). The transition, however, occurs considerably later in the season (August, September instead of May, June, July) so that the spring warming occurs in water with low salinities typical of winter. In October S_1 is observed to be greater than S_2. However, S_1 and S_2 were calculated from 36 and 33 October measurements, respectively, and a close examination reveals that this inequality is consistent with the data and not due to a few erratic measurements. Furthermore, the ΔS anomaly in October ($\Delta S < O$) was detected at every station except G (see Table D.2). The average October ΔS, from Table D.2 is $-0.103‰$ with a standard deviation of 0.06%, and shows a marked trend for lower values with greater distance from the straits, a trend that was found to be a general feature at all seasons.

The salinities at the straits (interpolated as mentioned above) are not included in Fig. D.22 because of their erratic variations, combined with the fact that they

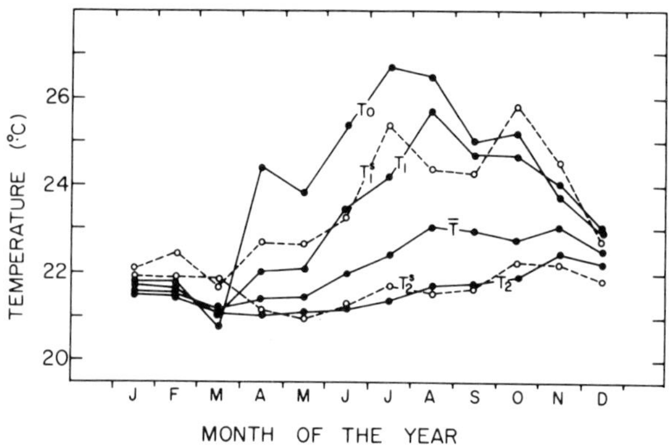

Fig. D.21. Annual variation of average surface temperature (T_0); average temperature of the uppermost 300 m ($\bar{T}_1$); temperature of the upper layer (T_1, T_1^s); lower layer T_2, T_2^s in the Gulf of Aqaba and in the Straits of Tiran, based on DCPE (1974–1977) and Snapir (1967) (Paldor and Anati 1979)

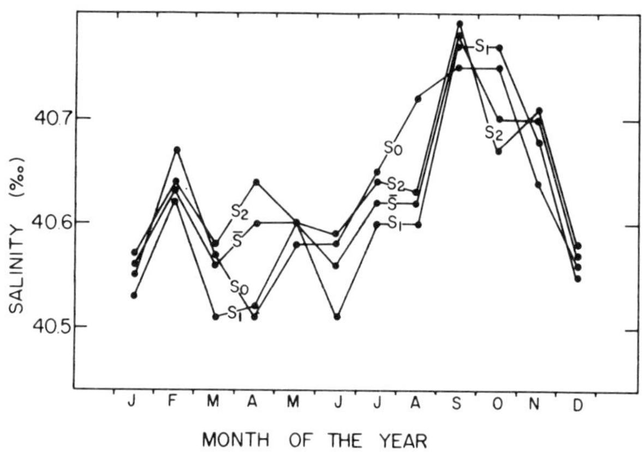

Fig. D.22. Annual variation of the surface salinity S_0, the average salinity of the uppermost 300 m, $\bar{S}$, the salinity of the upper layer S_1, and the lower layer S_2 (Paldor and Anati 1979)

are estimated rather than measured. High values (S_2^s 41‰ in August) and low values (S_1^s 40.3‰ in March) would appear in the diagram (see Table D.1).

The highest rate of change of salinity coincides with the highest surface temperatures. Although this does not take into account wind speed and air moisture, it is consistent with the fact that evaporation is highest when the saturation vapor pressure of the air at sea level is highest.

To observe the annual changes in the type of water of the upper mass (T_1, S_1), observations at all the stations for any given months are plotted on a common T-S diagram (Fig. D.23). The spring warming occurs at a lower salinity than the autumn cooling, so that the trajectory in the T-S space takes the shape of a crescent. August is definitely the warmest month and March the coldest. The area of each polygon can be taken as a measure of the horizontal inhomogeneity within the Gulf; it appears from Fig. D.23 that there is not much seasonal change in this

Table D.1. Monthly variation of temperature T and salinity S in the Gulf of Aqaba (Paldor and Anati 1979)

Months:	J	F	M	A	M	J
$\bar{T}$	21.57	21.81	21.10	21.43	21.43	22.01
T_0	21.79	21.51	20.75	24.39	23.82	25.28
T_1	21.68	21.59	21.06	22.05	22.05	23.39
T_2	21.51	21.46	21.13	21.01	21.05	21.17
T_1^s	22.16	22.45	21.73	22.70	22.68	23.21
T_2^s	21.83	21.84	21.90	21.38	21.10	21.32
$\bar{S}$	40.56	40.63	40.56	40.60	40.60	40.56
S_0	40.55	40.67	40.57	40.51	40.58	40.58
S_1	40.53	40.62	40.51	40.52	40.60	40.51
S_2	40.57	40.64	40.58	40.64	40.60	40.59
S_1^s	40.39	40.45	40.30	40.38	40.50	40.46
S_s^s	40.50	40.55	40.47	40.50	40.60	40.57
ΔS	0.11	0.10	0.17	0.12	0.10	0.11

Months:	J	A	S	O	N	D
$\bar{T}$	22.44	23.05	22.95	22.88	23.01	22.46
T_0	26.68	26.53	24.99	25.21	23.77	22.90
T_1	24.18	25.69	24.68	24.66	23.69	22.99
T_2	21.34	21.66	21.67	21.86	22.28	22.16
T_1^s	25.41	24.91	24.90	25.77	24.52	22.94
T_2^s	21.75	21.60	21.68	22.32	22.24	21.97
$\bar{S}$	40.62	40.62	40.78	40.70	40.70	40.57
S_0	40.65	40.72	40.75	40.75	40.64	40.56
S_1	40.60	40.60	40.77	40.77	40.68	40.55
S_2	40.64	40.63	40.79	40.67	40.71	40.58
S_1^s	40.38	40.81	40.67	40.59	40.53	40.43
S_2^s	40.54	41.03	40.68	40.67	40.65	40.56
ΔS	0.16	0.22	0.10	0.08	0.12	0.13

Overbars indicate averages in the upper 300 m over the whole basin. Subscripts 0, 1, 2, indicate surface values, upper 150 m averages, and 150 to 300 m averages. Superscript s indicates values at the Straits of Tiran

parameter, except for July and August. This may be an indication that the warm, summer Red Sea water undergoes enhanced evaporation and cooling, so that the north end is all the more saline and colder than the south end of the Gulf of Aqaba.

The crescent shape of the yearly trajectory in the T-S space is even clearer when single stations are plotted separately; Fig. D.24 compares the north Red Sea proper (Station H) with the north of the Gulf (Station A), displaying the narrower range of annual variation at A. The overall salinity difference of about ΔS 0.3% as seen in the figure, implies a **time lag**

$$\tau = \Delta D\, S/ES$$

between H and A along the surface. With a depth of $D = 150$ m and an evaporation rate of $E = 1$ cm d^{-1}, a time lag of $\tau = 4$ months is calculated.

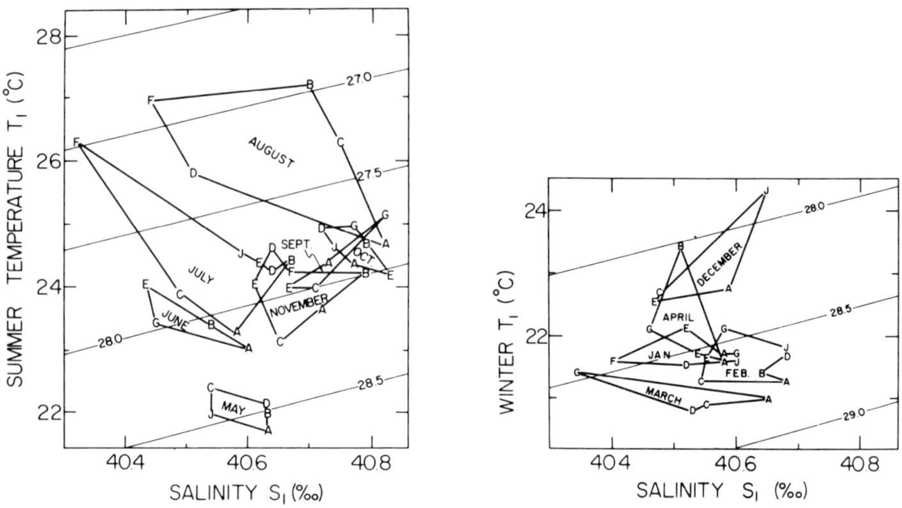

Fig. D.23. T-S diagram of the upper water mass (T_1, S_1), for summer and winter in the Gulf of Aqaba. *A-J* DCPE stations; *thin diagonal lines:* lines of constant σt; *thick lines* define a closed polygon for each month around the types of water found at each site. Note the low temperature in the north (Station A) in the early summer (May to August) and the high salinity there in winter (January to April) (Paldor and Anati 1979)

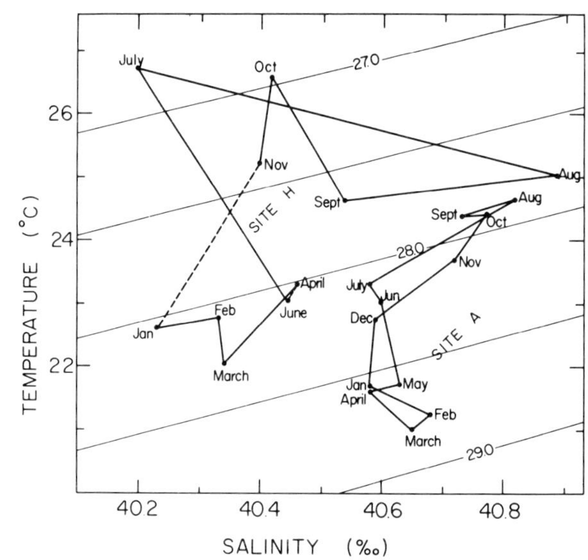

Fig. D.24. T-S diagram comparing the annual evolution of the upper water mass (T_1, S_1) within the Gulf of Aqaba (Station A) with that just outside the Gulf (Station H). *Thin diagonal lines:* lines of constant σt. Note the typical "crescent" shape and the tendency for lower temperatures and higher salinities at Station A. The type of water for Station H in August is based on one station only and is probably exceptionally high in salinity. No measurements at Station H are available for February or May (Paldor and Anati 1979)

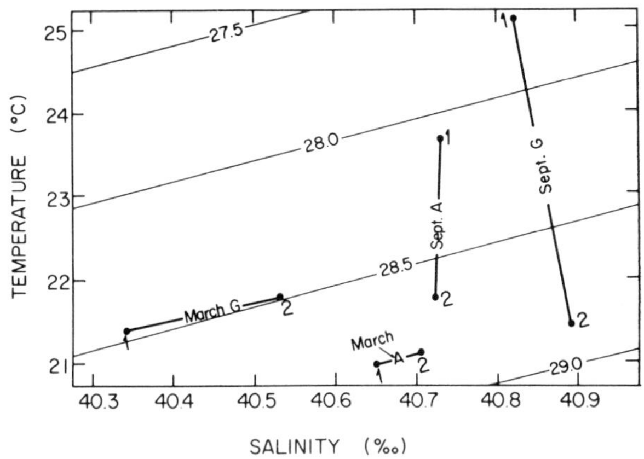

Fig. D.25. Water types in the upper *(1)* and lower *(2)* layers in the Gulf of Aqaba. Station A is at the northern end of the Gulf, Station G at the southern end (near the Strait of Tiran). The upper layer values are computed from the standard depths 0, 50 and 100 m, and the lower layer values from the standard depths 150, 200 and 300 m. (Based on DCPE data, 1974–1977) (Paldor and Anati 1979)

Since τ is less than about ½ of the residence time it would follow that the latter is in the upper waters twice as long as that calculated from the 1974/1975 survey alone.[3]

An independent check is provided by the heat-balance equation.

$$\tau = D\,\Delta T\,C_p/H,$$

which, considering a heat flux $H = -114\ \mathrm{W\ m^{-2}}$, derived from meteorological observations, gives an overall temperature difference of

$$\Delta T = -1.88\ ^{\circ}\mathrm{C}$$

(in agreement with Fig. D.24).

The 3 years of data show again that summer temperatures in the water column prevail over 5 to 6 months, April to October, while a typical winter period is relatively short (January, February, and March). The peaks in winter or summer temperature occur at different times depending on the depth, with a delay in the deeper layers. The temperatures in the north are consistently lower than in the south. However, Station B is an exception to this rule with its occasional high surface temperatures, particularly in the spring (max. 29.2 °C in May).

The period of high summer salinity is shorter (2 to 3 months) and occurs later in the season than that of the high summer temperature. The narrow overall range of salinities is illustrated in Fig. D.23. Considering the small size of the Gulf or, equivalently, the short residence time of its water, this narrow range is not surprising. The highest salinities were at Station A; from among 17 measurements with

3 On the other hand, a model based on a synoptic CTD survey, carried out by J. Erez, B. Krumgalz and A. Hecht in February 1982 in the Gulf of Aqaba and in the northern Red Sea, indicates a time lag between stations H and A in the order of $\tau = 72$ **days** for a layer of 100 m depth

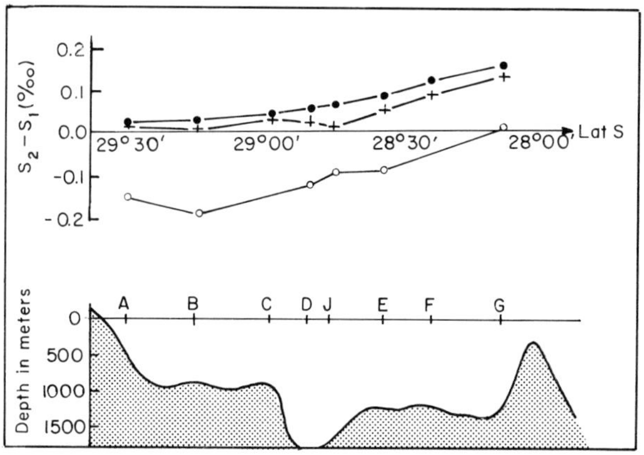

Fig. D.26. Variation of $\Delta S = S_2 - S_1$ along the main axis of the Gulf of Aqaba (roughly north–south). *Solid circles:* average for February, March, April; *crosses:* average for June, July, August, September; *open circles:* October; *shaded area:* bottom relief. (Based on DCPE data, 1974–1977) (Paldor and Anati 1979)

salinity about 41‰, 16 were at Station A; the only exception (41.02‰) was at Station E. The highest salinity was at Station A in August 1975 (41.57‰) and the lowest at Station B on November 1974 (40.03‰).

Figure D.25 displays on a T-S diagram the points (T_1, S_1) and (T_2, S_2) for north and south, winter and summer. Station A (north) is vertically more homogeneous than Station G (south), both in winter and summer, as would be expected from circulation models of semi-closed basins in arid zones. Note also the smaller overall ranges of temperature and salinity at Station A than at Station G and the changing character of the stratification between winter and summer: the winter column is mainly salinity-stratified, while the summer column is mainly temperature-stratified (see above).

From the two-layer thermohaline circulation model, the difference in salinity between lower and upper layers, $\Delta S = S_2 - S_1$, is expected to be (1) positive (evaporation exceeds precipitation), (2) decreasing toward the north (further away from the straits), (3) decreasing roughly linearly (uniform evaporation rate), and (4) nearly vanishing at the northern end (where the deep water is presumably formed).

To compare these features with observations, ΔS was computed for the months of February, March, and April, with missing sites interpolated, and then averaged over the 3 months. The same procedure was applied to the months of June, July, August, and September (Fig. D.26). The salinity difference ΔS does indeed display the expected properties. [The as yet unexplained negative ΔS values in October (Table D.2) are added for comparison.]

As a result of the thermohaline circulation pattern in the Gulf of Aqaba, as determined by climatic conditions and the presence of a narrow sill at Tiran, the Gulf is actually filled by **upper Red Sea waters.** These waters undergo seasonal modifications within the Gulf in accordance with wind stress, evaporation rates, heat flux, as well as by distance from the straits and from the shores. Thus, the

Table D.2. The salinity of the upper layer, S_1, the salinity of the lower layer, S_2, and their differences $\Delta S = S_2 - S_1$ for six DCPE stations in the Gulf of Aqaba in the month of October (Paldor and Anati 1979)

DCPE station	S_1	S_2	$\Delta S = S_2 - S_1$
A	40.770	40.623	−0.147
B	40.788	40.604	−0.184
D	40.722	40.604	−0.118
J	40.742	40.647	−0.094
E	40.832	40.744	−0.089
G	40.767	40.780	+0.013

nutrient regime in the Gulf is dependent upon the inflowing Red Sea waters which by themselves represent upper Gulf of Aden waters flowing in through the Straits of Bab-el-Mandeb. These facts, as well as the rapid exchange of waters over the sill of Tiran and their short residence time within the Gulf, have far-reaching ecological implications.

References: Anati (1974, 1976, 1980a, b), Anati et al. (1977), Assaf and Anati (1974), Assaf and Hecht (1974), Assaf and Kessler (1976), Deacon (1952), Hecht and Anati (1983), Hottinger (1977b), Klinker et al. (1975, 1976), Luksch (1901), Mergner and Schuhmacher (1974), Mohamed (1940), Morcos (1970), Murray et al. (1984), Oren (1962), Paldor and Anati (1979), Reiss (1977), Reiss et al. (1976, 1977), Shemesh (1980), Shemesh et al. (1981).

D.5 Nutrients

The range of phosphate, nitrate and silicate values as determined from 300 DCPE water samples collected during 1974/1975 in the Gulf of Aqaba and in the northernmost Red Sea is summarized in Table D.3, according to the two major seasons (April–October and January–March).

Generally, the nutrient values obtained at Station H in the Red Sea proper (Fig. D.9) are well within the range of those previously recorded from the northern Red Sea. These values reflect the continuous decrease in nutrient concentration in the Red Sea from south to north, a decrease which is particularly dramatic in the upper 100 m north of latitude 22° N, where nutrient concentration (in summer) is about 1/5–1/10 of that in the southern Red Sea, near the Straits of Bab-el-Mandeb.

There is no deep-water upwelling in the northernmost Red Sea which could lead to an appreciable enrichment in nutrients of the upper waters in the Gulf of Aqaba. However, a very short period (May–June) of northern Red Sea mid-water upwelling seems to be reflected in somewhat higher productivity during July–August in the southern and middle Gulf of Aqaba (see Chapt. D.4).

Table D.3. Maximum and minimum values (μg at l^{-1}) of nutrients in the Gulf of Aqaba and northernmost Red Sea, 1974–1975 (Klinker et al. 1978)

Nutrient and season	Stations A–G (Gulf of Aqaba)								Station H (Red Sea)					
	Maximum	Station	Depth (m)	Month	Minimum	Station	Depth (m)	Month	Maximum	Depth (m)	Month	Minimum	Depth (m)	Month
PO_4–P														
Summer	0.61	E	300	Oct.	0.03	A	100	Apr.						
						B	100	June						
						G	100	Apr.						
Winter	0.99	F	600	Jan.	0.03	A	150	Jan.	1.98	600	Mar.	0.08	50	Jan.
						E	50	Feb.						
NO_3–N														
Summer	7.19	B	600	Apr.	0.07	B	100	Apr.						
Winter	5.67	D	1,000	Jan.	0.09	A	200	Feb.	16.29	800	Jan.	0.46	50	Mar.
						A	300	Feb.						
Si														
Summer	4.90	A	500	Apr.	0.03	A	100	May						
						G	150	May						
						G	0	Apr.						
Winter	6.46	A	400	Feb.	0.06	B	100	Feb.	14.80	1,300	Jan.	0.11	700	Jan.

Except the episodic summer inflow into the Gulf of slightly richer in nutrients (and dense) Red Sea waters at still depth there is no connection between the deep waters of the Red Sea and the Gulf of Aqaba. Thus, most of the waters flowing through the Straits of Tiran into the Gulf are initially nutrient-depleted **Red Sea upper waters.** As a result, nutrient content within the Gulf of Aqaba is generally very low (cf. also Table D.3).

The average concentrations of phosphate, nitrate, and silicate in the Gulf of Aqaba during the two major seasons in 1974/1975 are shown in Figs. D.27–D.29 and D.31–D.33. (The water density distribution, as well as the circulation pattern and oxygen levels, are shown again for comparison in Figs. D.30 and D.34.)

Phosphate concentration is generally low. In the northern sector the upper 200 m water mass is similar with regard to phosphate content both in summer and winter. In the southern sector higher values of phosphate are present in winter. The central deep waters in the northern sector in summer are richer in phosphate than in winter.

Nitrate values are also low within the Gulf of Aqaba throughout the year. The upper 300–400 m have similar average values in summer and in winter in the northern sector, while higher values are present between 200 and 300 m in the southern sector. Generally, the deep waters are richer in nitrate in summer than in winter. (Ammonia levels reach less than 2 $\mu g\ l^{-1}$ and are, therefore, below those detectable by standard methods.)

With regard to **silicate**, the upper 400 m in the northern sector have a similar content throughout the year. In the southern sector, the upper 100 m have similar silicate content in summer like the upper waters in the north, while the deeper, 100–400 m waters of the southern sector are richer in silicate. In winter the whole water column down to 400 m in the southern sector is richer in silicate than in the northern one. No appreciable changes in silicate content occur in the deep water mass, except during the diatom bloom in March.

The nutrient isopleths, especially of phosphate and nitrate, indicate a close correlation with the circulation pattern, both seasonally and geographically.

The boundary between the very nutrient-poor upper waters and the relatively richer, main deep water mass is situated in summer at approximately 150–200 m depth in the southern sector and between 350 and 550 m in the northern one. This boundary closely parallels the trend of the 28.9 σt isopycnal, the 40.7% isohaline and 4.6 ml l^{-1} oxygen isopleth. The relatively high nutrient content of the deep waters of the Gulf immediately inside the sill during summer is probably connected with the apparently transient, deep inflow of Red Sea waters at sill depth (see Chap. D.3).

During winter, the difference between the southern and the northern sector is more evident, the upper waters being much richer in nutrients in the more stratified southern sector than in the nearly homopycnal northern one. Data obtained at Stations D and E indicate that at the same depth nutrient content is higher toward the Arabian side than the Sinai side of the Gulf (cf. Chap. D.3). Generally, the disposition of nutrient isopleths strongly suggests a spinning movement of the deep water mass, as also indicated by temperature and oxygen values (see Chap. D.4). Horizontal advection of aging deep waters from the central part of the Gulf is indicated by total inorganic dissolved carbon, oxygen content, primary produc-

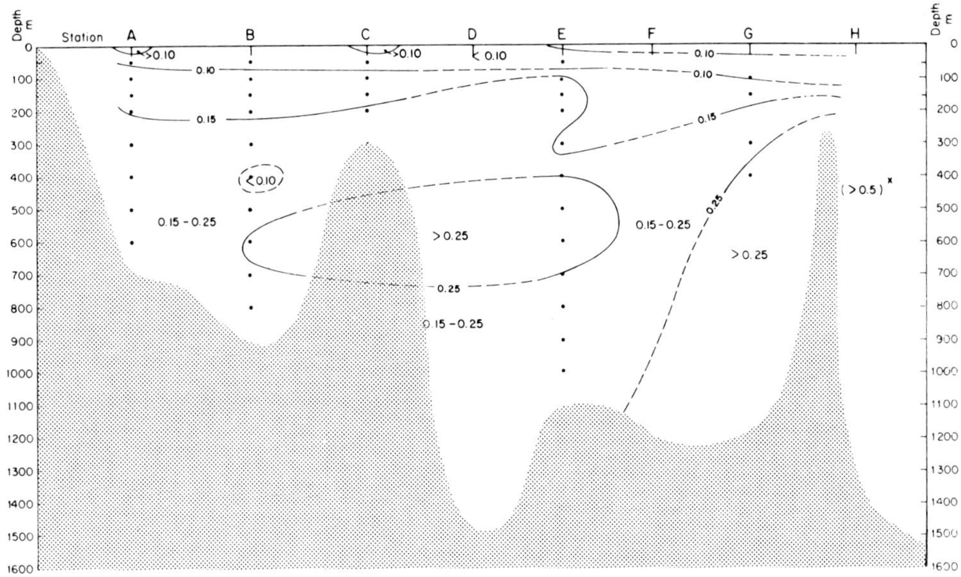

Fig. D.27. Averaged isopleths of PO_4-P (μg at l^{-1}) summer 1974. *Shaded area:* simplified bottom profile along station to station track (Klinker et al. 1978)

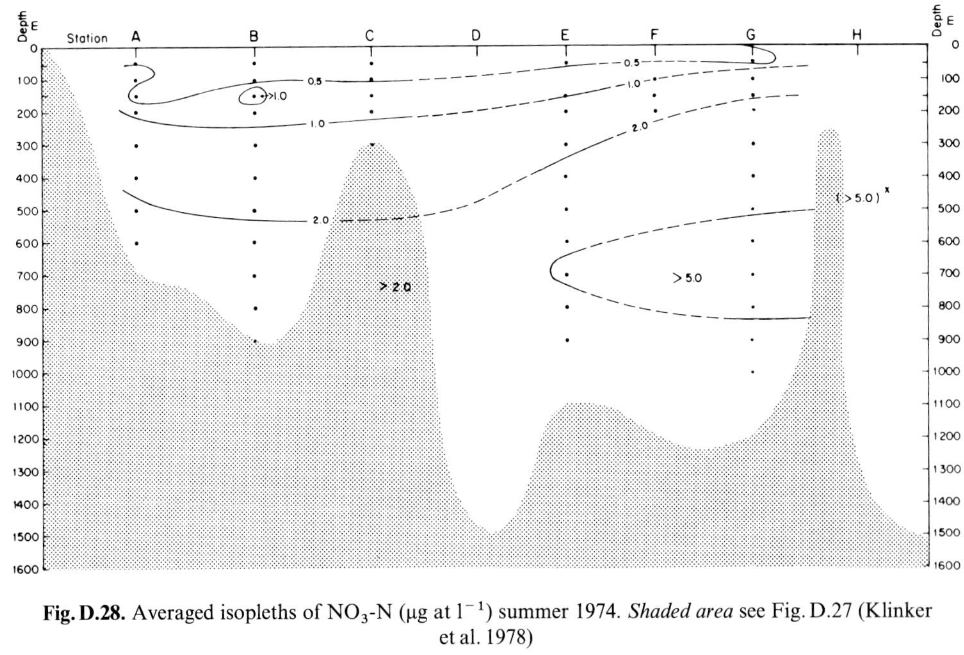

Fig. D.28. Averaged isopleths of NO_3-N (μg at l^{-1}) summer 1974. *Shaded area* see Fig. D.27 (Klinker et al. 1978)

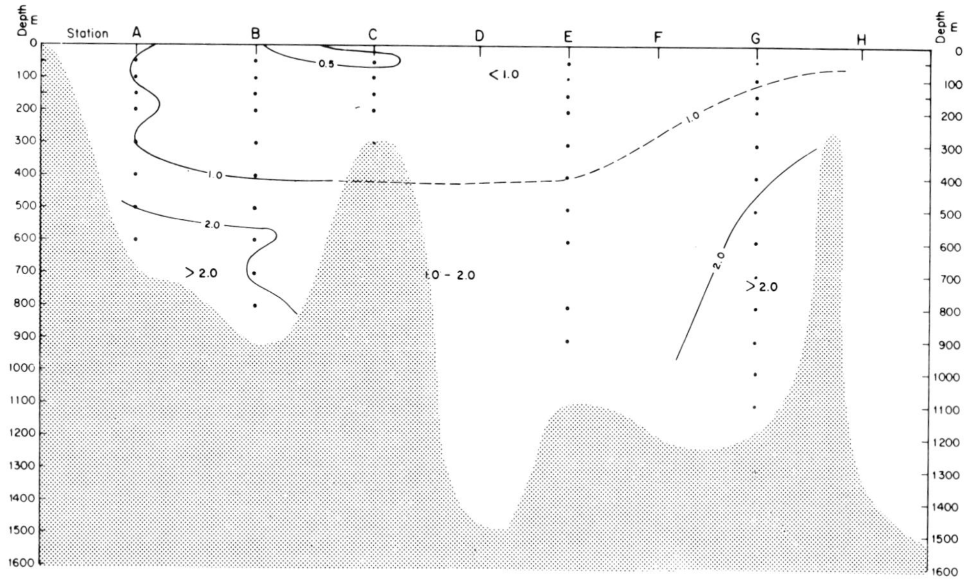

Fig. D.29. Averaged isopleths of SiO_4-Si (µg at l^{-1}) summer 1974. *Shaded area* see Fig. D.27 (Klinker et al. 1978)

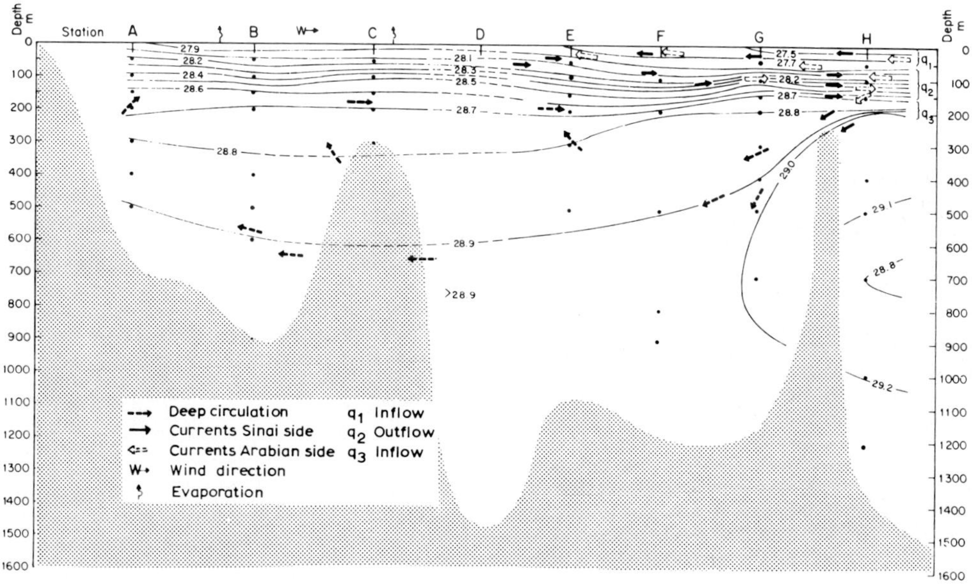

Fig. D.30. Circulation pattern during summer 1974. *Shaded area* see Fig. D.27 (Klinker et al. 1976)

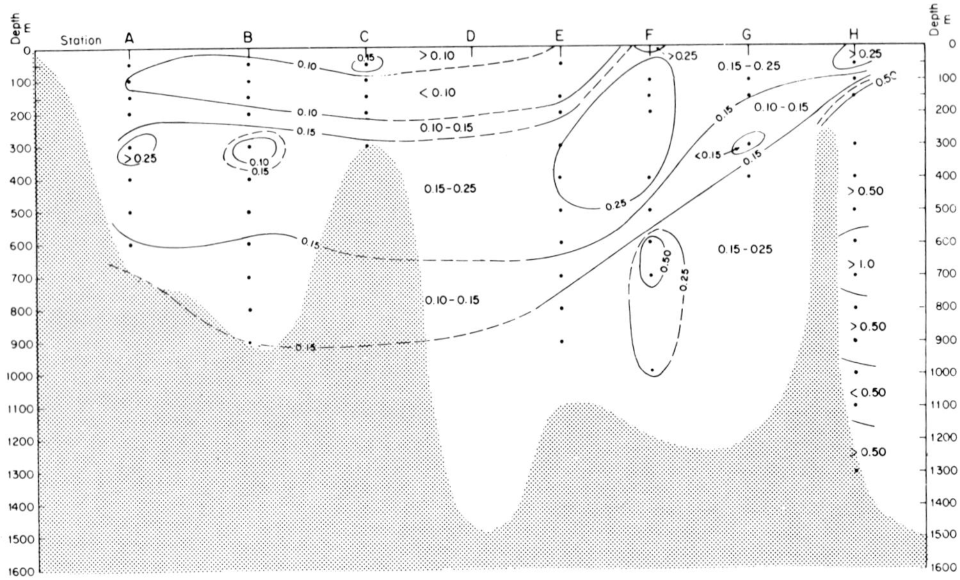

Fig. D.31. Averaged isopleths of PO_4-P (µg at l^{-1}) winter 1975. *Shaded area* see Fig. D.27. (Klinker et al. 1978)

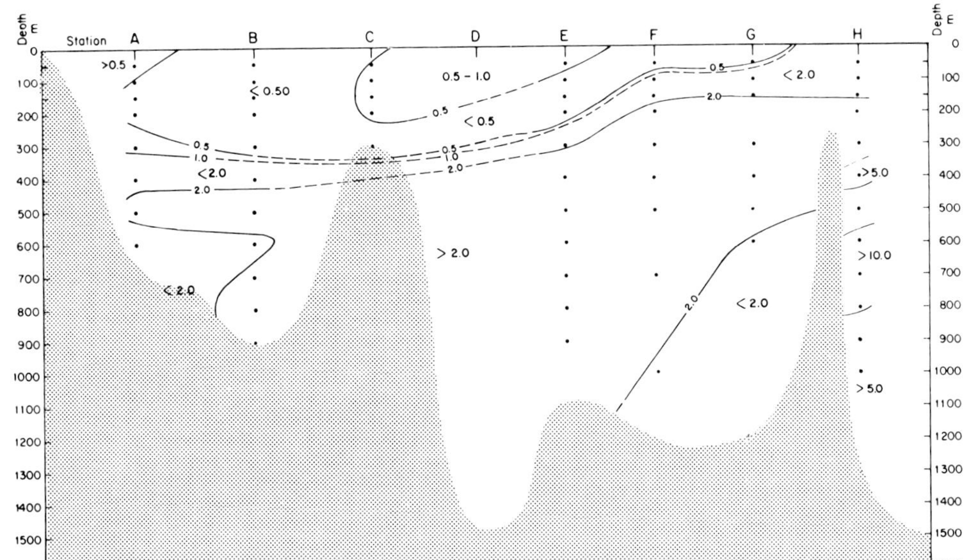

Fig. D.32. Averaged isopleths of NO_3-N (µg at l^{-1}) winter 1975. *Shaded area* see Fig. D.27 (Klinker et al. 1978)

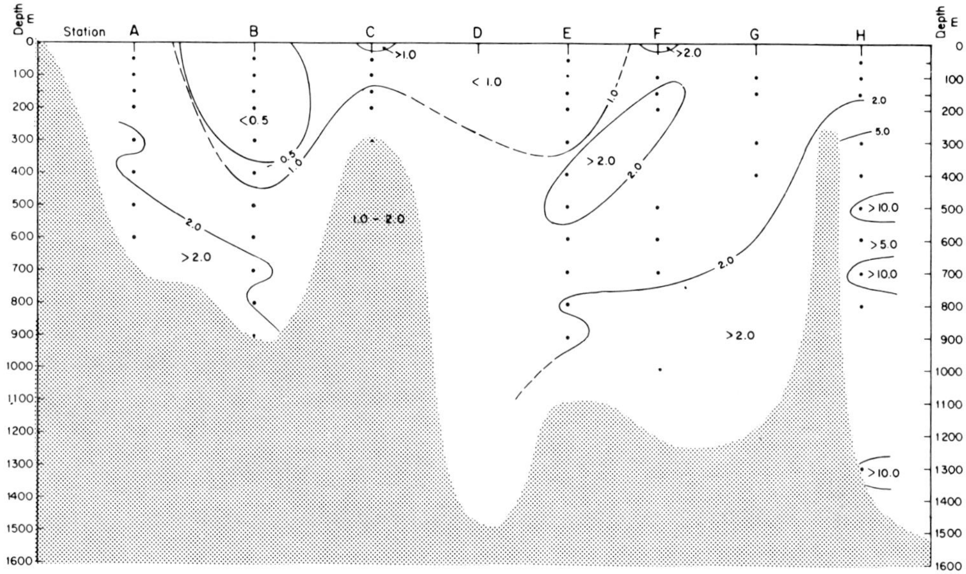

Fig. D.33. Averaged isopleths of SiO_4-Si (µg at l^{-1}) winter 1975. *Shaded area* see Fig. D.27 (Klinker et al. 1976)

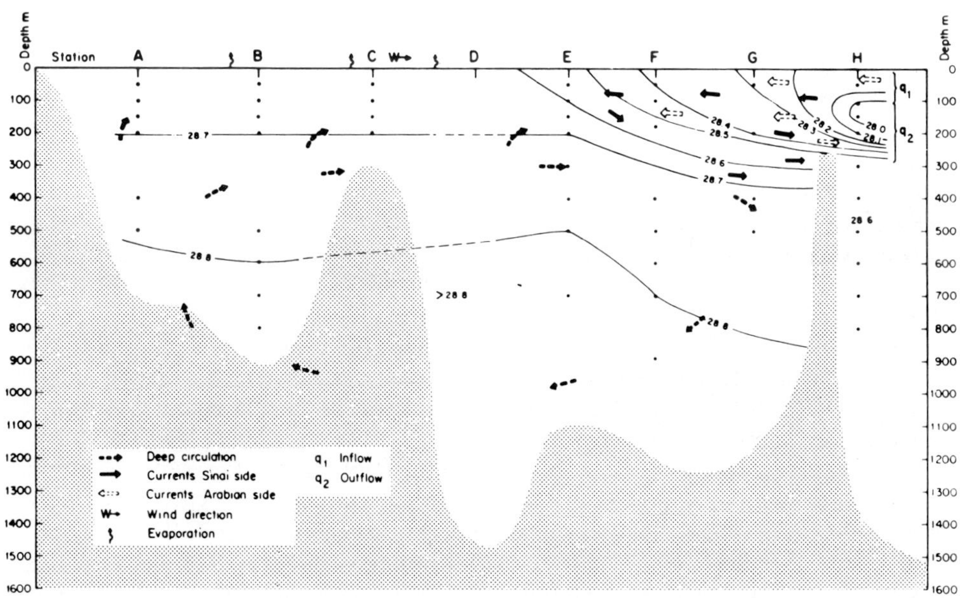

Fig. D.34. Circulation pattern during winter 1975 (Klinker et al. 1978)

tion and $\delta^{13}C$ of the deep waters at Station A. The rise of nutrient isopleths at the northern end of the Gulf follows that of isohalines and of oxygen isopleths, indicating upwelling in the north.

The differences between the waters with short residence time in the Gulf of Aqaba and those with long residence time in the Red Sea are well expressed by the nutrient and oxygen content of the deep water masses. While in the northernmost Red Sea nutrient content of the deep waters is relatively high, the highest values occurring with strong oxygen minima (see Figs. D.27–D.34), the deep waters of the Gulf of Aqaba are poor in nutrients and the oxygen content is not less than 3.75 $ml\,l^{-1}$.

The influence of nutrient content of the waters in the Gulf of either the effluents from the harbor towns of Elat and Aqaba or of the activity of the abundant higher plants in the coastal (0–70 m depth) "*Halophila* lawns" is not yet known in detail. Episodic extremely high phosphate values (more than 2.5 $\mu g\,l^{-1}$) throughout the column down to 600 m depth at Station A are definitely attributable to air-borne particulate matter brought into the sea from the phosphate fertilizer-loading facilities in Aqaba and Elat; these values are not considered here.

The seasonality of nutrient distribution in the upper 200 m is exemplified by data obtained at Station A during 1974/1975 (Figs. D.35 and D.36). (Note that upwelling influences this northernmost station.) The stratification within the upper water mass in summer is evident both in nitrates and phosphate content with high values in deeper layers at the end of the stratification period. A more homogeneous distribution is observed during winter, especially during the homothermal mixing stage in January–February (see also Fig. D.37). Three major depth boundaries within the upper waters can be recognized – dependent on season – by nutrient distribution, viz. at about 50 m, 100 m, and 150–170 m.

The seasonal depth profile of the ratio nitrates/phosphate is shown in Fig. D.38. The major season-dependent depth zones are again recognizable. Before the onset of winter mixing, the N/P ratio above 100 m is not greater than 8, while below 100 m it reaches 20. Ratios are much higher throughout the column at the beginning of the winter, especially below 100 m depth in December–January. In February–March high N/P ratios occur above 100 m. A prominent increase in the N/P ratio between 75 and 120 m is observed in June (cf. also Figs. D.35 and D.36).

There is no published information on the **details** of nutrient distribution in the uppermost 25 m of the water column which are, however, of interest to various aspects of primary production, photoinhibition, pollution effects, etc. Data on near-shore nutrient conditions are also very scanty. Measurements carried out in May 1976 in the upper 3 m water in the coral reef of the "Japanese Gardens", about 5 km south of the town of Elat; in the reef of the so-called "Fjord" at Marsa Murach, about 20 km south of Elat; and in the hypersaline Mangrove of Shurat Arwashiye north of Nabq, about 170 km south of Elat, indicate extreme nutrient-poor conditions (Table D.4) of the same order of magnitude as in the open surface waters. Phosphate and nitrate values obtained in May 1973, between the MBL and Wadi Taba, in coral reef and "*Halophila* lawn" areas, at depths between 12 and 70 m, are equally low and correspond to those obtained in the shallow reef and mangrove areas mentioned above (see also p. 167).

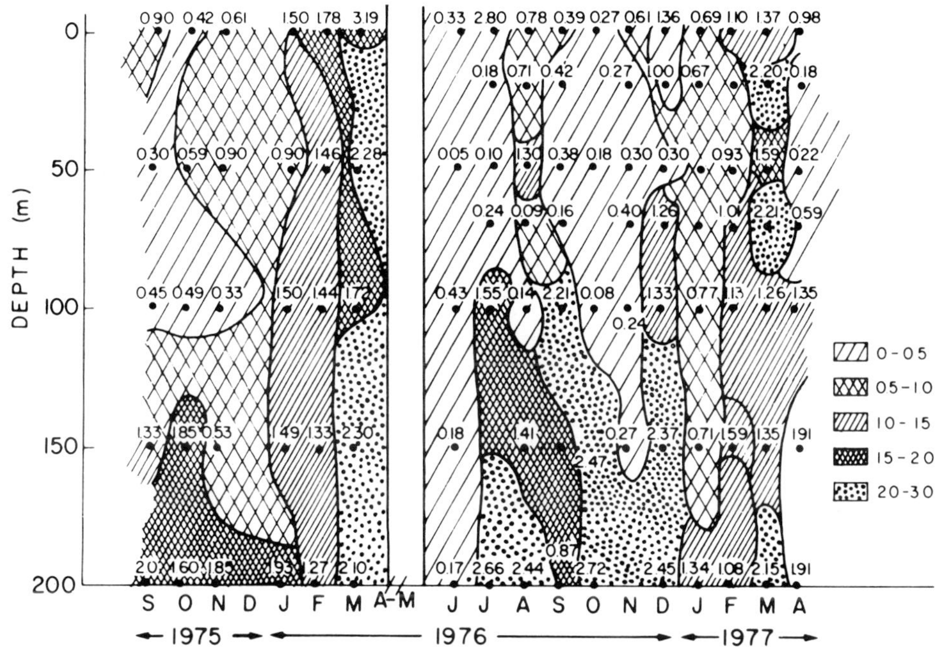

Fig. D.35. Seasonal and depth profile of nitrates (μg at l^{-1} NO_3+NO_2-N) in the upper 200 m at Station A during 1975–1977 (Levanon-Spanier et al. 1979)

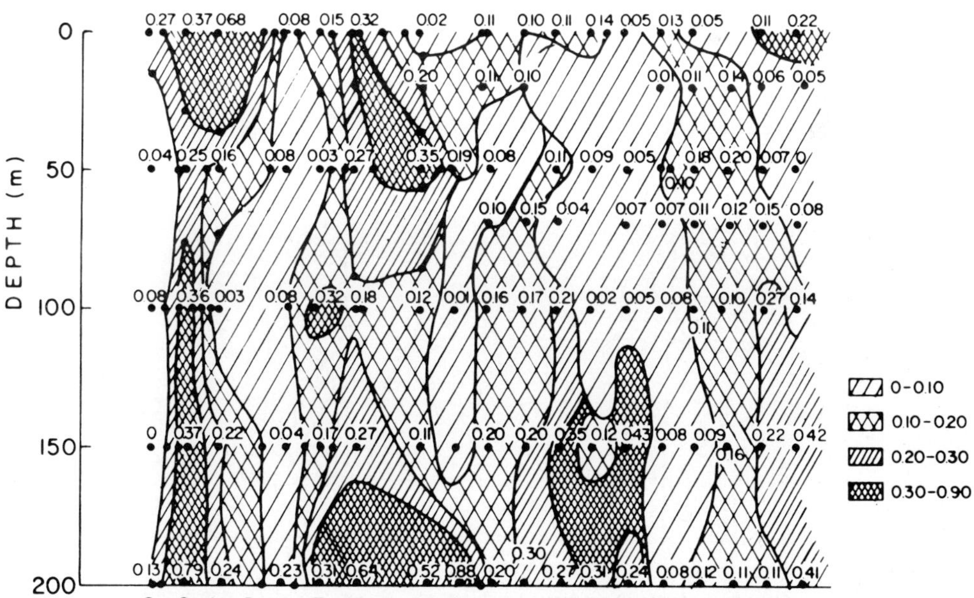

Fig. D.36. Seasonal and depth profile of phosphate (μg at l^{-1} PO_4-P) in the upper 200 m at Station A during 1975–1977 (Levanon-Spanier et al. 1979)

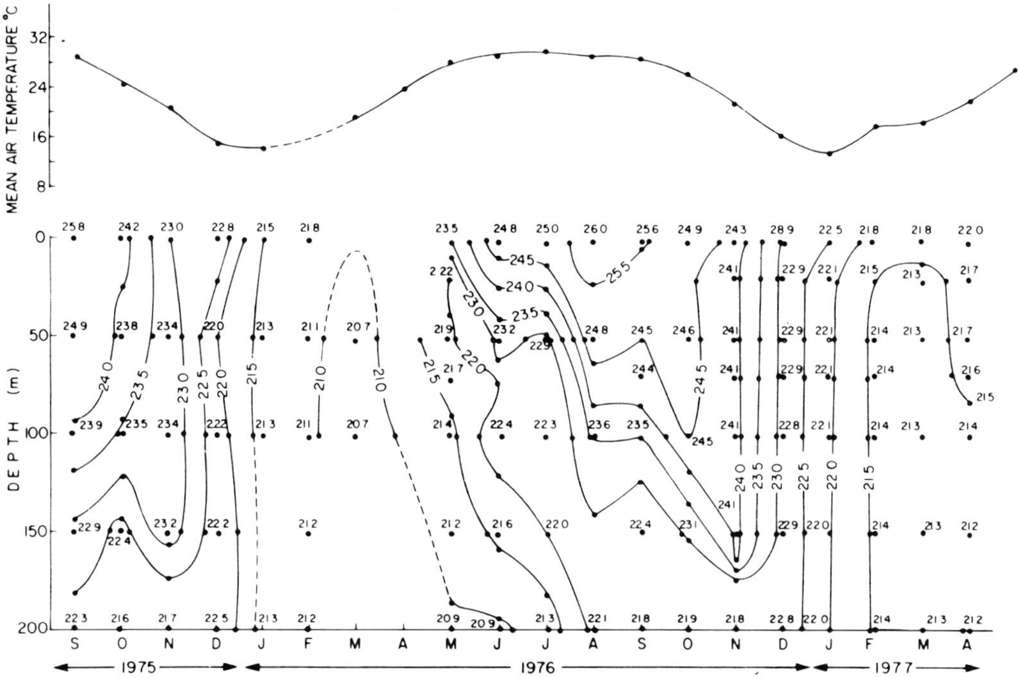

Fig. D.37. Seasonal and depth variations of temperature (°C) in the upper 200 m at Station A during 1975–1977 (Levanon-Spanier et al. 1979)

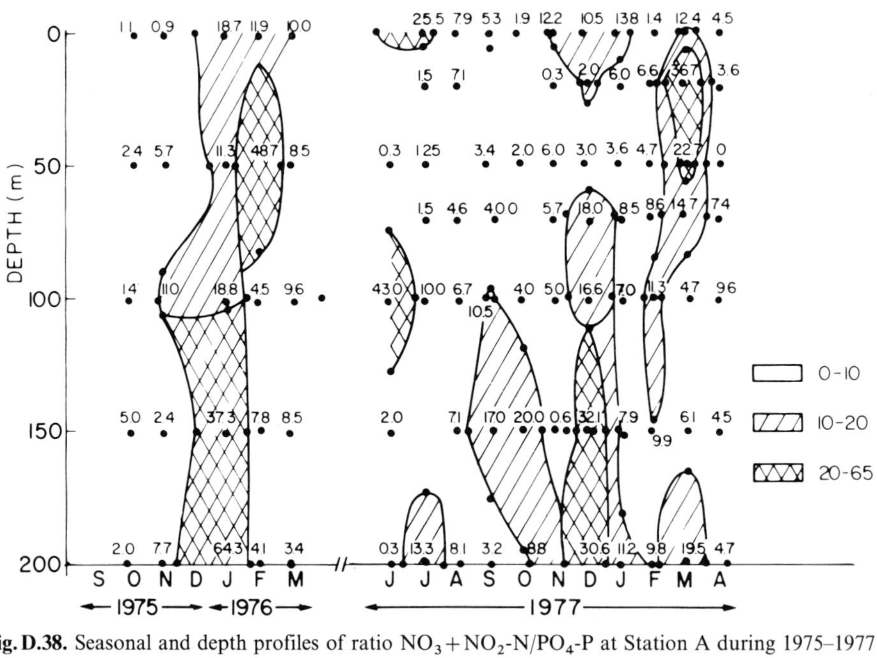

Fig. D.38. Seasonal and depth profiles of ratio NO_3+NO_2-N/PO_4-P at Station A during 1975–1977 (Levanon-Spanier et al. 1979)

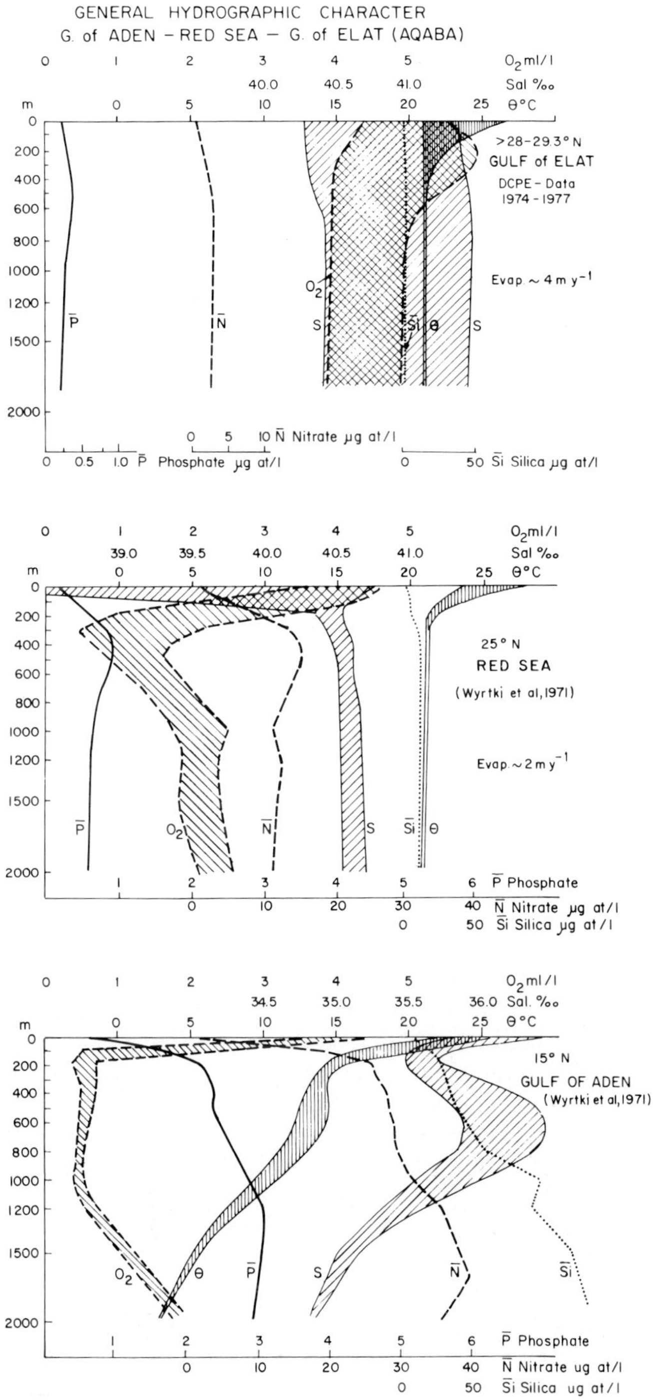

Fig. D.39. Generalized comparison of hydrographic characters and nutrient levels of the Gulf of Aden, Red Sea and Gulf of Aqaba (Halicz and Reiss 1981)

Table D.4. Nutrient levels in reef and mangrove areas, May 1976 (Sournia 1977)

Nutrient	Coral reef: "Japanese Gardens"	Coral reef: the "Fjord"	Mangrove pool near Nabq
N–NO_3, µg at. l^{-1}	$0.15 \pm$ [a]	0.36 ± 0.10	$0.22 \pm$ [a]
N–NO_2, µg at. l^{-1}	0.02 ± 0.02	0.02 ± 0.02	$0.01 \pm$ [a]
P–PO_4, µg at. l^{-1}	0.08 ± 0.05	0.03 ± 0.02	$0.25 \pm$ [a]
Si–SiO_4, µg at. l^{-1}	2.55 ± 2.53	1.17 ± 0.25	1.29 ± 0.33

[a] Confidence level not determined

The decrease in nutrient content along the stretch Gulf of Aden–Red Sea–Gulf of Aqaba is shown schematically – together with the changes in hydrographic characters – in Fig. D.39. While nutrient levels in the Gulf of Aden and in the southern Red Sea are those of productive waters, nutrient concentration in the northernmost Red Sea and in the Gulf of Aqaba are those of **oligotrophic waters.**

References: Halicz and Reiss (1981), Klinker et al. (1978), Krey and Babenerd (1976), Levanon-Spanier et al. (1979), Neumann and McGill (1962), Oren (1962), Reiss (1977), Reiss et al. (1976, 1977 a), Shemesh (1980), Shemesh et al. (1981), Sournia (1977), Wyrtki (1971).

D.6 Primary Production

Depth profiles of **primary production** and **chlorophyll** a in the upper 200 m at different DCPE stations (Fig. D.9) and of different seasons are shown in Fig. D.40.

The values indicate the differences between the low productive summer season (stratification) and the more productive winter period (mixing), as well as between the low productive northern sector of the Gulf of Aqaba on one hand, and the relatively more productive southern sector of the Gulf and northernmost Red Sea proper, on the other. This pattern thus closely follows that of the circulation-determined nutrient distribution. Figures D.41 and D.42 show the seasonality of chlorophyll a and primary production respectively in the upper 200 m at DCPE-Station A during 1975/1977.

Chlorophyll a concentrations range between 0.02 and 0.45 mg m^{-3} and chlorophyll a is observed to a depth of at least 170 m. During autumn and winter (November–March) the highest chlorophyll a concentrations (0.25–0.4 mg m^{-3}) are nearly evenly distributed throughout the water column down to 170 m. During summer (April–October) the lowest values (0.02–0.05 mg m^{-3}) were observed, although during May–June a high concentration (0.39 mg m^{-3}) was observed between 50 and 100 m. Generally, the depth boundaries of chlorophyll a follow those indicated by the nutrients.

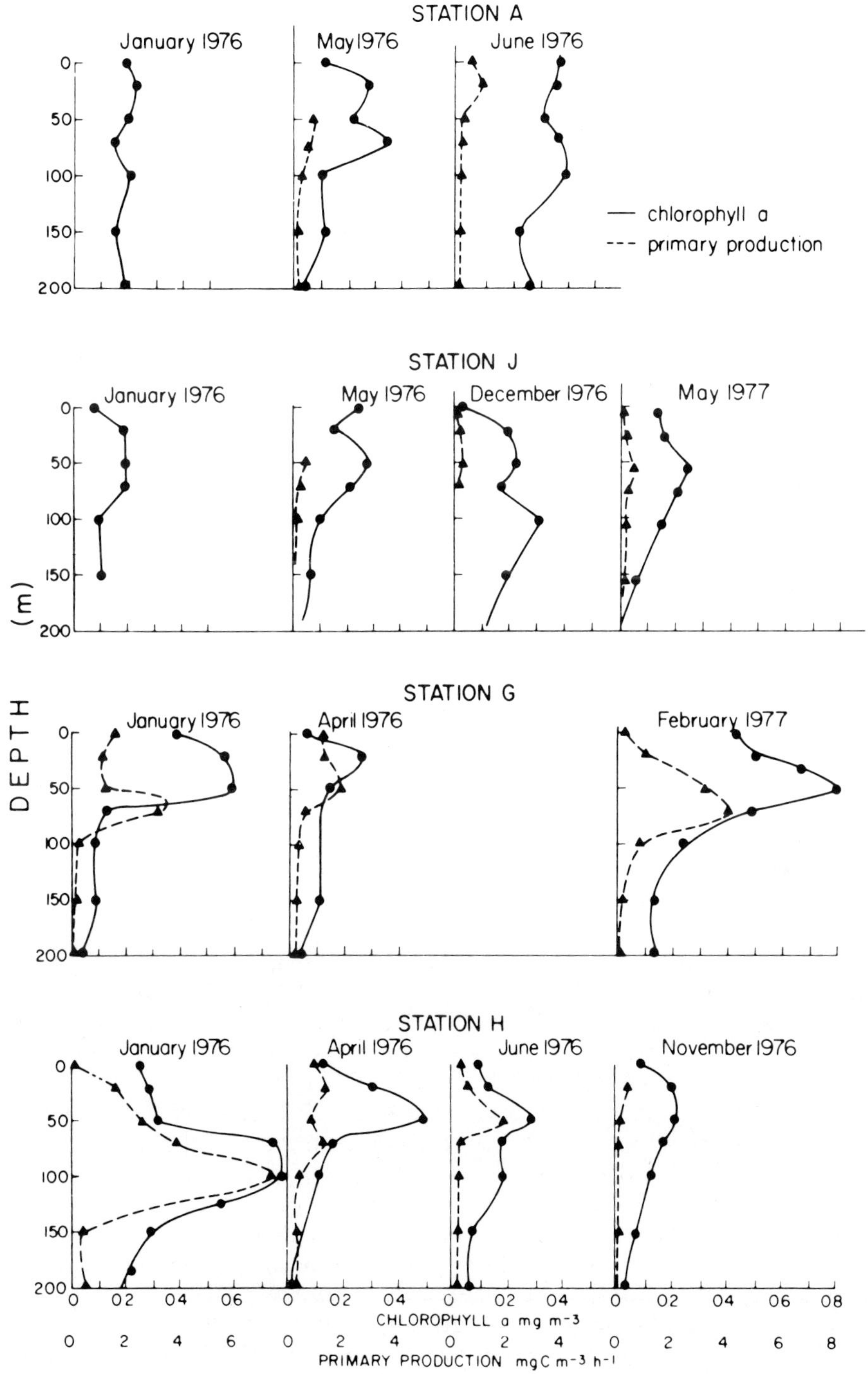

Fig. D.40. Depth profiles of chlorophyll a (mg m^{-3}) and primary production (mg C m^{-3} h^{-1}) at DCPE-Stations A, J, G, and H, Gulf of Aqaba during 1976–1977 (Levanon-Spanier et al. 1979)

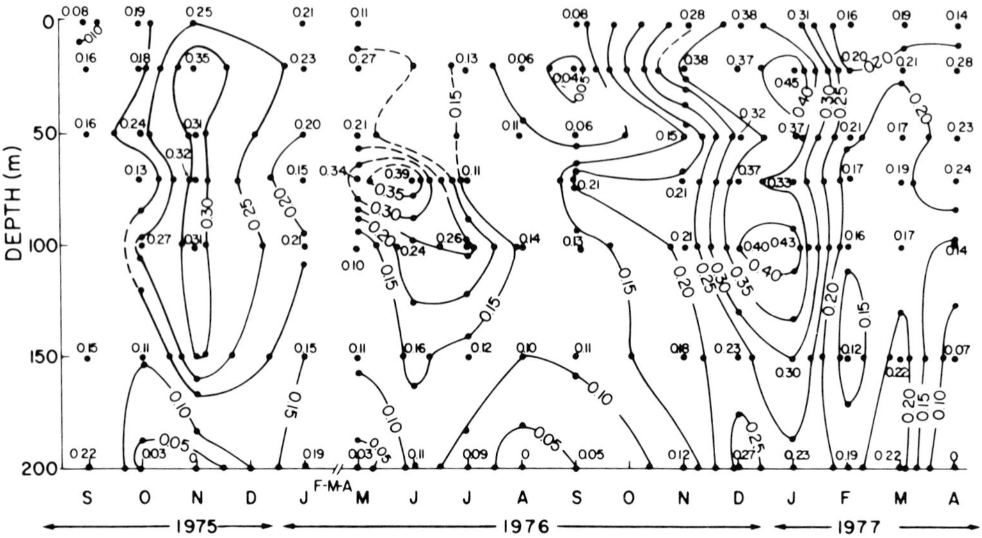

Fig. D.41. Seasonal depth profile of chlorophyll a (mg m^{-3}) at DCPE-Station A during 1975–1977 (Levanon-Spanier et al. 1979)

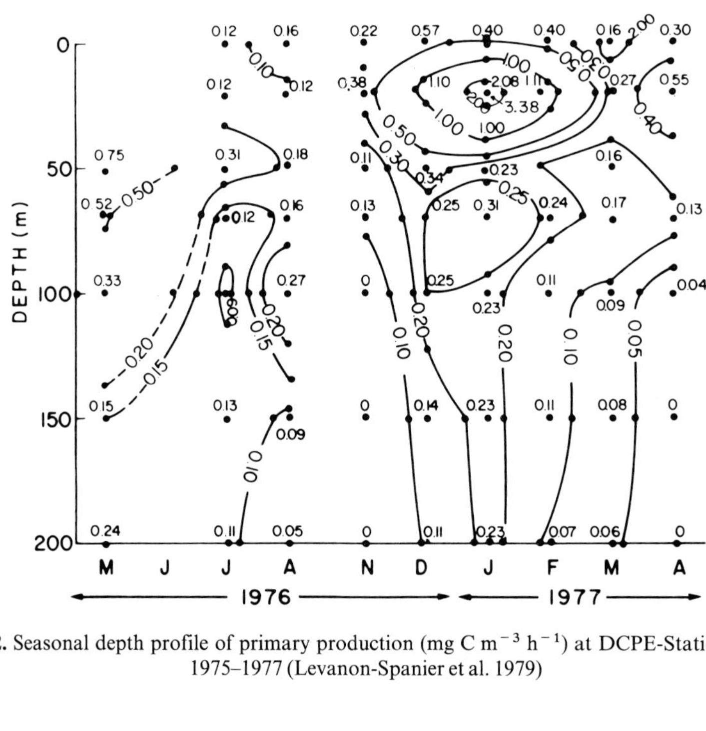

Fig. D.42. Seasonal depth profile of primary production (mg C m^{-3} h^{-1}) at DCPE-Station A during 1975–1977 (Levanon-Spanier et al. 1979)

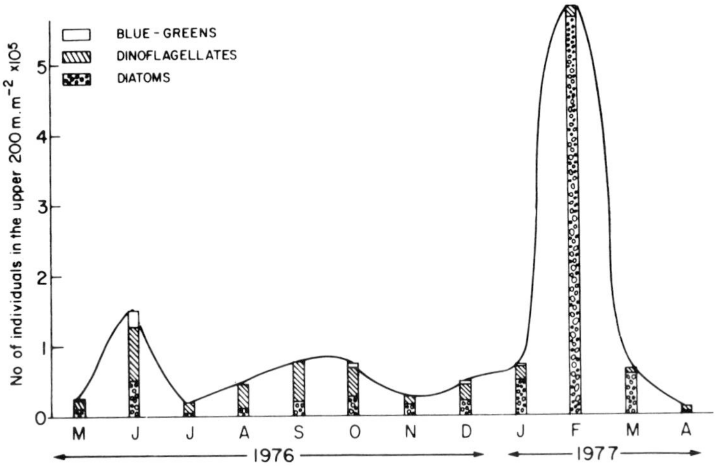

Fig. D.43. Phytoplankton abundance (individuals m^{-2}) in the upper 200 m at DCPE-Station A during 1976–1977 (Levanon-Spanier et al. 1979)

The range of primary production is between 0.05 and 3.38 mg C m^{-3} h^{-1}. The highest value was observed in January at 20 m depth.[4] During summer, low values were measured (0.05–0.27 mg C m^{-3} h^{-1}), although higher values (up to 0.75 mg C m^{-3} h^{-1}) were obtained in May-June between 50 and 100 m (see seasonality of chlorophyll a above).

Figure D.43 shows the abundance of microphytoplankton >65 μ in the upper 200 m at DCPE-Station A during 1975/1977. Figure D.44 summarizes the abundance of microphytoplankton (and tintinninds) in the upper 600 m at Station A during the years 1974–1975. Figure D.45 is a summary of the monthly fluctuations of nutrient distribution, primary production and chlorophyll a in the upper 200 m at Station A during 1975–1977, compared to the fluctuations in abundance of microphytoplankton >65 μ and nannoplankton <65 μ during the same period.

The depth integrals (Fig. D.45c, d) show the seasonality of chlorophyll a standing crop and primary production in the upper 200 m. The range is 16.2–69.8 mg m^{-2} chlorophyll a and 19.7–111.8 mg C m^{-2} day^{-1}. The most productive period is December–March, although another peak is evident in May–June. While during the winter mixing period productivity takes place throughout the upper 200 m and especially above 50 m depth, during summer stratification productivity is concentrated below 30 m (cf. Figs. D.40–D.42).

The winter peaks of primary production and chlorophyll a occur in December–January, hence earlier than the major abundance peaks of diatoms and dinoflagellates (February–March) (see Fig. D.45e, f). Thus, these microplankton groups do not account for the total chlorophyll a and production levels. On the

4 This high value is erroneously indicated by Levanon-Spanier et al. (1979) in their Fig. 1 as present at 20 m in November; the value at that point is 0.38 (see op. cit., text and their Fig. 10)

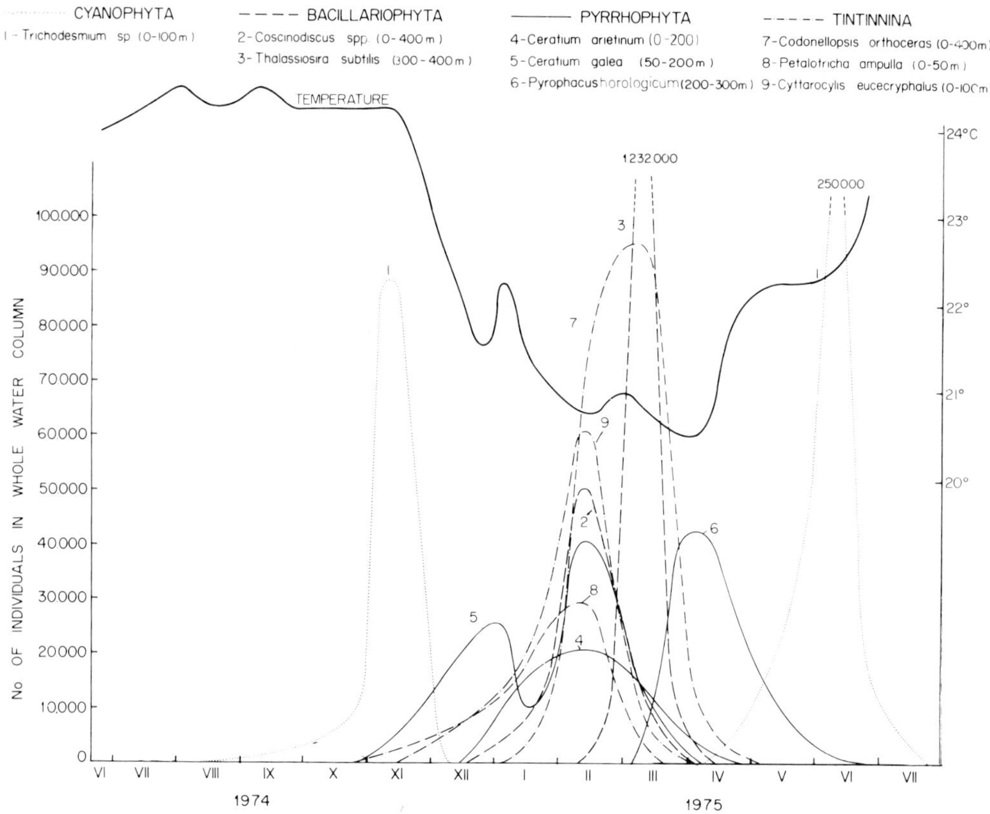

Fig. D.44. Seasonal succession and abundance of principal microplankton constituents between 0 and 600 m depth at DCPE-Station A during 1974–1975. (Highest concentrations at depth intervals indicated in paratheses in key) (Kimor and Golandsky 1977)

other hand, peak abundances of coccolithophorids (Fig. D.45h) are observed during December, thereby indicating that calcareous nannoplankton (which is lost from plankton nets during sampling) plays a major role in primary production in the Gulf of Aqaba. Nothing is known as yet on the quantitative role in primary production levels of other nanno- and picoplanktic organisms, such as naked monads, coccoid cells a few microns in diameter, and various colorless filaments which form in oligotrophic seas (e.g., Sargasso) together with coccolithophores a biomass perhaps equivalent to that of the microphytoplankton (Platt et al. 1983). All these organisms are not enumerated in 65 μ samples and are probably lost in SEM preparations for calcareous nannoplankton. The symbionts of planktic foraminifera are by themselves important primary producers. Laboratory experiments (J. Erez, pers. comm.) indicate photosynthetic rates per foraminiferal individual *(G. sacculifer)* which at about 150 individuals per 10 m^3 may contribute about 5% of the total primary production in the Gulf of Aqaba. Thus, the high primary production and chlorophyll a values in January and February (Fig. D.45) may be due in part to pico- and nannoplankters and foraminif-

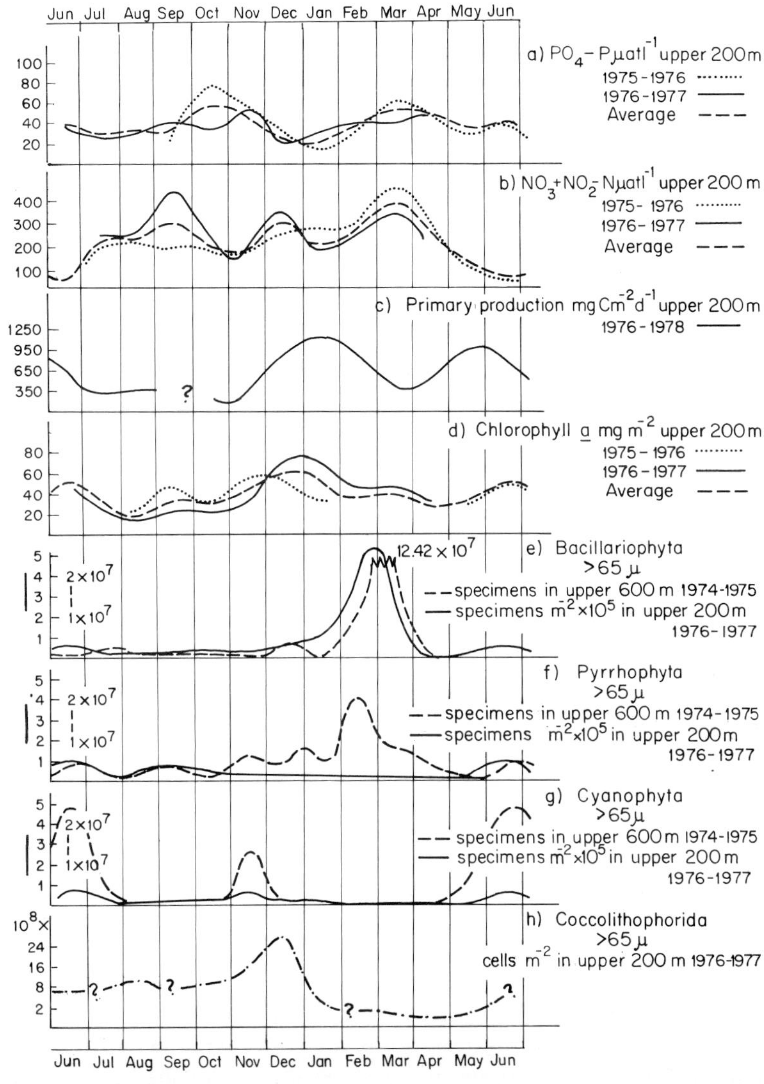

Fig. D.45. Seasonal succession of nutrients, primary production, chlorophyll a, microphytoplankton and calcareous nannoplankton at DCPE-Station A during 1974–1977. Microphytoplankton counts refer to stratified vertical hauls by Villefranche nets 65 μ with 58 cm net diameter. Nannoplankton counts refer to stratified filtered water samples. (Data from Kimor and Golandsky 1977, 1981, Klinker et al. 1978, Levanon-Spanier et al. 1979, Winter et al. 1979)

eral symbionts, not accounted for in the plankton study in the Gulf of Aqaba. Summer (May–June) peaks of primary production and chlorophyll a are accounted for mainly by abundance of blue-green algae (Figs. D.45 g–h and compare Figs. D.43 and D.44).

The seasonal pattern is also observed from depth profiles of the assimilation number mg C h^{-1} (mg chl a)$^{-1}$ (Fig. D.46). The highest specific activity (4–8)

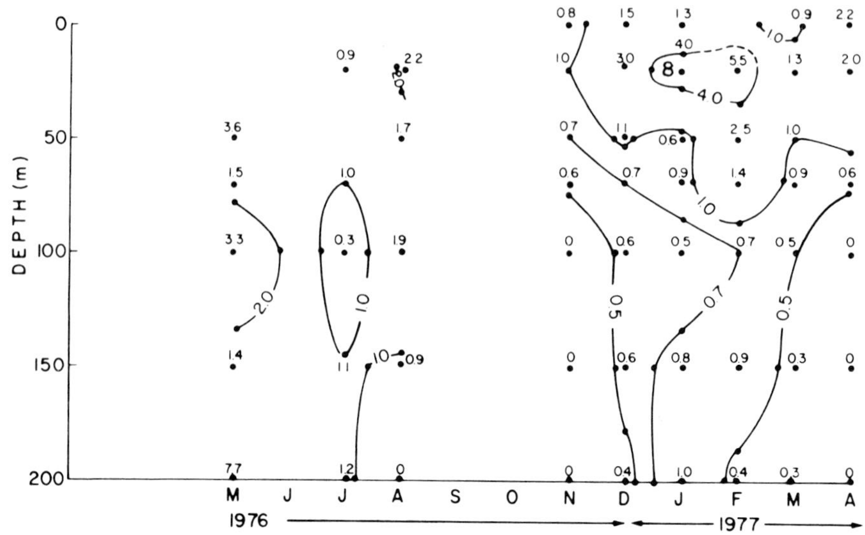

Fig. D.46. Seasonal depth profile of assimilation number [mg C h^{-1} (mg chl a)$^{-1}$] at DCPE-Station A during 1976–1977. (After Levanon-Spanier et al. 1979)

takes place in winter in the upper 50 m (see Figs. D.41 and D.42). During summer the assimilation number is low (but higher in May–June) and is characteristic of a **nutrient-limited phytoplankton.** Low assimilation numbers despite abundance of microphytoplankton at depths greater than 100 m during winter 1977 may be due in part to accumulations of nonactive phytoplankton at these levels.

The seasonal depth integrals of nitrates and phosphate, shown in Fig. D.45 a, b and compared with primary production and chlorophyll a in Figs. D.45 c, d, indicate that the increase of nitrates in the upper 200 m is followed by an increase in both primary production and chlorophyll a levels. This suggests that **nitrogen is the major limiting nutrient** and its supply during winter mixing to the upper waters enables phytoplankton to bloom. In June, and October–November, when the lowest nitrate concentration is observed in the water, the cyanobacterium *Trichodesmium* is abundant. However, this blue-green alga is known to be capable of fixing atmospheric nitrogen and its mass occurrence in the absence of combined nitrogen in surface waters has also been reported from the Sargasso Sea.

The seasonal depth distribution of the N/P ratio in the water (Fig. D.47) supports the suggestion that N is limiting. Before the winter mixing, the N/P ratio in the upper waters is not greater than B, while below 100 m it reaches 20. Immediately before the microphytoplankton bloom, N/P ratios (by atoms) increase to 18 in the upper waters and to 30 below 100 m. With the onset of the microphytoplankton bloom a considerable decrease in the ratio is observed throughout the water column, with a renewed increase following the peak of the bloom. It is noteworthy that the greatest abundance of coccolithophorids in December correlates with a high N/P ratio (see Chap. E).

Transects from DCPE-Station A toward the reef examined in March 1977 (Fig. D.48) indicate an increase of primary production and chlorophyll a values

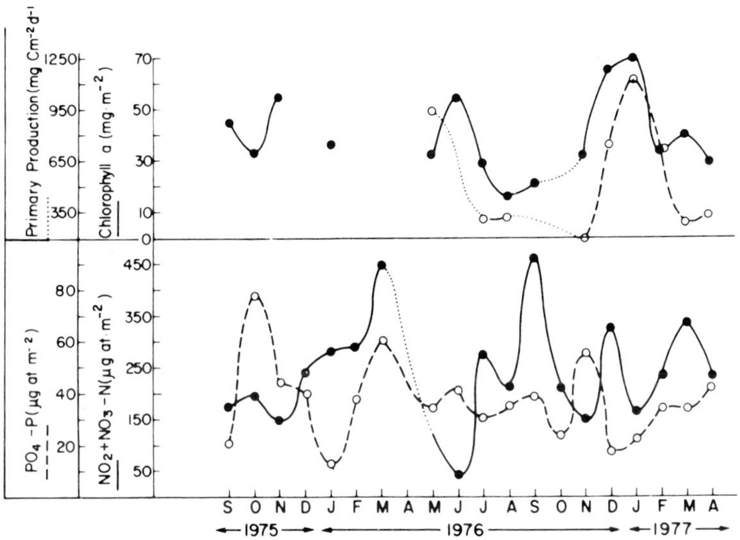

Fig. D.47. Seasonal profile of ratio $NO_3 + NO_2$-N/PO_4-P and primary production/chlorophyll a at DCPE-Station A during 1975–1977 (Levanon-Spanier et al. 1979)

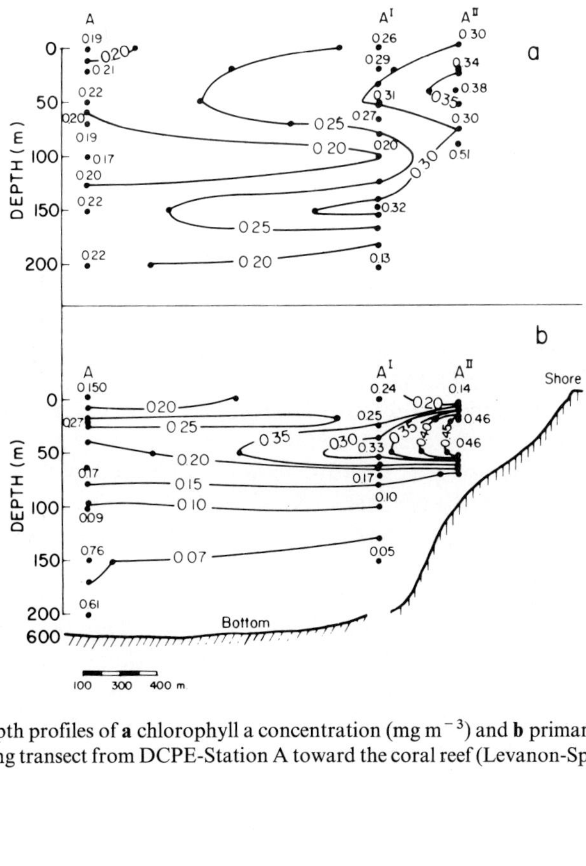

Fig. D.48 a, b. Depth profiles of **a** chlorophyll a concentration (mg m^{-3}) and **b** primary production (mg C m^{-3} h^{-1}) along transect from DCPE-Station A toward the coral reef (Levanon-Spanier et al. 1979)

in the reef area, though values are still very low. Measurements in May 1976 of primary production, chlorophyll a, phytoplankton and nutrient concentration in coral reef areas and mangroves (Table D.5) indicate extremely low levels of primary production in the free waters (comparable to those obtained in the open Gulf), nannoplankton (coccolithophorids and naked flagellates) making up the bulk of the phytoplankton.

Measurements of photosynthesis and respiration by coral tips in situ indicate high metabolic rates in the coral reef. Oxygen concentration in mangrove pools ranges between 1 and 11 mg O_2 l^{-1}, indicating high **gross benthic** production. Mangrove sediments harbor large crops of blue-green algae and especially of benthic, symbiont-bearing Foraminifera. Algal-foraminiferal symbiosis may account for much of the plant biomass found in the mangrove sands.

High benthic productivity in the coastal ecosystems is probably due to such energetic mechanisms as rapid recycling of nutrients or appreciable fixation of molecular nitrogen, facilitating a high benthic biomass in a greatly nutrient-impoverished environment.

Average primary production and chlorophyll a levels in the Gulf of Aqaba agree with those in the northern Red Sea (north of approximately lat 20° N). They are about 1/5–1/10 of those in the southern Red Sea (see Chap. D.5) and are comparable to other oligotrophic water bodies, including oceanic gyre centers (Table D.6).

Primary production is higher in winter than in summer. During the winter mixing period, the southern, more stratified sector is much more productive (and nutrient-richer) than the well-mixed northern one. Production values in the southern sector are very similar to those immediately outside the sill of Tiran, in the Red Sea proper (see Fig. D.40). As pointed out above (Chap. D.2) oxygen content in the waters of the Gulf of Aqaba is very high (3.7 ml l^{-1} min to 6.1 ml l^{-1} max). Outside the sill of Tiran oxygen content drops rapidly below sill depth to minima of 1.4 ml l^{-1} in deep waters.

In the absence of any significant land-derived apport of nutrients, production in **the Gulf of Aqaba is entirely dependent upon nutrient supply from the Red Sea.** These nutrients accumulate during summer (April–October) below the pycnocline and are supplied by turbulent diffusion to the upper waters with the breakdown of the thermocline, starting in November and, occurring mainly during the winter (January–March) mixing.

Continuous skimming of nutrients from south to north leads to the northern sector of the Gulf of Aqaba being generally less productive than the southern sector. The higher productivity at DCPE-Station A observed in May–June is probably due to upwelling at the northern end of the Gulf under permanent northerly wind stress (see Chap. D.3) and to the activity of *Trichodesmium*, which is a major primary producer in summer.

A study of the carbonate system, dissolved oxygen and carbon isotopes in the waters of the northern Gulf of Aqaba (DCPE-Station A) indicates that there are no significant changes in chemical or isotopic composition between day and night samples and that, therefore, effects of primary production on this system are negligible. The major factor controlling the $\delta^{13}C$ in the uppermost waters is the process of isotopic equilibrium with the atmosphere. Deviations from equilibrium,

Table D.5. Summary of observations, May 2–28, 1976 (mean and confidence interval at 0.05 level) (Sournia 1977)

	Coral reef "Japanese Gardens"	Coral reef the "Fjord"	Mangrove pool near Nabq
Nutrients			
N–NO_3, μg at. l^{-1}	0.15±	0.36±0.10	0.22± [a]
N–NO_2, μg at. l^{-1}	0.02±0.02	0.02±0.02	0.01± [a]
P–PO_4, μg at. l^{-1}	0.08±0.05	0.03±0.02	0.25± [a]
Si–SiO_4, μg at. l^{-1}	2.55±2.53	1.17±0.25	1.29±0.33
Total particle content dry weight, mg l^{-1}	0.42±0.16	0.41± [a]	7.10± [a]
Phytoplankton abundance cells $l^{-1} 10^3$	48±25	70±26	? [b]
Chlorophyll content of phytoplankton μg chl. a l^{-1}	0.27±0.11	0.23± [a]	0.31±0.06
Primary production of phytoplankton			
^{14}C light uptake, μg C l^{-1} day $^{-1}$ or (1 m deep station), mg C m^{-2} day^{-1}	10.2 ±4.5	29.9±6.3	8.6±1.18
^{14}C dark uptake, μg C l^{-1} day^{-1}	0.9±0.2	2.0± [a]	? [b]
Chlorophyll content of sediments			
μg chl. a (g dry weight)$^{-1}$	3.5±1.1	4.1±3.4	2.9±2.0
μg pheophytin (g dry weight)$^{-1}$	0.4±0.3	0.6±0.4	0.5±0.5

[a] In some cases, the confidence interval could not be estimated, due to the scarcity and/or non-normality of the data
[b] No data

Table D.6. Comparison between chlorophyll a and primary production in the Gulf of Aqaba and in other marine oligotrophic water bodies (Levanon-Spanier et al. 1979)

Region	Depth of photic zone (m)	Chlorophyll *a*		Primary production	
		($mg\,m^{-3}$)	($mg\,m^{-2}$)	($mg\,Cm^{-2}d^{-1}$)	($g\,Cm^{-2}y^{-1}$)
Gulf of Elat (North)	170	0.03–0.45	16.2–69.8	19.7–111.8	76.2[a]–84.6[b]
Sargasso Sea, off Bermuda		0.05–1.00		50 –830	72 –77
N. Pacific, central gyre	150	0.04–0.23		144 –294	
N. Pacific, off Hawaii	130	0.03–0.40	27.58 (average)	82.0 (average)	86.7 (gross)

[a] December–February
[b] March–November

i.e., enrichment in $\delta^{13}C$ in the uppermost water layer (about 10 m; cf. Chap. D.3) where primary production is lower due to photoinhibition, are caused by rapid evaporation, mixing and fast gas exchange when the water–atmosphere diffusion layer is removed during high winds and turbulence, and not by photosynthesis.

During winter (December–March) the water mass over 600 m (surface to bottom) at Station A is homogeneous, its average characteristics being temperature

$T = 21.1\,°C$, dissolved oxygen $O_2 = 0.225\,mmol\,l^{-1}$, alkalinity $CA = 2.37\,mEq\,l^{-1}$, total dissolved inorganic carbon $TC = 2.06\,mmol\,l^{-1}$ and $\delta\,^{13}C$ 2% PDB. During summer, the upper water mass above the thermocline (180–200 m), characterized by an average of $T = 24.6\,°C$, $O_2 = 0.225\,mmol\,l^{-1}$, $CA = 2.37\,mEq\,l^{-1}$, $TC = 2.1\,mmol\,l^{-1}$ and 1.9% $\delta\,^{13}C$, is separated from a deep water mass, characterized by an average $T = 21\,°C$, $CA = 2.37\,mEq\,l^{-1}$, with less oxygen ($O_2 = 0.210\,mmol\,l^{-1}$), more $TC = 2.3\,mmol\,l^{-1}$ and lower 1.8% $\delta\,^{13}C$. Although the decrease in O_2 and $\delta\,^{13}C$ and the increase of TC with depth during summer is explained by processes of oxidation of organic matter in deep waters (and not by dissolution of calcium carbonate) note the high oxygen, the pH between 8.17–8.24, and the alkalinity, as well as the presence of abundant aragonitic pteropods in sediments (see below). Calculations show that the primary production is not sufficient to account for the shift in $\delta\,^{13}C$ between winter and summer values of the deep water within 2 months (the time of the formation of the thermocline). Instead, the changes in the **deep water** at Station A are attributed to advection of aging waters from the central deep-water core of the Gulf (see Chap. D.3). Deep-water ventilation to the atmosphere takes place in the north during winter mixing.

References: Krey and Babenerd (1976), Levanon-Spanier et al. (1979), Patzert (1972), Platt et al. (1983), Shemesh (1981), Shemesh et al. (1981), Sournia (1977).

D.7 Composition of Plankton

The ranges of $>65\,\mu$ and $200\,\mu$–$330\,\mu$ plankton biomass concentrations, given in terms of displacement volume (DV ml 1,000 m^{-3}) as derived from 1,900 stratified hauls at DCPE stations during 1974–1975, are given in Table D.7. It should be noted that, although the $>65\,\mu$ biomass is composed mainly of phytoplankton, it contains also significant numbers of zooplankters like tintinnids and foraminifera, as well as juvenile stages of copepods and various larvae.

Table D.8 compares the mean values of biomass in the Gulf of Aqaba with the rather scanty data from the Red Sea proper and the Gulf of Aden. It can be seen that, in general, zooplankton biomass values in the Gulf of Aqaba are similar to those from the northern Red Sea and much lower than the values obtained in the middle and southern Red Sea or in the Gulf of Aden. On the other hand, 65 μ microplankton biomass in the upper 100 m in the Gulf of Aqaba is similar to that in the middle Red Sea and lower than in the northern Red Sea in summer. Between 100 and 500 m, the microplankton 65 μ biomass is much higher during the winter mixing period in the Gulf of Aqaba than in the more stratified Red Sea.

Mean annual values of biomass in the upper 300 m water column both in the Gulf of Aqaba and in the northernmost Red Sea are low and similar to those in the oligotrophic southeastern North Atlantic Central water mass. The seasonal succession of biomass and abundance of major components in the upper 600 m at DCPE-Station A during 1974–1975 are shown in Fig. D.49. The curves for 65 μ samples (mainly phytoplankton) and for 200–330 μ (mainly zooplankton) seem to indicate grazing-dependent opposite trends (Fig. D.49 e).

Table D.7. Maximum and minimum values of biomass (displacement volume ml 1,000 m^{-3}) in the Gulf of Aqaba, 1974–1975 (Klinker et al. 1978)

Net size and season	Stations A–G (Gulf of Aqaba)								Station H (Red Sea)					
	Maximum	Station	Depth (m)	Month	Minimum	Station	Depth (m)	Month	Maximum	Depth (m)	Month	Minimum	Depth (m)	Month
65 µm net														
Summer	504.5	A	0–50	June	3.1	E	500–600	Oct.	153.9	0–50	Aug.	8.5	400–500	Aug.
Fall	160.0	F	50–100	Nov.	3.6	A	400–500	Nov.	144.0	0–50	Nov.	10.7	500–600	Nov.
Winter	249.1	A	0–50	Mar.	6.4	G	200–300	Feb.	146.2	50–100	Jan.	3.8	400–500	Jan.
200 µm net														
Summer	392.9	A	0–50	Oct.	3.5	B F	400–500	Oct. Aug.	106.1	50–100	Aug.	4.0	500–600	Aug.
Fall	230.0	C	0–50	Dec.	3.9	A	500–600	Nov.	277.7	50–100	Nov.	8.6	500–600	Nov.
Winter	144.8	A	0–50	Jan.	3.1	A D	400–500	Mar. Jan.	40.0	0–50	Jan.	3.0	100–200	Jan.
330 µm net														
Summer	596.5	A	0–50	June	1.6	A	400–500	July	65.5	50–100	Aug.	27.8	100–200	Aug.
Fall	153.4	G	50–100	Dec.	3.3	F	200–300 300–400	Nov.	104.0	0–50	Nov.	4.0	500–600	Nov.
Winter	83.0	C	0–50	Mar.	3.5	A	400–500	Feb.	72.0	0–50	Mar.	3.1	400–500	Jan.

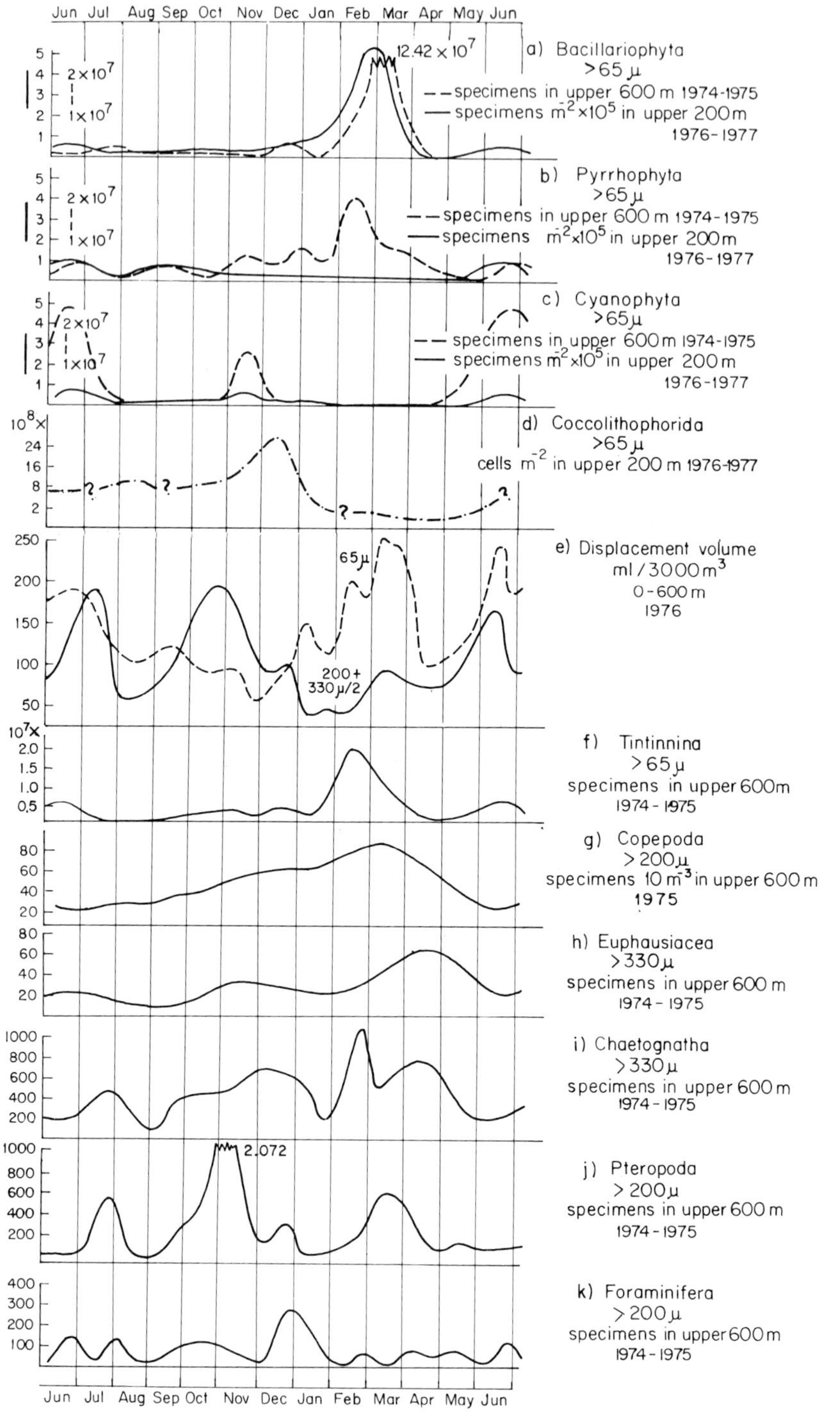

Fig. D.49. Seasonal succession of biomass and abundance of principal microphytoplankton and microzooplankton in the upper 600 m at DCPE-Station A during 1974–1975. (Data from A. Almogi-Labin, O. Adar, Y. Shapiro, J. Klinker, C. Kropach, in Reiss et al. 1977a, b, Kimor and Golandsky 1977, S. Prado-Por 1983). (All zooplankton data except for copepods, derived from total hauls)

Table D.8. Comparison of mean biomass values in Gulf of Aqaba with those from Red Sea proper and Gulf of Aden (Klinker et al. 1978)

Season and depth (m)	Mesh size of nets (µm)	Gulf of Aqaba	Red Sea: Station H	Red Sea: North to 23 °N	Red Sea: Middle to 15 °N	Red Sea: South	Gulf of Aden
Summer	65						
0–100		116[a]	98[a]	32[b,e]/42[b,e]	113[b,e]		
100–200		39[a]	42[a]		21[b,e]		
200–500		21[a]	46[a]		8[b,e]		
Summer	> 180						
0–100		82[a]	51[a]	< 100[a,d]/43[b,e]	< 300[a,d]/59[b,e]	300–500[a,d]	< 200[a,d]
100–200		29[a]	43[a]				457[b,f]
200–500		20[a]	23[a]				
Winter	> 180						
0–100		49[a]	37[a]	47[b,c]	81[b,c]	105[b,c]	571[b,c]
100–200		24[a]	15[a]	9[b,c]	14[b,c]	38[b,c]	102[b,c} 420[b,f]
200–500		18[a]	8[a]	10[b,c]	17[b,c]	19[b,c]	34[b,c]

[a] Displacement volume (ml 1,000 m^{-3}) DCPE. [b] Wet weight (mg m^{-3}).
Data from: [c] Delalo (1966). [d] Ponomareva (1968). [e] Gordeyeva (1970). [f] Gapishko (1971)

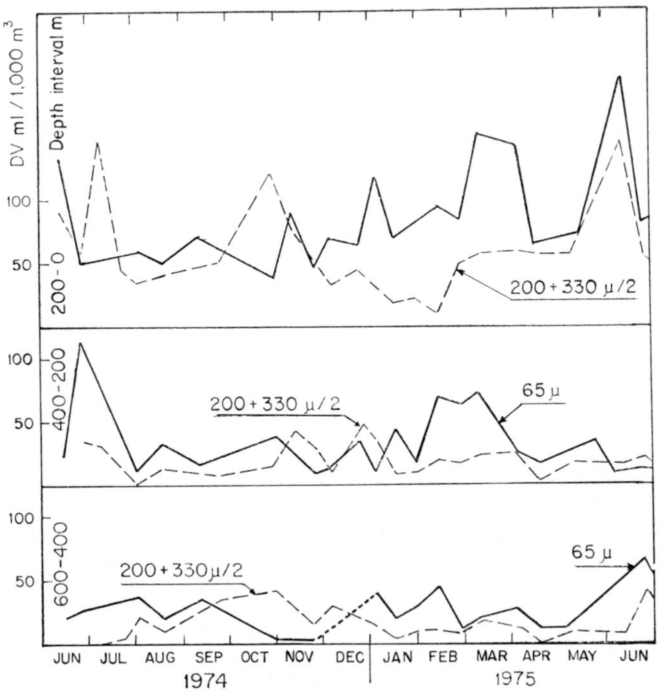

Fig. D.50. Depth distribution in the upper 600 m of biomass (displacement volume DV) at DCPE-Station A during 1974–1975 (Reiss et al. 1977 a, b)

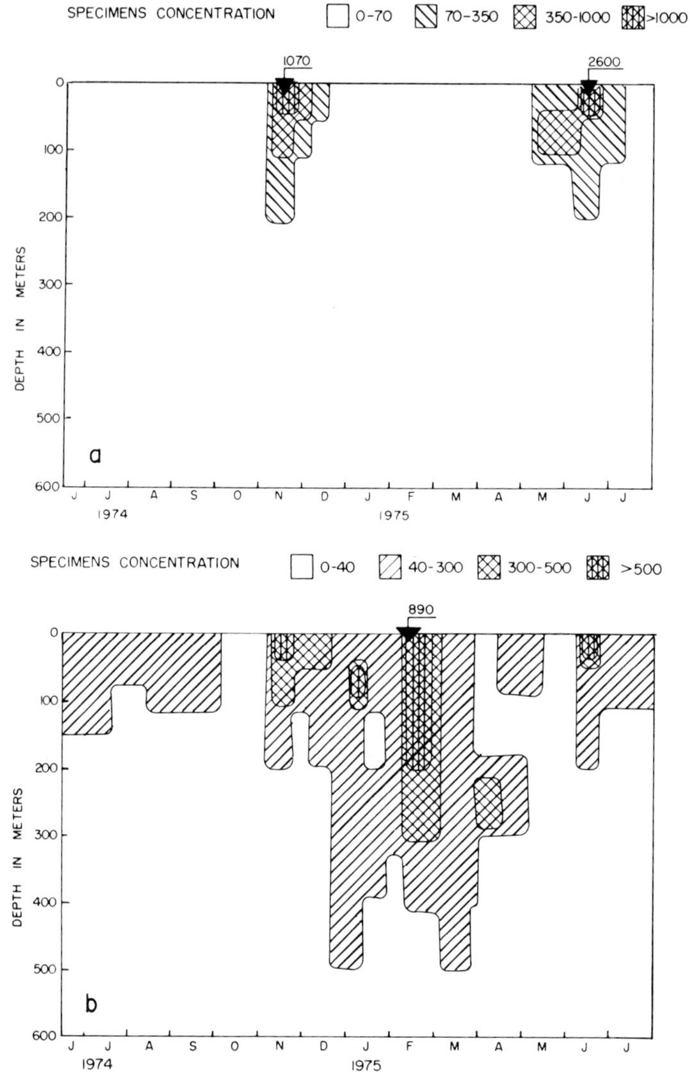

Fig. D.51 a–d. Seasonal and bathymetric distribution of major phytoplankton groups (**a** Cyanophyta, **b** Pyrrhophyta, **c** Bacillariophyta) and tintinnids (**d**) in the upper 600 m at DCPE-Station A during 1974–1975 (Kimor and Golandsky 1977)

Comparison with Fig. D.49 a–c, f shows that the major components of the 65 μ biomass are diatoms as well as tintinnids and dinoflagellates during winter, and blue-green algae and tintinnids during summer and fall. Figure D.49 g–k shows that pteropods and chaetognaths are mainly responsible for high 200 μ biomass values in fall (October–November) and in summer (July), as well as – with the addition of euphausiids and copepods – for the high peaks during February–March. The high 200 μ biomass value in June 1975 is not accounted for by any of the groups shown in Fig. D.49; it may be due to a particular abundance of echino-

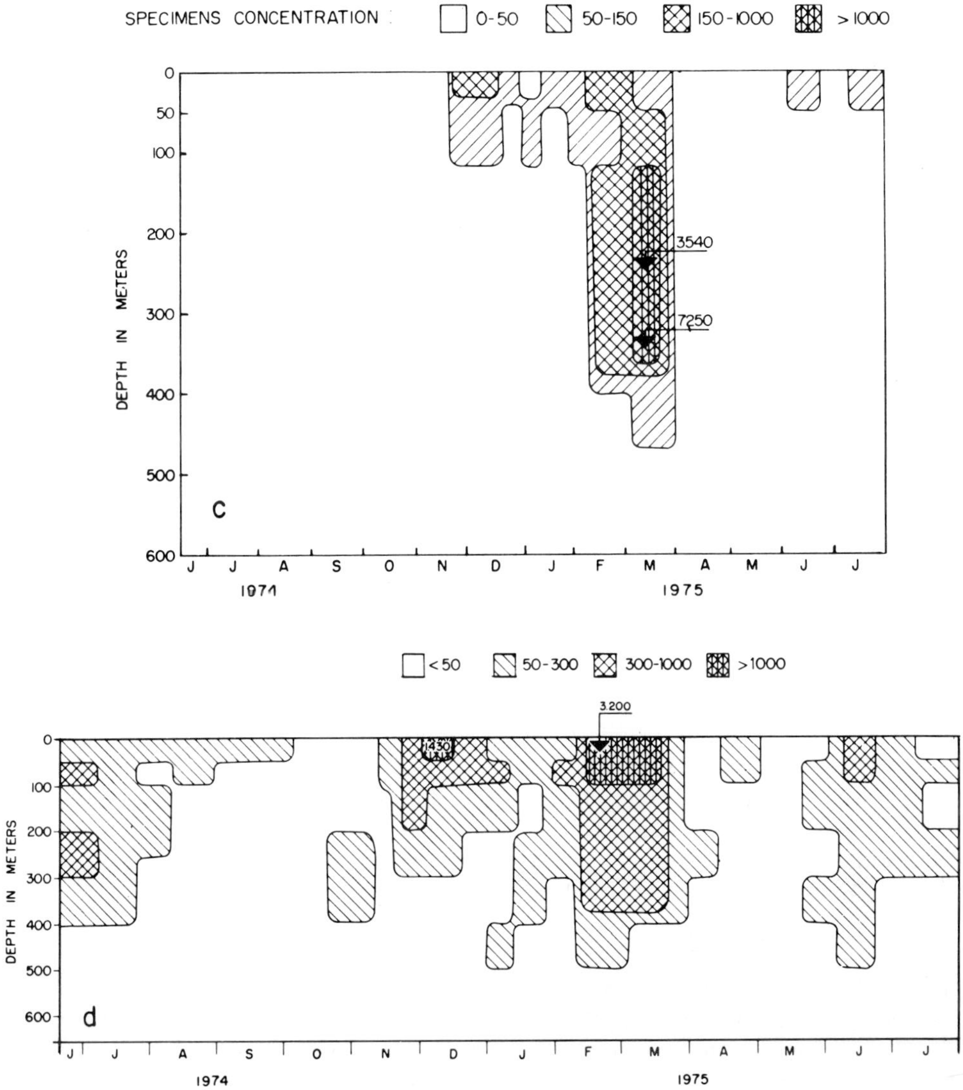

Fig. D.51 c, d

derm veliger stages observed at that time. The planktic foraminifera seem to have little influence on bulk biomass.

The distribution of biomass by depth intervals at DCPE-Station A during 1974–1975 is given in Fig. D.50. The latter figure shows that, although the highest biomass values occur within the upper 200 m, there are still appreciable concentrations of both phyto- and zooplankton down to 600 m depth.

The seasonal and bathymetric distribution of the major phytoplankton groups and of the tintinnids during 1974–1975 at DCPE-Station A is summarized

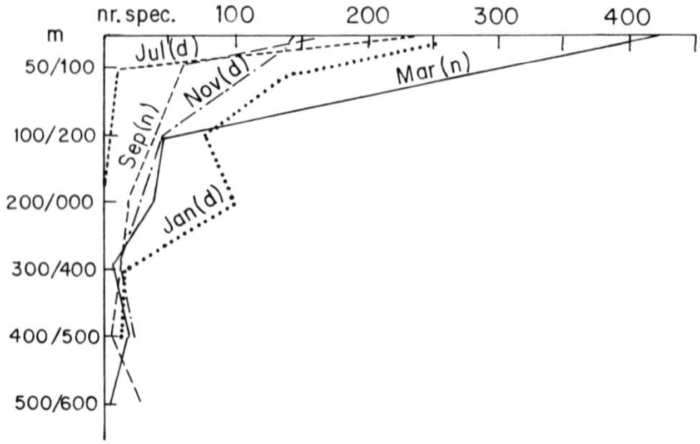

Fig. D.52. Seasonal and depth distribution of (calanoid) copepods in the upper 600 m at DCPE-Station A during 1975. (Data from Prado-Por 1983)

in Fig. D.51 a–d. The distribution of copepods is given in Fig. D.52, while that of coccolithophores is shown in Fig. D.53. (The seasonal depth distribution of pteropods and planktic foraminifera will be discussed in detail in Chap. E.)

Generally during summer there is a decrease in numbers of phytoplankton organisms (unicellular and colonial) with depth, particularly below 200 m. In the upper 100 m a minor peak of dinoflagellates and a major one of blue-green algae occur in May–June (cf. Primary Production). The November peak of blue-green algae occurs down to 200 m depth. Dinoflagellates and diatoms occur in great abundance down to depths of 500 m during February–March. High concentrations of dinoflagellates occur in April between 200–300 m.

Tintinnids occur down to depths of 500 m and are particularly abundant between surface and 200 m in December, between surface and 400 m depth in February–March, as well as to a lesser extent between 50 and 300 m in June.

The largest number of (mostly herbivorous and omnivorous) copepods occur during the winter period mostly in the upper 200 m, although appreciable numbers are still present down to 400 m. A noteworthy copepod peak is observable in June–July in the upper 100 m.

The abundance peak of the nannoplanktic coccolithophorids is in December throughout the upper 400 m water column. Although abundance is much lower in early summer, a slightly greater concentration is observable in May, in the upper 50 m and a more pronounced one in August, in the upper 100 m.

In comparing the abundance patterns of the different 65 μ microplankton groups with the biomass data, one must be aware of the size differences of individuals in the different groups and within the latter. Thus, e.g., the April mass occurrence between 200 and 300 m of dinoflagellates is due to newly emerged cysts of *Pyrophacus,* the peak abundance of diatoms between 200 and 400 m in March is due to a minute *Thalassiosira,* while the cyanophyte population is dominated by *Trichodesmium* trichomes with about 400 cells per filament.

Copepods (mostly calanoids) and pteropods are generally larger than 200 μ, while planktic foraminifera occur both in the 200–330 μ and in the 65 μ samples. Euphausiaceans and chaetognaths are large and occur mostly in 330 μ samples.

Despite these differences, the contribution of the major groups (except the coccolithophores) to the biomass is discernable from Fig. D.49.

The seasonal and bathymetric distribution of the micro- and nannoplankton is largely controlled by the nutrient regime as determined by the circulation pattern (supply of nutrients from the Red Sea; summer stratification in the Gulf of Aqaba and winter homothermy and mixing; upwelling at the northern end of the Gulf). The seasonal inverse correlation between nutrients, especially nitrates, on one hand and microphytoplankton biomass and primary production on the other, has been noted above. A considerable drop in phosphate content between 200 and 300 m at Station A in spring has been observed at the time of proliferation of dinoflagellates at that depth. The bloom of diatoms in March is probably responsible for an observed drop by nearly half in silicate concentration from that observed in January.

Although the seasonal and bathymetric changes in temperature of the warm waters in the Gulf of Aqaba are less than 7 °C, the influence of temperature on seasonality of thermosensitive groups of plankton is exemplified by the tintinnids which are abundant below 23 °C and by *Trichodesmium* whose blooms occur at the beginning of the fall cooling and immediately prior to the major summer warming (May–June) of the upper waters (see Chaps. D.4–D.6). In addition to increased nutrient supply to the upper waters, lowered temperatures may be responsible for the diatom winter bloom.

Changes in salinity are very small (about 1.5‰) (see Chap. D.4) and apparently do not affect the seasonality or bathymetry of the plankton groups as a whole.

Density values are high throughout the column and change only within about 2 σt, mainly as a function of temperature. On the other hand, the existence of a pycnocline – even if weak – and especially of its breakdown in winter seems to affect seasonality and bathymetry of the plankton.

An ecological factor of crucial importance to planktic (and benthic) organisms in the Gulf of Aqaba is light, i.e., surface irradiance, depth of photic layer and light quality, to be briefly discussed below.

References: Almogi-Labin (1982), Gorsky (1975), Kimor (1973), Kimor and Golandsky (1977, 1981), Klinker et al. (1978), Levanon-Spanier et al. (1979), Mikkelsen (1973), Prado-Por (1983), Reiss (1979), Reiss et al. (1974, 1976), Winter et al. (1979).

D.8 Light

The depth of the euphotic layer has considerable bearing on primary producers, as well as on zooplankters directly dependent on them. It is noteworthy in this connection that in the Gulf of Aqaba the food chain is "shortened." Tintin-

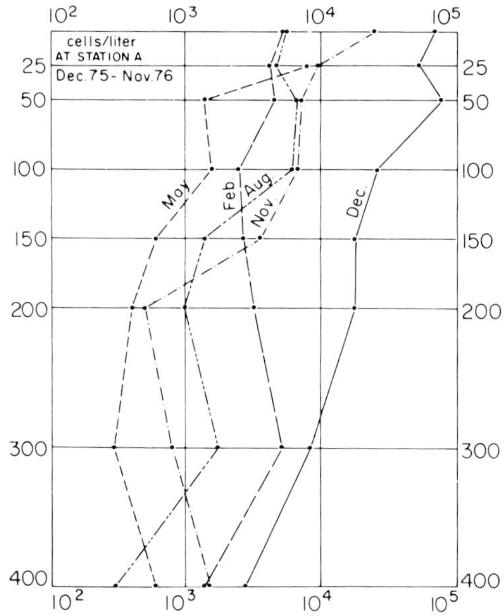

Fig. D.53. Abundance of coccospheres per liter of water in the upper 400 m at DCPE-Station A during 1975–1976 (Winter et al. 1978)

nids feed on minute diatoms and nannoplankton, the great majority of copepods are herbivores and omnivores, the pteropods in the Gulf are micro- and nannoplankton feeders and the euphausiaceans are mostly herbivores. The chaetognaths and planktic foraminifera feed apparently mainly on copepods.

The depth to which the major primary production takes place in the Gulf extends at least to 170 m. Photoinhibition is observed between surface and about 20 m, especially in summer, as indicated by the quantitative depth distribution of the major phytoplankton groups (Figs. D.51 a–c and D.53). Enough light of appropriate quality must be present to enable blooms and reproduction at depths of more than 400 m. Moreover, among the dinoflagellates and the diatoms, different species are restricted to different depths, probably in response to their light requirements ("shade flora") (see Fig. D.44). The major coccolithophore blooms occur in the whole 400-m column during the lowest insolation rates. Certain tintinnids sink to greater depths during summer, probably as a photonegative response. The majority of the planktic foraminifera in the Gulf of Aqaba are symbiont-bearers and hence light-dependent. Vertical migration of euphausiaceans and siphonophores as a response to moonlight has been observed in the Gulf. It is also important to note that the depth range in the Gulf of Aqaba of hermatypic reef corals is about 110 m, of coralline algae about 130 m, "*Halophila* lawns" flourish down to depths of about 70 m and symbiont-bearing, benthic Foraminifera are abundant down to depths of 150 m. It may be of interest to compare the situation in the Gulf of Aqaba with that in the strongly illuminated and transparent Eastern Mediterranean in summer, where phytoplankton maxima were observed always at about 100 m, i.e., below the 1% surface light level (between 50 and 70 m), and where dinophyceans in active division, as well as prasinophytes and symbiont-bearing acantharians, were found in depths of 1,000 m.

The strong illumination of the waters in the Gulf of Aqaba is due to geographical position, climate, and circulation pattern.

Insolation of the area is very high, strongest in June and weakest in December (see Chap. D.1). As pointed out above, cloudiness is very low and nearly restricted to the winter months. Apport of abiogenic material in suspension is relatively insignificant and restricted to transport by winter wadi floods, while aeolian transport of particles from the Arava mainly affects the northernmost Gulf. Plankton biomass is low (see above) and the base of the weak summer thermo- and pycnocline is situated at about 200 m. This situation results in strong illumination and deep light penetration in the transparent waters of the Gulf of Aqaba.

Although only relatively few light studies in the Gulf have been carried out up to now, their results are very significant. Observations from the GEO-submersible indicated that in November 1982 surface wave action could be discerned through the domeport at depths of about 120 m, while at 200 m light penetration was sufficient to enable reading and writing in the submersible. Transparency of the water is greatly reduced in February–March (cloudiness, phytoplankton abundance, abiogenic matter in suspension) when surface visibility through the domeport war restricted to 40–50 m.

Penetration measurements of photosynthetic active radiation carried out with a Lambda photometer from the GEO-submersible in October–November 1981 and March 1982 near the MBL, Elat, are summarized in Fig. D.54. The 1% level of surface light (2,000–2,080 Å^{x}E at air–water interface) is at about 100 m and the 0.02% level at 200 m (cf. visibility in the domeport above). Differences between the curves at the dates indicated are small. It should be noted, however, that these dates coincide apparently with relatively low production levels (see Fig. D.45).

The diurnal changes in light penetration observed from the GEO-submersible at one station in the northern Gulf of Aqaba are shown in Fig. D.55. The changes in deeper waters in the early afternoon are particularly interesting and the position of the 1% level between 80 and 100 m is also noteworthy. Light measurements carried out at DCPE-Station A with a photovoltaic cell (and continuous surface comparative monitoring) from aboard the R. V. Arnona in May 1982 are summarized in Fig. D.56 a–c. The transmission curve (Fig. D.56 a) indicates the position of the 1% level of downward surface irradiance at about 80 m (see also above) with important attenuation boundaries for red light at about 20 m and between 40 and 50 m (cf. Primary Production Boundaries above). It should be noted that the measurements were carried out during the early afternoon (cf. Fig. D.55) on a day with small clouds and at the beginning of a relatively high production period (mainly cyanophytes; Figs. D.51 c and D.45, and see above). The ratio downward/upward irradiance (Fig. D.56 b) indicates considerable backscattering of light (particularly blue and green) below 80 m, where red light is strongly attenuated. The red light attenuation is connected with phytoplankton activity, while the backscattering trend is probably due to zooplankton present below 80 m. The oxygen measurements (Fig. D.56 c) support such an interpretation. (Lower values in the uppermost 25 m are probably due to photoinhibition.)

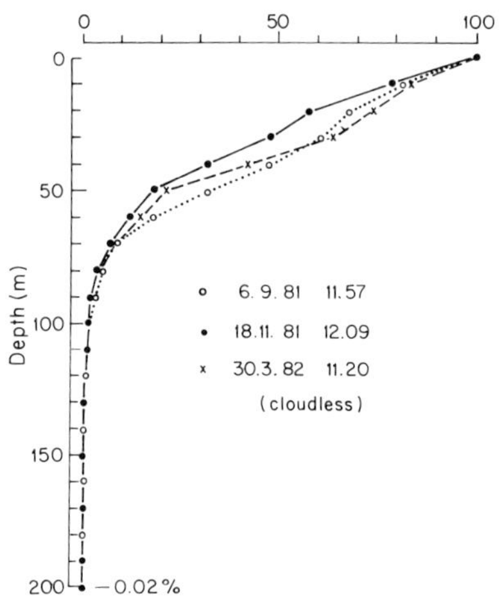

Fig. D.54

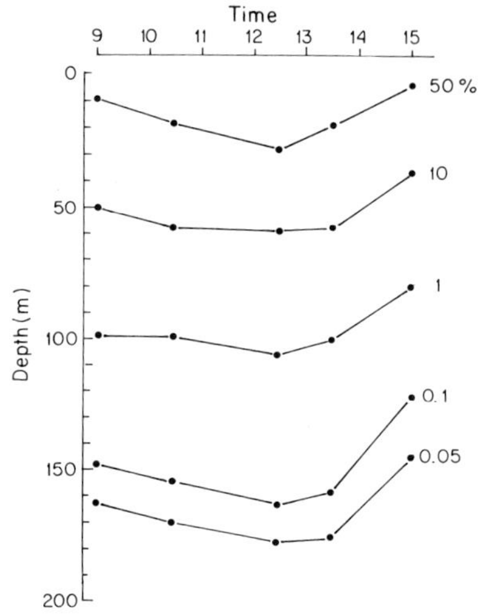

Fig. D.55

Fig. D.54. Light penetration (% of surface value) measured at noon on cloudless days in different months

Fig. D.55. Diurnal change in light penetration. Time relates to start of measurement. Surface values between 09.00 and 14.42 h at air/water interface 2,000–2,080 μE. (Figures D.54 and D.55 from Fricke and Schuhmacher, in press)

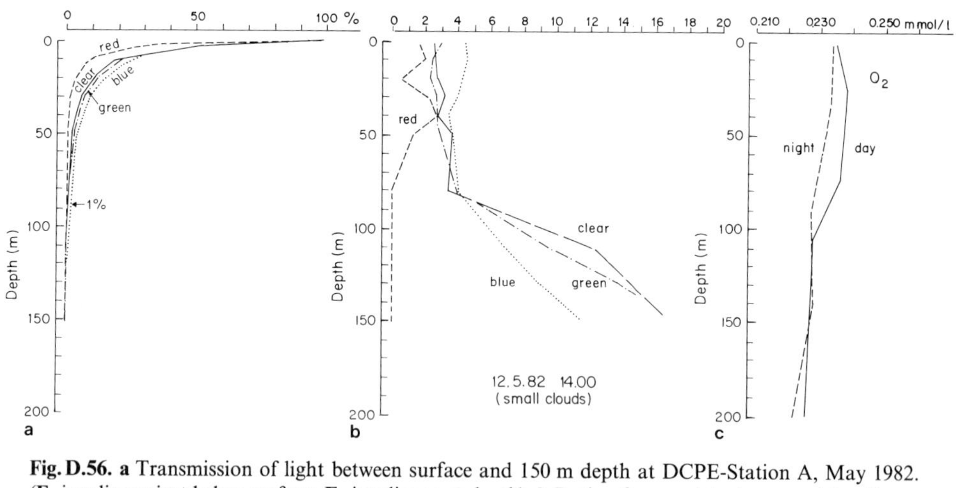

Fig. D.56. a Transmission of light between surface and 150 m depth at DCPE-Station A, May 1982. (E_o irradiance just below surface; E_d irradiance at depth). **b** Ratio of upward to downward irradiance between surface and 150 m depth at DCPE-Station A, May 1982. Note that although general trend at depth is valid, the absolute values may be exaggerated due to the sensitivity range of the instrument. **c** Oxygen content (mmol l^{-1}) between surface and 150 m depth at DCPE-Station A, May 1982 (Erez and Peguine, unpubl.)

Comparison of attenuation coefficients (clear light) in the Gulf of Aqaba with those characteristic of water types defined by Jerlov (1968) shows the upper 20 m at Station A in May to lie between types II and III, while the waters below 20 m lie between types II and I, i.e., between the Sargasso Sea and Eastern Mediterranean on one hand and the Red Sea on the other.

References: Almogi-Labin (1982), Avraham (1981), Erez and Peguine (unpubl.), Fricke and Hottinger (1983), Fricke and Schuhmacher (1984), Gorsky (1975), Jerlov (1968), Kimor and Golandsky (1977, 1981), Kimor and Wood (1975), Klinker et al. (1978), Krey and Babenerd (1976), Levanon-Spanier et al. (1979), Oren (1962), Patzert (1972), Prado-Por (1983), Reiss (1979), Reiss et al. (1974, 1976, 1977a), Shemesh (1980), Shemesh et al. (1981), Sournia (1977), Winter (1982a), Winter et al. (1979).

D.9 Characteristics of Water Masses

In summary, the **arid climatic conditions** and the existence of the **sill at Tiran** determine the **circulation pattern** in the – because of geological history – deep Gulf of Aqaba. As a consequence of the circulation, **nutrient content** of the waters and **primary production** are low, the Gulf of Aqaba being **oligotrophic** during most of the year and comparable to other marine oligotrophic water bodies in the world, especially **oceanic gyre centers.** The depth of the **euphotic zone** in the Gulf reaches 170 m, i.e., a depth determined for other oligotrophic subtropical seas, although primary producers are prolific in the **subphotic zone** to depths of at least 400 m. On the other hand, the Gulf of Aqaba differs from permanently stratified, typically oligotrophic seas by a short **winter mixing period,** when production over the upper 200 m would characterize the Gulf temporarily as a moderately productive water body.

The generally oligotrophic character of the desert-enclosed Gulf of Aqaba is also expressed in the **micro-** and **nannoplankton** with significant similarities to oceanic gyre centers, like the Sargasso Sea. The south–north changes in biological water mass characteristics from the highly productive Gulf of Aden, with similarities to equatorial areas through the Red Sea to the oligotrophic Gulf of Aqaba, with many similarities to gyre centers, are of considerable relevance to considerations of the ecology of micropaleontologically important groups of organisms.

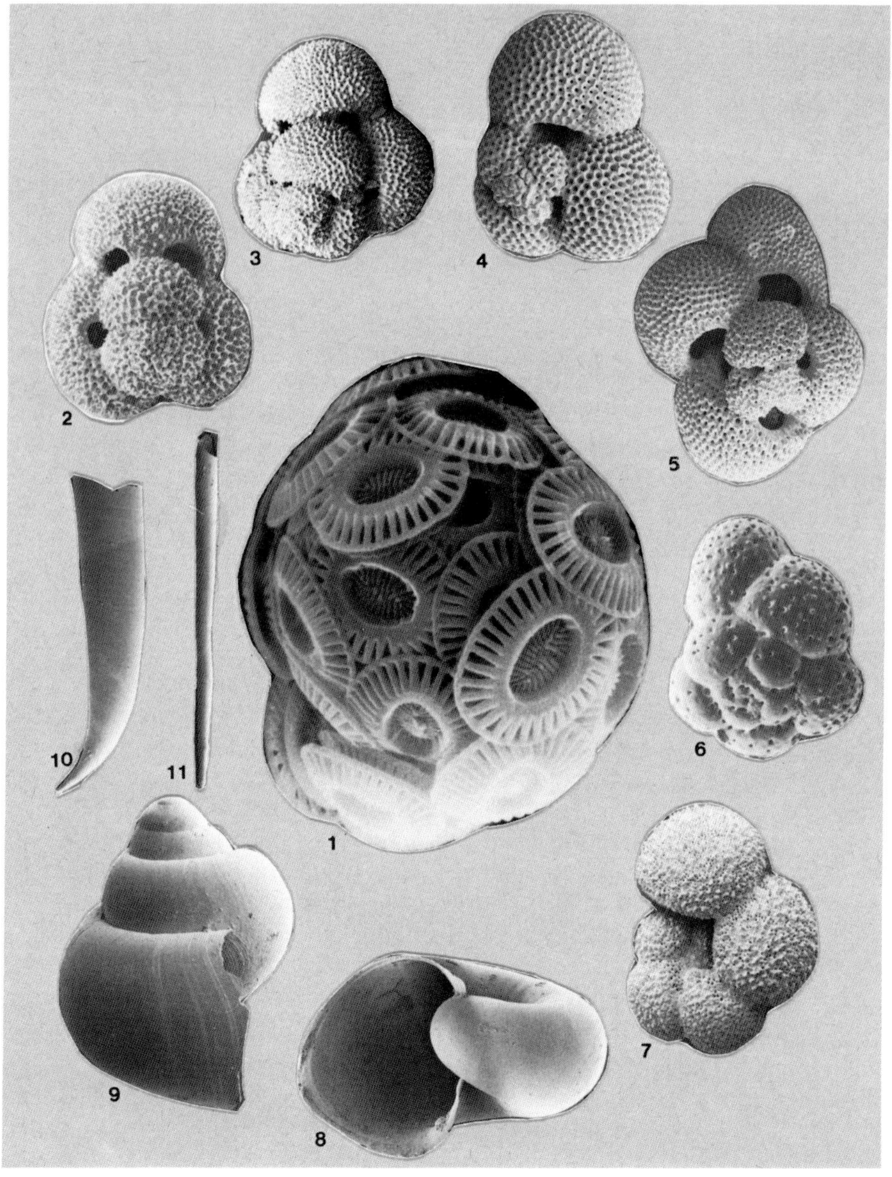

Plate E.1. Major calcareous planktic organisms from the Gulf of Aqaba

1	*Emiliania huxleyi* (Lohmann) (Coccolithophore)	Natural size:	30 μm
2–3	*Globigerinoides ruber* (d'Orbigny)	Natural size:	2:460 μm 3:350 μm
4–5	*Globigerinoides sacculifer* (Brady)	Natural size:	4:500 μm 5:700 μm
6	*Turborotalita clarkei* (Rögl and Bolli)	Natural size:	135 μm
7	*Globigerinella siphonifera* (d'Orbigny) (*2–7* Foraminifera)	Natural size:	550 μm
8	*Limacina inflata* (d'Orbigny)	Natural size:	1000 μm
9	*Limacina trochiformis* (d'Orbigny)	Natural size:	800 μm
10	*Creseis virgula virgula* (Rang)	Natural size:	2500 μm
11	*Creseis acicula* (Rang) (*8–11* Pteropoda)	Natural size:	6000 μm

E. Shell Producers in the Water Column

E.1 Calcareous Plankton (Plate E.1)

The key to the reconstruction of ancient pelagic environments lies largely in the sedimentary record of the shells of planktic organisms. Most important in this respect are the shells of planktic Foraminifera and Coccolithophorida. They form in places, together with shells of Pteropoda, the bulk of the present-day calcareous oozes, covering about 1/3 of the Earth's surface. Their Mesozoic-Cenozoic fossil record is probably the most complete and best-documented among marine groups.

The usefulness of biogenic constituents as paleoceanographic indicators depends upon knowledge of the factors influencing the distribution of the living organisms producing the shells in present-day seas, of those determining the preservation on the sea floor of their remains, as well as upon an understanding of shell morphology as related to evolution on one hand and to changes in the ecosystem on the other.

However, despite the intensive research carried out on various aspects of calcareous planktic groups (see the latest comprehensive reviews by Gartner 1981, Berger 1981, Vincent and Berger 1981), very little is known on their biology and nearly nothing on their functional shell morphology. In the absence of first-order criteria, distribution patterns of calcareous planktic organisms must be correlated empirically with ecological factors. Interpretations of such correlations are facilitated by studying environments in which the range of variability of many ecological factors is minimal.

However, most of the research carried out pertains to open-ocean environments where pronounced gradients of physical and chemical factors occur in season and depth, where calcium carbonate dissolution usually occurs and where low rates of sedimentation are common.

The narrow ranges of bathymetric and seasonal variations of physicochemical water mass characteristics in the Gulf of Aqaba make the latter of particular interest to studies of calcareous plankton. In addition, no $CaCO_3$ dissolution takes place in the fairly deep Gulf of Aqaba, as indicated by the excellent preservation of aragonitic shells in the sediment and by the high degree of calcite saturation ($\Omega = 5$–6). Finally, the desert-enclosed Gulf of Aqaba lies at the extreme end of an environmental gradient which runs from the highly productive Indian Ocean and Gulf of Aden through the increasingly isolated, highly saline and oligotrophic Red Sea.

The bathymetric and seasonal distribution patterns of living coccolithophores, planktic foraminifera and pteropods, as well as the composition of their accumulations in Recent sediments of the Gulf of Aqaba provide information of relevance not only to the regional paleoceanographic reconstruction (Chap. H), but also to the interpretation of ecological and paleoecological relations of these organisms in the world oceans. Evidence from deep-sea cores raised in the Gulf and laboratory experiments have also helped to elucidate the ecological tolerance limits of various widely distributed calcareous planktic organisms.

E.2 Coccolithophorida

Although studies on the living coccolithophores in the Gulf of Aqaba, carried out between 1972 and 1977, are still relatively restricted, their results are meaningful with regard to the ecology and paleoecology of coccolithophores both in the Gulf and in other oceanic environments.

The **standing crop** of coccolithophorids shows a unimodal yearly distribution within the upper 400 m water column. There is a general decrease in cells with depth, particularly below 150 m in summer and fall, and below 300 m in winter. Cell counts are highest within the upper 50 m, although during the winter mixing period cells are more evenly distributed throughout the column (see Fig. D.53).

The greatest abundance is observed in late fall: 16.8×10^3 cells l^{-1} in surface waters and 27×10^8 cells in the unit column ($100\ cm^2 \times 200$ m) in December 1975 at DCPE-Station A. Cell density decreases rapidly during the later winter and increases gradually from a low in spring in the upper 100 m, but not at depth. In August 1976, 5.5×10^3 cells l^{-1} were counted at the surface and 8.45×10^8 coccolithophores were contained in the unit column.

The **composition** of the calcareous nannoflora in the Gulf of Aqaba is given in Table E.1.

Table E.1. List of coccolithophorid species in the Gulf of Aqaba (Winter 1982a, Winter et al. 1979)

Coccolithaceae Kamptner (1928)
Calcidiscus leptoporus (Murray and Blackman 1890)
Crenalithus? sp.
Emiliania huxleyi (Lohmann 1902)
Gephyrocapsa ericsoni McIntyre and Bé (1967)
G. oceanica Kamptner (1943)
G. cf. *G. ornata* Heimdal (1973)
G. protohuxleyi McIntyre (1970)
G. oceanica-E. huxleyi dimorphic
Genus? sp.?
Oolithotus fragilis (Lohmann) subsp. *cavum* Okada and McIntyre (1970)
Umbilicosphaera sibogae (Weber-Van Bosse 1901)

Table E.1 (continued)

Helicosphaeraceae Black (1971), emend. Jafar and Martini (1975)

Helicosphaera carteri (Wallich 1977)

Rhabdosphaeraceae Lemmermann (in Brandt and Apstein 1908)

Acanthoica cf. *A. acathifera* Lohmann (1913)
A. aculeata Kamptner (1941)
Anthosphaera quatricornu (Schiller 1914)
A. robusta (Lohmann 1902)
Discosphaera tubifera (Murray and Blackman 1898)
Rhabdosphaera clavigera Murray and Blackman (1898)
R. longistylis Schiller (1925)
Umbellosphaera irregularis Paasche (1955)
U. tenuis (Kamptner 1937)

Syracosphaeraceae Lemmermann (in Brandt and Apstein 1908)

Anoplosolenia brasiliensis (Lohmann 1919)
Calciopappus caudatus Gaarder and Ramsfjell (1954)
Halopappus cf. *H. adriaticus* Schiller (1914)
Ophiaster formosus Gran (1912)
O. hydroideus (Lohmann 1903)
Michaelsarsia sp. Borsetti and Cati (1976)
Syracosphaera corolla Lecal (1966)
S. corrugis Okada and McIntyre (1977)
S. elatensis Winter (1979)
S. exigua Okada and McIntyre (1977)
S. mediterranea Lohmann (1902)
S. pirus Halldal and Markali (1955)
S. prolongata Gran ex Lohmann (1913)
S. pulchra Lohmann (1902)
S. rotula Okada and McIntyre (1977)
S. variabilis (Halldal and Markali 1955)
S. sp. A

Calyptrosphaeraceae Boudreaux and Hay (1969)

Calyptrosphaera catillifera (Kamptner 1937)
C. oblonga Lohmann (1902)
C. pirus Kamptner (1941)
C. sp. A
Corisphaera arethusae Kamptner (1941)
C. aff. *C. gracilis* Kamptner (1937)
Helladosphaera cornifera (Schiller 1913)
Homozygosphaera quadriperforata (Kamptner 1937)
H. wettsteini (Kamptner 1937)
Sphaerocalyptra bannockii Borsetti and Cati (1976)
S. hasleana (Gaarder 1962)
S. quadridentata (Schiller 1913)
Zygosphaera divergens Halldal and Markali (1955)
Thoracosphaera cf. T. *heimi* (Lohmann 1919)
Thoracosphaera? sp. A

All species were illustrated by Winter et al. (1978, 1979), Winter (1982a) and Mikkelsen (in Reiss 1974)

Among the 53 species identified, 37 have been recorded from both the Atlantic and Pacific Oceans, three from the Pacific only and two exclusively from the Mediterranean Sea. Seven species seem to be endemic to the Gulf of Aqaba. The distribution of three others is unknown. A record of *Gephyrocapsa caribbeanica* (Boudreaux and Hay) by Mikkelsen (1973) refers probably to a misidentification.

Calcidiscus leptoporus, Discosphaera tubifera and *Umbellosphaera irregularis,* common in the Pacific and Atlantic Oceans, are very rare in the Gulf of Aqaba. Conspicuous is the absence in the deep Gulf of prominent deep-water species such as *Florisphaera profunda* Okada and Honjo and *Thoracosphaera flabellata* Halldal and Markali which are common in the Pacific and Atlantic, and present in the central and southern Red Sea proper.

There is **no distinct depth distribution** of the various species in the Gulf of Aqaba and, therefore, deep-water living assemblages reflect the composition of those found in surface waters. All species occurring in the water column occur in the Recent sediments of the Gulf, although delicate holococcoliths are rare (see also Chap. H).

Seasonality of coccolithophore assemblages is discernable in the Gulf, especially by the relative abundance of component species. The seasonal relative abundance of coccolithophore species in surface waters during 1975–1976 is shown in Fig. E.1 and the dominance of species with more than 10% of the assemblage in the water column is given in Table E.2.

The "winter" (December–February) assemblage is dominated by *Emiliania huxleyi* and *Gephyrocapsa ericsoni*, as well as by frequent *Anthospharea quatricornu,* spineproducing *Syracosphaera* spp. and scapholite coccolithophores. The

Table E.2. List of major coccolithophore species (more than 10% of the population) occurring between surface and 400 m depth in the Gulf of Aqaba in order of abundance (Winter et al. 1979)

November 20, 1975	**November 26, 1976**
G. ericsoni	*E. huxleyi*
E. huxleyi	*G. ericsoni*
December 9, 1975	**February 2, 1976**
G. ericsoni	*E. huxleyi*
E. huxleyi	*G. ericsoni*
Spine producing Syracophaeraceae	*G. oceanica*
	Spine-producing Syracophaeraceae
May 3, 1976	**August 9, 1976**
E. huxleyi	*E. huxleyi*
G. ericsoni	*R. longistylis*
U. sibogae	*U. tenuis*
	H. cornifera
	G. protohuxleyi

"summer" (August) assemblage is characterized by a strong reduction in specimens of *Gephyrocapsa*, by a profusion of holococcolith coccospheres and by the frequency of *Syracosphaera*, *Umbellosphaera tenuis* and *Rhabdosphaera longistylis*. A "fall" (November) assemblage is generally dominated by *Gephyrocapsa*. In 1976 this assemblage had more affinities in species composition with the summer assemblage than that in November 1975.

Compound **species diversity** in surface waters of the Gulf of Aqaba has been determined by the method used by Honjo and Okada (1974) for coccolithophores in the Pacific Ocean according to the formula:

$$H' \text{ (natural bels)} = \sum_{i=1}^{s} P_i \ln P_i \text{-} (S\text{-}1)/2N,$$

where P_i is the proportion of the ith species in the population, S is the number of species and N the number of specimens in the sample.

The highest diversity, 2.47 natural bels, was recorded in the summer and the lowest, 1.41 natural bels, in the "transitional" November assemblages. However, the high summer diversity is due to the abundance of holococcolithophores which contributed 12 out of 36 species occurring during that period and to the decrease in dominance of *Emiliania huxleyi* and *Gephyrocapsa ericsoni* (see also Fig. E.1).

A comparison of coccolithophore standing crop and diversity with surface water temperature, regional insolation, nutrients, primary production and chlorophyll a is shown in Fig. E.2. It can be seen that, while coccolithophore diversity correlates positively with water temperature, standing crop shows an inverse correlation with insolation rates. Figure E.2 also shows that, as pointed out above (Chaps. D.6 and D.7), coccolithophores are major primary producers in the Gulf of Aqaba especially during early winter.

The winter abundance of coccolithophores in surface waters of the Gulf of Aqaba is comparable with the highest summer–fall cell densities in the equatorial Pacific Ocean zone and with the greatest winter abundance in the subtropical and northernmost tropical zones of the western North Atlantic (Honjo and Okada 1974, Okada and Honjo 1973, Okada and McIntyre 1979). The August cell density in the Gulf of Aqaba is comparable with that during summer and early fall in the North Pacific transitional zone and with that in late summer in the North Atlantic subtropical zones. High standing crops during autumn and winter and scarcity of coccolithophores during the warmest months are also characteristic of the subtropical North Atlantic central water mass (op. cit.).

The highest diversity in the Gulf of Aqaba (August) is comparable with that in the region of the northern border between the central and transitional zones of the North Pacific, and is even higher than in the equatorial and southernmost central zones of the N. Pacific (note, however, the number of holococcolith species in the Gulf of Aqaba mentioned above). The number of species identified in the Gulf of Aqaba is comparable with that in the central zone (5°–30° lat N) of the North Pacific and in the subtropical, northernmost tropical and transitional zones of the North Atlantic.

The depth distribution of species in the photic zone, known from the subtropical-tropical North Pacific, has not been observed either in the Gulf of Aqaba or in the North Atlantic Ocean.

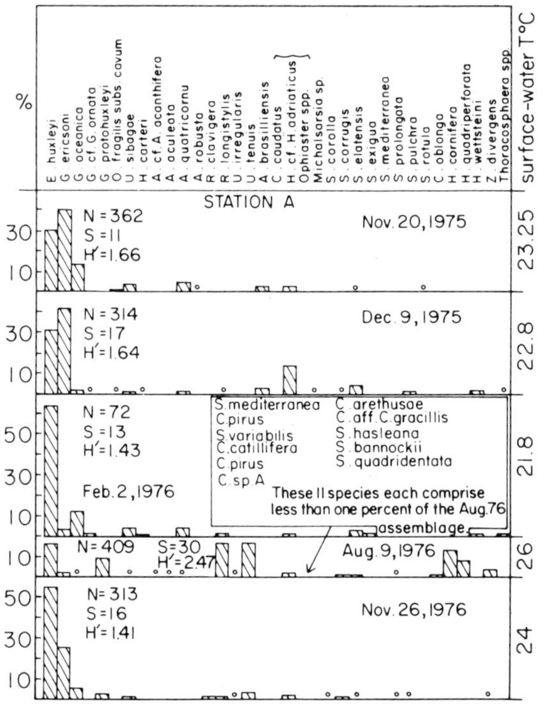

Fig. E.1. Percentage distribution of surface coccolithophore species in the Gulf of Aqaba DCPE-Station A. Note change in community structure and diversity with increase in temperature in the August assemblage. *N* number of specimens counted in the SEM; *S* number of species present; *H′* diversity index in natural bels. O = < 1% (Winter et al. 1979)

Comparison of the **community structure** of the coccolithophore assemblages in the Gulf of Aqaba with those in other regions (including the Red Sea) shows that (despite the absence of deep-water species, the rarity of such species as, e.g., *Umbellosphaera irregularis* and the presence of *Gephyrocapsa protohuxleyi*) the Aqaba flora is closely comparable – especially in the dominance of *Emiliania huxleyi, G. ericsoni,* and *G. oceanica* during most of the year – to the floras occurring at the southernmost border (8°–10° lat N) of the central north zone of the Pacific Ocean and in the subtropical central and northernmost tropical zones (20°–40° lat N) of the northwestern Atlantic Ocean.

The strong reduction in abundance of *E. huxleyi, G. ericsoni,* and *G. oceanica* together with the diversification of the assemblage and increase in frequency of *Umbellosphaera tenuis* at high summer temperatures in the Gulf of Aqaba is also observed in the subtropical, northernmost tropical and transitional zones of the N. Atlantic. It appears that the greatest general similarity of the Aqaba assemblages is with those occurring in the subtropical southwestern N. Atlantic Central Water Mass (Sargasso Sea) and at the southern border of the North West Pacific Central Water Mass with the equatorial one.

Like the Gulf of Aqaba, these highly insolated regions are characterized by very warm, highly oxygenated, oligotrophic and clear waters, stratified during

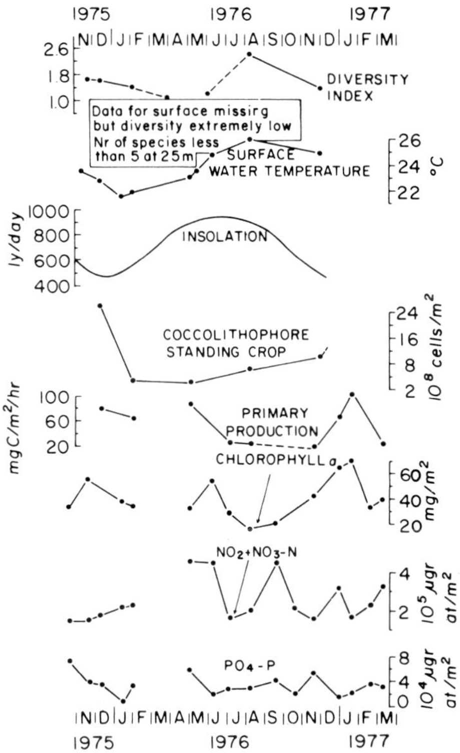

Fig. E.2. Comparison of coccolithophore species diversity and standing crop to physical, biological and chemical parameters in the Gulf of Aqaba. Diversity is given in natural bels. Insolation is undepleted solar radiation for latitude 30° N. Graphs of parameters, except for diversity, temperature and insolation are derived by integration of in situ values in a photic column of 200 m. Primary production, chlorophyll and nutrients from Levanon-Spanier et al. (1979) (Winter et al. 1979)

most of the year. Seasonal fluctuations in temperature, salinity, oxygen and nutrient content in the **upper waters** parallel those in the Gulf of Aqaba (Table E.3; see Chap. D). Replenishment of nutrients within the upper waters occurs in these regions during winter cooling.

Table E.3. Comparison of physical, chemical and nutrient values in upper 150 m in the North Atlantic central water mass (Bermuda), subtropical-tropical North Pacific Ocean (lat. 20°–30°N) and Gulf of Aqaba (Winter et al. 1979)

	Aqaba	North Atlantic	North Pacific
Temperature (°C)	20.50–27.30	19.00–27.70	20.00–27.50
PO_4-P (μg at l^{-1})	0.03– 0.32	0.02– 0.16	0.02– 0.60
NO_3+NO_2-N (μg at l^{-1})	0.07– 2.00	0.01– 1.50	
Oxygen (ml l^{-1})	4.00– 6.10	4,25– 5.50	4.00– 5.00
Salinity (‰)	40.26–41.63	36.10–36.70	34.50–35.50

Differences between the Gulf of Aqaba and the oceanic regions mentioned are the extremely high salinities and densities in the Gulf; the great depth (1,830 m) at which temperatures of above 20 °C occur in the Gulf (as compared with about 150 m in the central water masses); and the homothermal winter stage of the **whole column** of the Gulf in winter as compared with the permanent, deep-water stratification in the ocean.

As pointed out by various authors, light intensity and temperature are major ecological factors affecting coccolithophores. Strong insolation and deep light penetration in nutrient-poor, clear waters have photoinhibitive effects and may prevent large standing crops in summer.

The temperature ranges for the more common 38 species occurring in the Atlantic and Pacific Oceans, as given by Okada and McIntyre (1979), indicate that most of the species are eurythermal over a wide range, 12 species occur in waters warmer than 10°–12 °C, only *Syracosphaera borealis* being restricted to less than 12 °C and *Coccolithus pelagicus* to less than 15 °C. Of particular interest to the Aqaba assemblages are some stenothermal species, i.e., *Gephyrocapsa caribbeanica* with a maximum recorded temperature of 21 °C, *Calciopappus caudatus* with a maximum of 20 °C and *Umbellosphaera irregularis* with a minimum of 21 °C. The absence of *G. caribbeanica* and the presence of *U. irregularis* and *C. caudatus* in the Gulf of Aqaba fit exactly the minimum temperature in the Gulf.

The scarcity of *U. irregularis* and *Discosphaera tubifera* in the Gulf of Aqaba as compared with their abundance in the Atlantic and Pacific Oceans cannot be explained, however, by temperature, and this is also true of the absence in the Gulf (but presence in the Red Sea) of the deep-water species *Florisphaera profunda* and *Thoracosphaera flabellata*. The possibility is not excluded that the very high salinity (and density) in the Gulf of Aqaba and in the northernmost Red Sea present a filter barrier for certain coccolithophores and prevent others from flourishing. On the other hand, high salinity does not seem to have any influence on standing crop or diversity of the flora, and the same is true of seasonal salinity fluctuations in the open ocean.

The role of stratification (pycnocline) in the photic zone, together with light penetration, must also be considered as possibly influencing distribution of deep-water organisms. In the ocean, as well as in the central and southern Red Sea, a pycnocline within the upper 250 m is present even in winter, while in the Gulf of Aqaba the pycnocline nearly disappears during winter. On the other hand, both *F. profunda* and *T. flabellata* apparently prefer high temperatures in the open ocean similar to those in the Gulf of Aqaba during summer, when a pycnocline is developed.

It may also be argued that the absence or scarcity of major pelagic species is due to the proximity of the open waters of the Gulf to land and hence to its being "neritic" in character. The abundance in summer in the Gulf of holococcolithophores – recorded as indicative of near-shore assemblages – may support such a suggestion. On the other hand, tropical and subtropical waters described as "neritic" (e.g., the Red Sea) are dominated by *Gephyrocapsa oceanica,* which is not the case in the Gulf of Aqaba.

A closer comparison of the coccolithophore community structure during the seasons in the Gulf of Aqaba with those characteristic of different zones in the

Table E.4. Comparison of the distribution and abundances of the important coccolithophore species in the Gulf of Aqaba and in the Pacific (Okada and McIntyre 1977); note that the winter coccolithophore assemblage in Aqaba has far more "equatorial" elements than the summer assemblage (Winter et al. 1979)

Species	Percent composition in Aqaba	Occurrence in Pacific zones	Abundance in Pacific
Summer (August 9, 1976)[a]			
E. huxleyi	16	Equatorial-subarctic	Abundant
G. protohuxleyi	8.5	not present	
H. cornifera	12.5	Central-transitional	Occasional
H. quadriperforata	9	Equatorial-central	Rare to occasional
R. longistylis	17	Kuroshio current	Occasional
U. tenuis	17	Central-transitional	Common to abundant
L. divergens	5	Kuroshio current	Rare
Winter (November 20, 1975)			
A. brasiliensis	2.5	Equatorial Central-transitional	Common Occasional
A. quatricornu	5	Transitional	Rare
E. huxleyi	30	Equatorial-subarctic	Abundant
G. ericsoni	40	Equatorial-transitional	Occasional to common
G. oceanica	12.5	Equatorial Central-transitional	Abundant to common Common
O. fragilis subsp. *cavum*	1	Equatorial-central	Common to abundant
Ophiaster spp.	2	Equatorial-transitional	Occasional
U. sibogae	4	Equatorial-transitional	Common
H. wettsteini	1–9	Equatorial-transitional	Occasional to common

[a] Other minor components of summer holococcolithophore assemblages are not found in equatorial zones

Pacific Ocean reveals that the species composing the summer assemblages in the Gulf have much greater affinity to the central zones than the winter assemblages which contain far more "equatorial" elements (Table E.4). Of the six major summer species in the Gulf, four are not found in the Pacific equatorial zone.

On the other hand, the important winter species in the Gulf of Aqaba are all present in the equatorial regions. Of particular interest is *Homozygosphaera wettsteini* which is the only holococcolith-bearing species of the 13 identified so far occurring in winter, but not in the summer assemblages of the Gulf of Aqaba. This species is found in the equatorial to transitional zones of the Pacific. Further evidence of the affinities of the winter assemblage of the Gulf of the equatorial zones are the species characterized by scapholiths which are abundant both in the Aqaba winter and in the equatorial zones. Since the equatorial zones are productive regions, a fertility-dependent relationship is suggested between the seasonal composition of the Aqaba flora and its distribution in the open ocean.

A further indication for such a fertility dependence of coccolithophores is provided by the ratio *Emiliania huxleyi/Gephyrocapsa oceanica*. Figure E.3 shows the

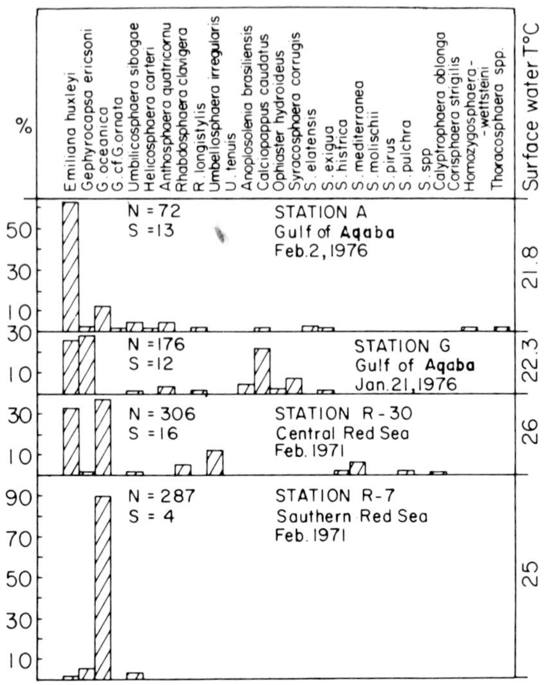

Fig. E.3. Comparison of percentage distribution of coccolithophore species in surface waters between the Gulf of Aqaba (1976) and the Red Sea (1971) during winter. *N* number of specimens counted; *S* number of species present. Red Sea data from Okada and Honjo (1975) (Winter et al. 1979)

percentage distribution of coccolithophorids in surface waters in the Gulf of Aqaba and in the Red Sea during winter. Figure E.4 shows the distribution of *E. huxleyi* and *G. oceanica* in core-top sediments and in living floras along the stretch Gulf of Aden–Red Sea–Gulf of Aqaba, as compared with changes in temperature, salinity, and phosphate content of the upper waters.

The dominance of *G. oceanica* over *E. huxleyi* in the Red Sea has been attributed to the "neritic" character of the latter (see, however, above). Considering the well-known euryhalinity of *E. huxleyi*, one would be tempted to correlate its northward-increasing dominance over *G. oceanica* in terms of the increasing salinity. However, stable isotope and fossil evidence obtained from deep-sea cores in the Gulf of Aqaba and in the Red Sea (see Winter 1982a, Winter et al. 1983, Chap. H) strongly indicate that *G. oceanica* is more euryhaline than *E. huxleyi* in very high salinity (>45‰) conditions. Laboratory experiments also indicate that *E. huxleyi* does not survive in salinities greater than 45‰. The change from an *E. huxleyi* and *G. ericsoni*-dominated assemblage in the southernmost central zone of the N. Pacific (lat. 8°–10° N) to a *G. oceanica*-dominated assemblage in the equatorial zone immediately to the south, without any appreciable change in (the oceanic) salinity is also noteworthy in this connection.

On the other hand, the gradual change from a *G. oceanica*-dominated assemblage in the Gulf of Aden to an *E. huxleyi*-dominated one in the Gulf of Aqaba

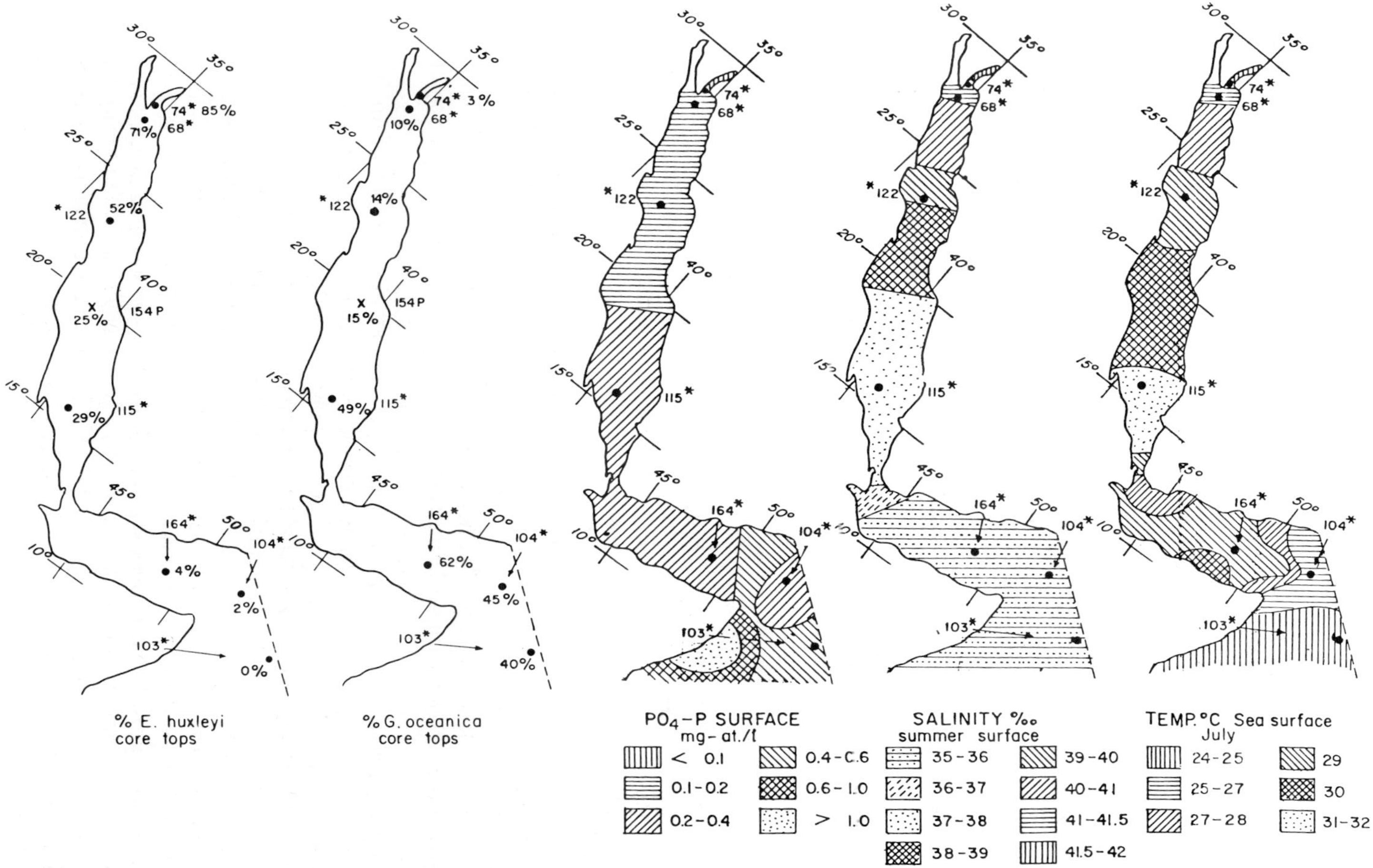

Fig. E.4. Distribution of *E. huxleyi* and *G. oceanica* and ecological parameters along the stretch: Gulf of Aqaba – Red Sea – Gulf of Aden. Based on core-top data from Atlantis II-cores (marked with *asterisk*; Winter 1982a), Red Sea cores RC 9–164, V 14–104 and C 61–158 (marked x) (McIntyre 1969) and on living floras betwen lat 21° N and 11° N (Okada and Honjo 1975) (Winter 1982a)

can be well correlated with the strong gradient of nutrients along this stretch (Fig. E.4), thus indicating a dependence of the *huxleyi/oceanica* ratio upon fertility of the sea. The importance of phosphate as a factor controlling the distribution of coccolithophores, especially of *G. oceanica*, in the Pacific Ocean has been shown by Roth and Berger (1975). Correlation of core-top data from the Indian Ocean (B. Molfino, unpubl.) and from the Pacific Ocean (Geitzenauer et al. 1976) with potential primary production and PO_4-P values in the water respectively, also suggests that nutrient content of the sea influences the distribution of *E. huxleyi* and *G. oceanica*. It appears that *E. huxleyi* is found in low to highly fertile regions of the sea, while *G. oceanica* replaces it in warm and productive waters, even if of very high salinity.

The northward decreasing abundance of *Calcidiscus leptoporus* from the Gulf of Aden toward the Gulf of Aqaba (Müller 1976) parallels that of its decrease from the equatorial zone to the southern central zone in the Pacific and may be thus equally dependent upon nutrient levels.

Considering the fact that neritic waters are generally more productive than open ocean ones, the view of the central-southern, more productive Red Sea as "neritic" in comparison with the oligotrophic Gulf of Aqaba is compatible with the nutrient-productivity pattern.

The causal relationship between abundance or presence/absence patterns of certain coccolithophore species on one hand, and nutrients on the other, is unknown at present. The efficiency of nutrient uptake is known in only a few species, among them *E. huxleyi*. Phagotrophy may be more widespread among the photoautotrophic coccolithophores (see, e.g., *Coccolithus pelagicus*) than generally assumed. *E. huxleyi* may also be able to use organic matter, which may explain its abundance in oligotrophic waters. Little is known as yet on influence of light in relation to nutrient-dependent photosynthetic rates in coccolithophores. Nevertheless, there seems to be a correlation between calcification and light as connected with photosynthetic rates, as well as between coccolith shape and light. Nutrient levels, through phyto- and zoo-plankton biomass, have considerable importance in influencing turbidity, hence illumination of the upper waters, and may thus indirectly determine the distribution and even morphologies of different species.

Some observations on the **morphology** of certain species in the Gulf of Aqaba are relevant to ecologial-biogeographic problems. Seasonal variation in morphology of the coccoliths of *E. huxleyi* is expressed by the central area which is solid toward its middle part by partial fusion of elements in winter specimens and cribrate to comb-like in summer specimens. Such a variation has been recorded from the Atlantic Ocean, where it is correlatable with temperature and latitude (McIntyre and Bé 1967). That (seasonal and regional) light intensive may be involved in influencing this variation can, however, not be excluded. Encrusted specimens are very rare in the living flora of the Gulf of Aqaba. Encrusted coccoliths with extreme shape modification occur, however, in Late Quaternary sediments in the Gulf and in the Red Sea proper (see Chap. H). They are found mainly in winter collections in the southern sector of the Gulf, at all depths. Malformed specimens like those reported from nitrogen-deficient, marginal (or neritic) environments are nearly absent in the Gulf of Aqaba, despite its low nutrient content.

The morphology of the placoliths of *Gephyrocapsa oceanica* in the Gulf of Aqaba places them in the second of the three groups described by Okada and McIntyre (1977). None of the specimens bears an additional bridge, common on specimens from the Pacific and Atlantic. Most of the coccospheres in the Gulf occur in clusters, a salient feature reported for phenotypes of *G. oceanica* which, however, do not occur in the Gulf of Aqaba. Numerous coccospheres collected from surface waters in winter exhibit apparent etching, i.e., separation of unit crystals along the seams, a feature previously reported for placoliths collected from great depth, but never from surface waters, supersaturated with regard to $CaCO_3$, like those in the Gulf. Possibly the etching is due to digestion of coccolithophores by larger organisms.

Rare dimorphic coccospheres with placoliths of both *E. huxleyi* and *G. oceanica* occur in winter in the southern sector of the Gulf. Although this dimorphism supports the close affinity of these species, the possibility is not excluded that in fact (seasonal?) dithecatism of *Emiliania* is involved.

Abundant living specimens of *Gephyrocapsa* which, from a morphological point of view, must be attributed to *G. protohuxleyi* – a species generally believed to be extinct since 75,000 years BP. – occur in summer and fall collections in the Gulf. Although it is possible that this species has persisted to the present day only in the Gulf of Aqaba (see also Chap. H), the specimens occurring in the Gulf may belong in fact to a "transitional" morphotype of a species "plexus" whose end members (co-occurring in the Gulf) are *Gephyrocapsa ericsoni* and *Emiliania huxleyi.*

In summary, the study of the coccolithophore assemblages in the Gulf of Aqaba and their empirical correlation with ecological factors in the Gulf itself, in the Red Sea and the Gulf of Aden, as well as in the Atlantic and Pacific Oceans, indicate that temperature is a major factor influencing composition, diversity, and seasonality of the assemblages; light intensity affects seasonality and bathymetry of standing crops; salinity and density may affect certain species, possibly preventing deep-water species to survive; and that nutrient levels influence seasonality of standing crop and the dominance of certain species – possibly because of their phagotrophic abilities – in nutrient-poor environments. The changes in coccolithophore assemblages along the stretch Gulf of Aden – Red Sea – Gulf of Aqaba closely parallel those from equatorial regions into central water masses of the ocean, thereby reflecting an environmental gradient in the warm ocean from highly productive to oligotrophic waters (see also Chap. B). Morphology of certain species is apparently temperature-dependent, although light levels may also play a role. No influence of nutrient levels on malformation (reported elsewhere) was observed in the Gulf of Aqaba.

References: Geitzenauer et al. (1976), Honjo and Okada (1974), McIntyre and Bé (1967), Mikkelsen (1973), Okada and Honjo (1973, 1975), Okada and McIntyre (1977, 1979), Reiss (1974), Roth and Berger (1975), Winter (1982 a), Winter et al. (1978, 1979, 1983).

E.3 Foraminiferida

The present-day assemblages of **planktic Foraminifera** in the Gulf of Aqaba and in the northernmost Red Sea were studied both in the living plankton and in bottom sediments. A total of 1,300 stratified vertical plankton hauls, collected in 1974–1976 between surface and 600 m depth, with 65 μ, 200 μ and 330 μ nets (58 cm diam.) at eight DCPE-stations, were examined. Several hundred surface sediment samples (obtained by grab and Willemoes chandelier) were examined qualitatively. Seventy-two sediment samples, as well as six deep-sea, core-top samples, were also studied quantitatively (see Fig. E.5, as well as Fig. D.9).

Since 65 μ plankton samples were scattered with regard to sampling date and location, and in order to facilitate comparison of data from the Gulf of Aqaba and the Red Sea with those from other seas, the data discussed here refer mostly to 1,068 stratified plankton samples obtained with 200 μ nets. The differences in composition of foraminiferal assemblages in 200 μ and 330 μ nets were found to be very small, and for this reason in some cases quantitative data from both net sizes were combined.

Fifteen species of planktic Foraminifera characteristic of warm waters, were recognized in plankton and sediment samples, i.e. (see Plate E.1):

x		2 (H)	1. *Globigerina bulloides* (d'Orbigny 1826)
x		2 (H)	2. *Globigerina calida* (Parker 1962)
s		2 (I)	3. *Globigerinella siphonifera* (d'Orbigny 1839)
	p	2 (H)	4. *Globigerinita glutinata* (Egger 1893)
	p	(H)	5. *Globigerinita* cf. *G. uvula* (Ehrenberg 1861)
	x	s	6. *Globigerinoides elongatus* (d'Orbigny 1826)
x	s	1 (L)	7. *Globigerinoides ruber* (d'Orbigny 1839)
x	s	1 (L)	8. *Globigerinoides sacculifer* (Brady 1877)
	x	1	9. *Globoturborotalita rubescens* (Hofker 1956)
	x	2 (I)	10. *Globoturborotalita tenella* (Parker 1958)
	x	2 (H)	11. *Hastigerina pelagica* (d'Orbigny 1839)
x	s	2 (H)	12. *Orbulina universa* (d'Orbigny 1839)
x	s		13. *Orbulina* sp. ("small")
			14. *Tenuitella anfracta* (Parker 1967)
x	(s)		15. *Turborotalita clarkei* (Rögl and Bolli 1973)

The species were illustrated and discussed by Reiss et al. (1974), but the generic position of some of them is changed here in accordance with recent taxonomic revisions (see Vincent and Berger 1981). The specimens reported earlier as *Globigerina quinqueloba* Natland from the Gulf of Aqaba and northernmost Red Sea, as well as from other warm-water regions, belong apparently all to *T. clarkei*.

Eleven species of those listed above and marked "x" are definitely spinose, two marked "p" are nonspinose, pustulate to pseudospinose, and one, *T. anfracta,* is microperforate-smooth. *T. clarkei*, with its widely scattered pores, is most probably spinose like other species of *Turborotalita*.

Three species (marked 1) are "shallow-water" (epipelagic) ones, recorded as living predominantly in the mixed layer (upper 50 m) of the world oceans and

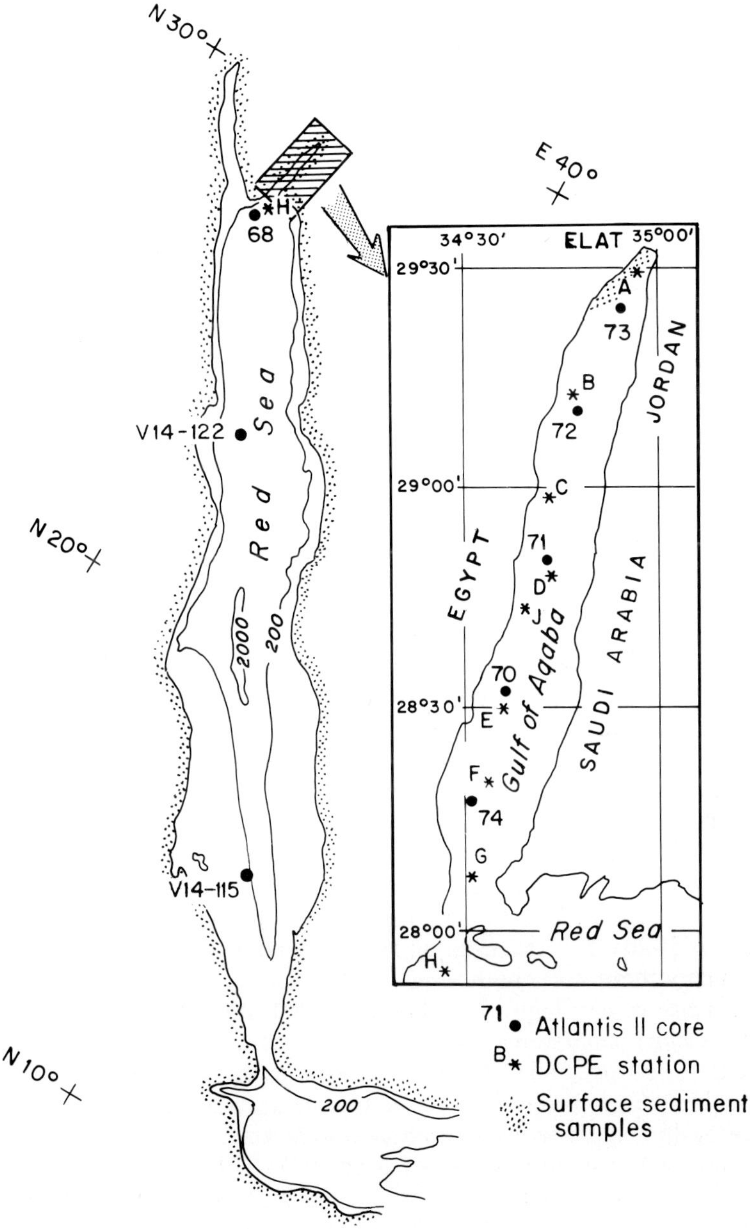

Fig. E.5. Location map of DCPE stations, Atlantis II deep sea cores and quantitatively examined surface sediment samples having low vertical diel migrations (L). Seven species (marked 2) are recorded as "intermediate" (shallow mesopelagic) species, living predominantly between 50 and 100 m. Among them, four species have high diel migration ranges (H), two intermediate ones (I), and one a low diel migration range (L). [For information on ecology and distribution of planktic Foraminifera, the reader is referred to the publications cited in Reiss et al. (1974) and especially to the comprehensive bibliography in Vincent and Berger (1981)].

Six out of the 15 species recorded (marked "s") are known to be associated with dinophycean sequestered endosymbionts, as well as with dynophycean commensals and parasitic swarmers. It is not known at present whether *T. clarkei* possesses symbionts, although other species of the genus *Turborotalita* do posses sequestered zooxanthellae. In the plankton of the Gulf of Aqaba *T. clarkei* has a whitish translucent appearance which is usually not characteristic of symbiont bearers.

The feeding habits of the different species are only incompletely known. *Hastigerina pelagica* and *Globigerinoides sacculifer* seem to be exclusively carnivorous (copepods and other crustacenas). *Orbulina universa, Globigerinella siphonifera* and *Globigerinoides ruber* seem to prefer a largely carnivorous diet, although they feed also on diatoms and phytoflagellates. *Globigerina bulloides* was reported to ingest exclusively certain diatoms and chlorophytes. The diet of the other species is completely unknown at present. Digestion of endosymbionts in *Hastigerina pelagica, Globigerinella siphonifera, Globigerinoides sacculifer,* and *G. ruber* from the Gulf of Aqaba has been suggested by Hansen (1975). Although artifacts of preparation for EM study might be involved in these observations, the possibility is not excluded that under special conditions (e.g., starvation) digestion of symbionts may actually take place.

The **standing crop** of planktic Foraminifera (specimens per 10 m^3) of the 600 m column at the different DCPE stations is given in Fig. E.6. It can be seen that at the same time of the year the standing crop in the northern sector (Stations A–C) is three times lower than that in the southern one (Stations E–G). Immediately outside the Straits of Tiran (Station H) the standing crop is three times higher than in the southern sector of the Gulf.

Dominance is very high in the **low-diversity** foraminiferal assemblage, three species – *Globigerinoides sacculifer, Globigerinoides ruber* (counted together with the rare *G. elongatus*), and *Globigerinella siphonifera* – making up, in varying proportions, between 96% and 99% of the specimens in 200–330 μ samples. In 65 μ samples the dominant species is *Turborotalita clarkei.*

The **seasonal variation** during the period 1974–1976 of the total standing crop and of the dominant species in the 600 m column, averaged over DCPE-Stations A–G in the Gulf of Aqaba, is shown in Fig. E.7. Although **long-range variation** is evident, several general trends can be discerned: June, September, and February are poor in foraminifera, while October to January, and April are periods of high foraminiferal standing crop (August 1975 is an exception, with high standing crop at Station D). *Globigerinoides sacculifer* is mainly responsible for this general distribution pattern. *G. ruber* is relatively frequent all year round, except during the winter season. *Globigerinella siphonifera,* on the other hand, is rare during the summer months.

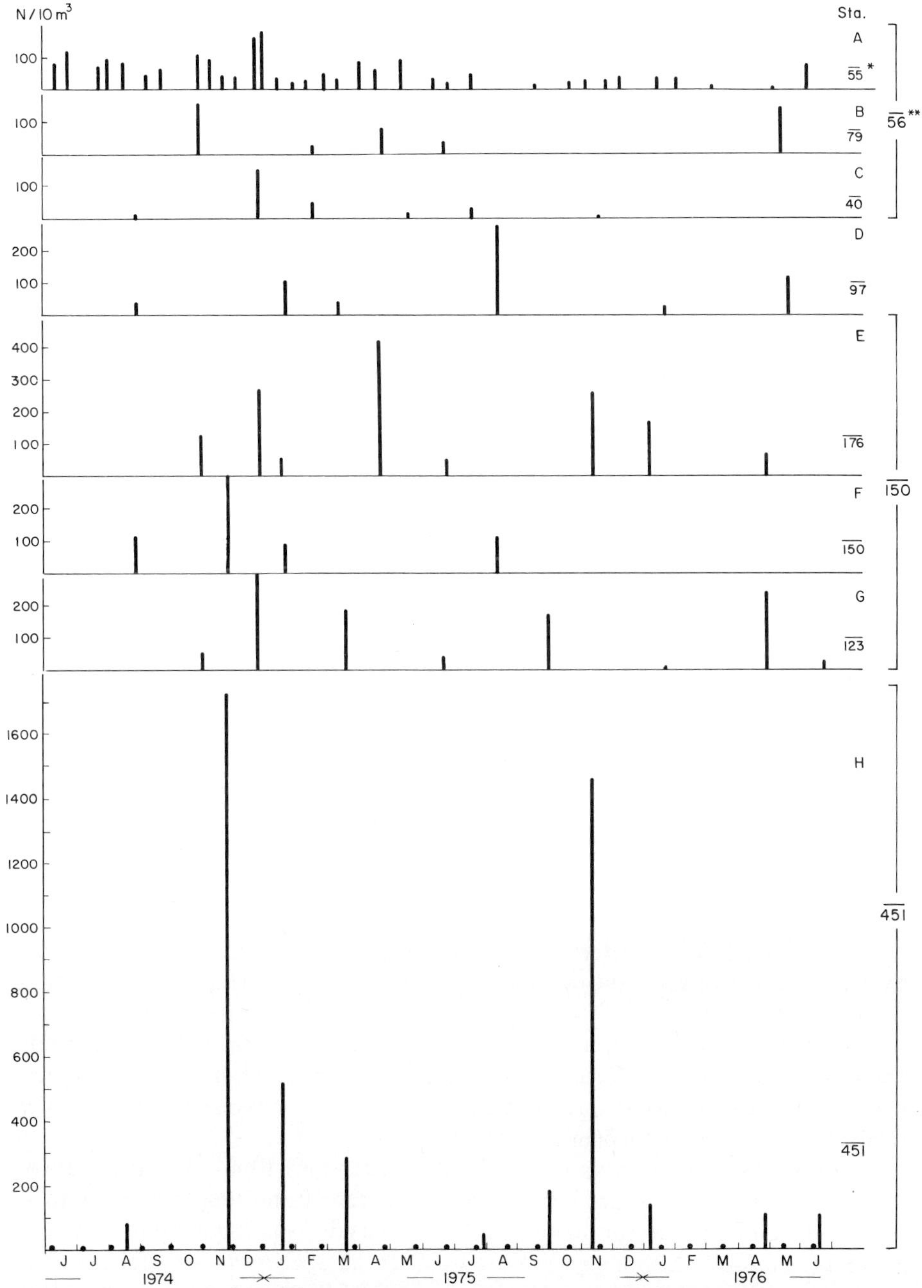

Fig. E.6. Standing crop of planktic foraminifera >200 μ in upper 600 m by station and sampling date. Yearly average for each station. Yearly average for northern and southern sectors of the Gulf of Aqaba, as well as for the northernmost Red Sea (Almogi-Labin 1981)

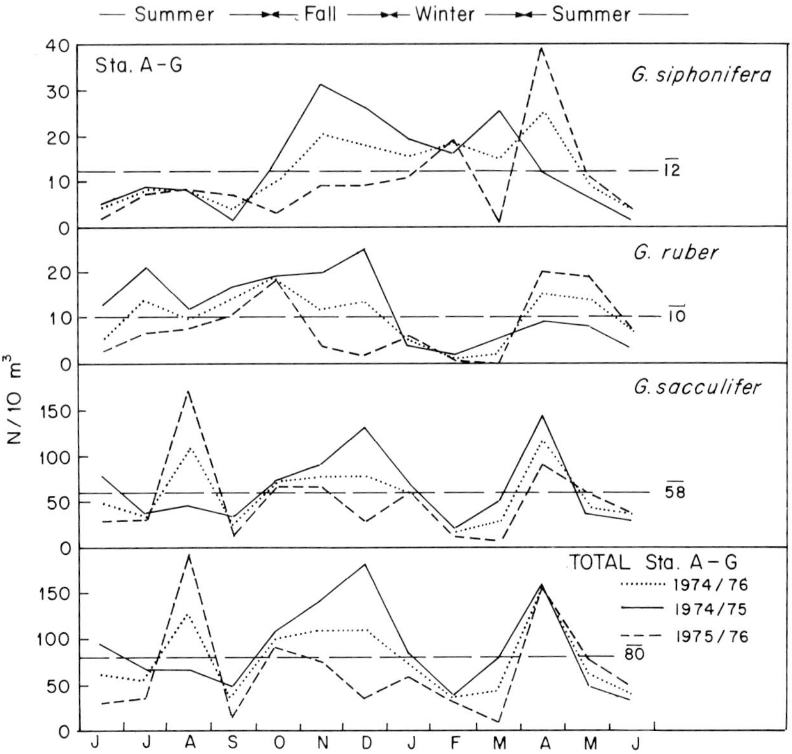

Fig. E.7. Monthly average standing crop of planktic Foraminifera >200 μ (total and dominant species) in the upper 600 m at DCPE-Stations A–G (Gulf of Aqaba) 1974–1976. Note change of scale from *G. siphonifera* and *G. ruber* to *G. sacculifer* and total (Almogi-Labin 1981)

The **seasonal depth pattern** of the total standing crop in the northern and southern sectors of the Golf of Aqaba, as well as in the northernmost Red Sea, is shown in Fig. E.8.

The general increase in abundance from north to south, indicated already above (Fig. E.6), is paralleled by a gradient of greater abundance to greater depths. High abundance in the uppermost 50 m is usually associated with greater richness of foraminifera in deeper waters.

Maximum concentrations (more than 50 specimens 10 m^{-3}) in the Gulf of Aqaba are found in the upper 100 m. Values between 10 and 50 specimens 10 m^{-3} occur in the northern sector down to 200 m, but as deep as 300 m in the southern sector of the Gulf. Although absolute numbers of specimens are higher in the northernmost Red Sea (Station H) the abundance-depth pattern there is similar to that in the southern sector of the Gulf.

Expectionally high standing crops were observed down to 400 m at the beginning of the homothermal stage in December 1974 both in the northern and southern sectors of the Gulf. High general abundance to great depth was observed at Station E in April 1975.

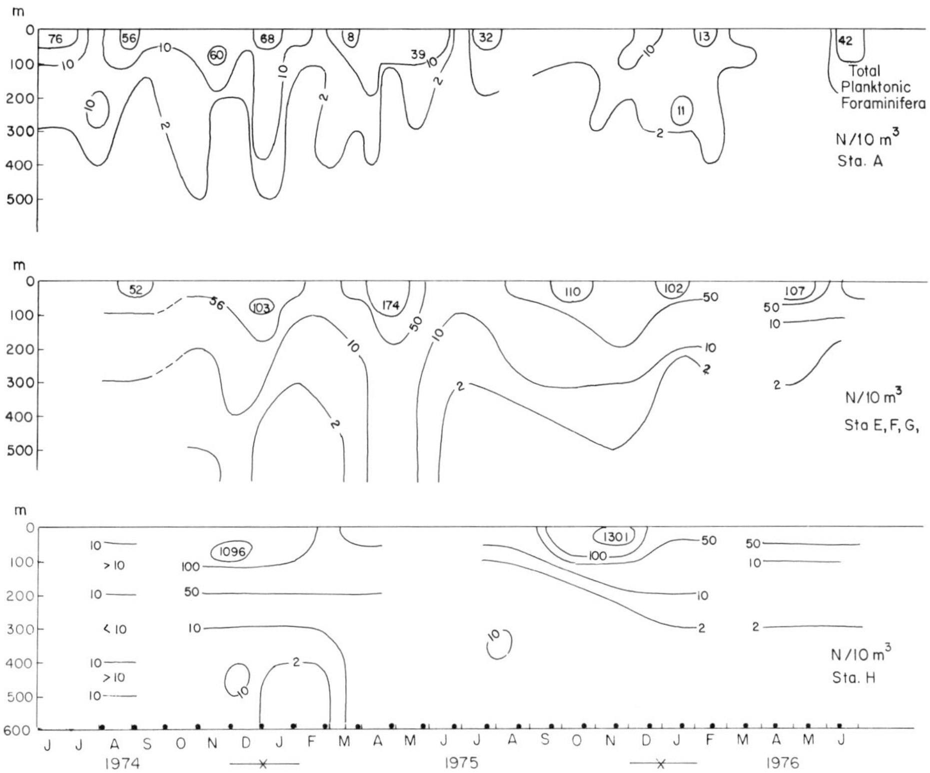

Fig. E.8. Depth-distribution of total standing crop of planktic Foraminifera in the northern and southern sectors of the Gulf of Aqaba and northernmost Red Sea, during 1974–1976 (Almogi-Labin 1981)

The distribution of **standing crop by stations** and **depth,** averaged in accordance to the main seasons, is shown in Fig. E.9. The general decrease in average standing crop from south to north can be observed, with the highest average values in the Red Sea immediately outside the Straits of Tiran. The highest abundance there in November–December is associated with a relative high abundance all over the Gulf of Aqaba in the upper 100 m in the northern sector, deepening to 200 m in the south. Similar high average abundance values are noticeable in January–March only in the southern sector of the Gulf, within the upper 50–100 m. During April–October, the upper 50 m in the southern and middle sectors are relatively rich in planktic Foraminifera. Figure E.10 shows the **seasonal depth distribution of the three major species** in the two sectors of the Gulf of Aqaba and outside the Straits of Tiran.

The influence of *Globigerinoides sacculifer* on the total standing crop (Fig. E.8) is evident, both within the Gulf of Aqaba and outside it in the Red Sea. While in the northern sector of the Gulf *G. sacculifer* occurs in relatively high numbers mainly within the upper 50 m, intermediate values are found down to 200 m. In the southern sector of the Gulf relatively high abundances of *G. sacculifer* occur

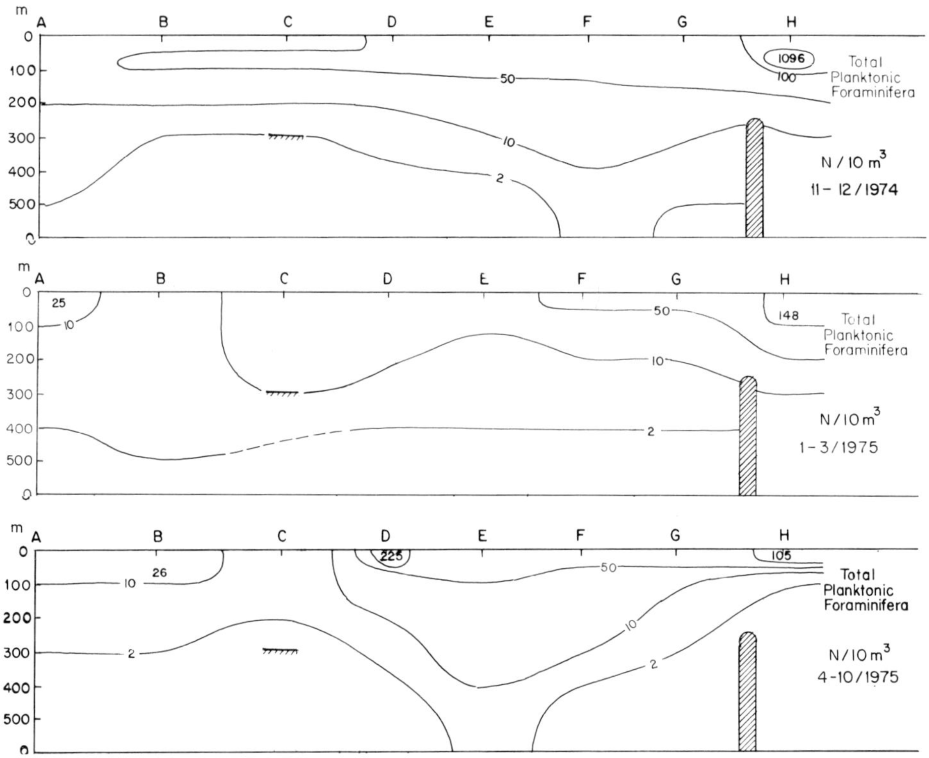

Fig. E.9. Average seasonal standing grop of planktic Foraminifera in the Gulf of Aqaba and in the northernmost Red Sea (Almogi-Labin 1981)

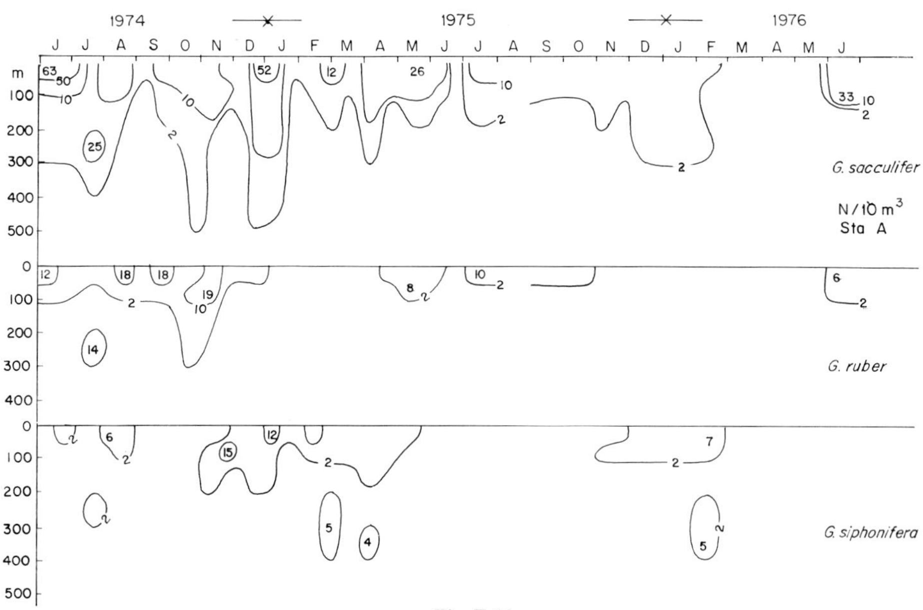

Fig. E.10

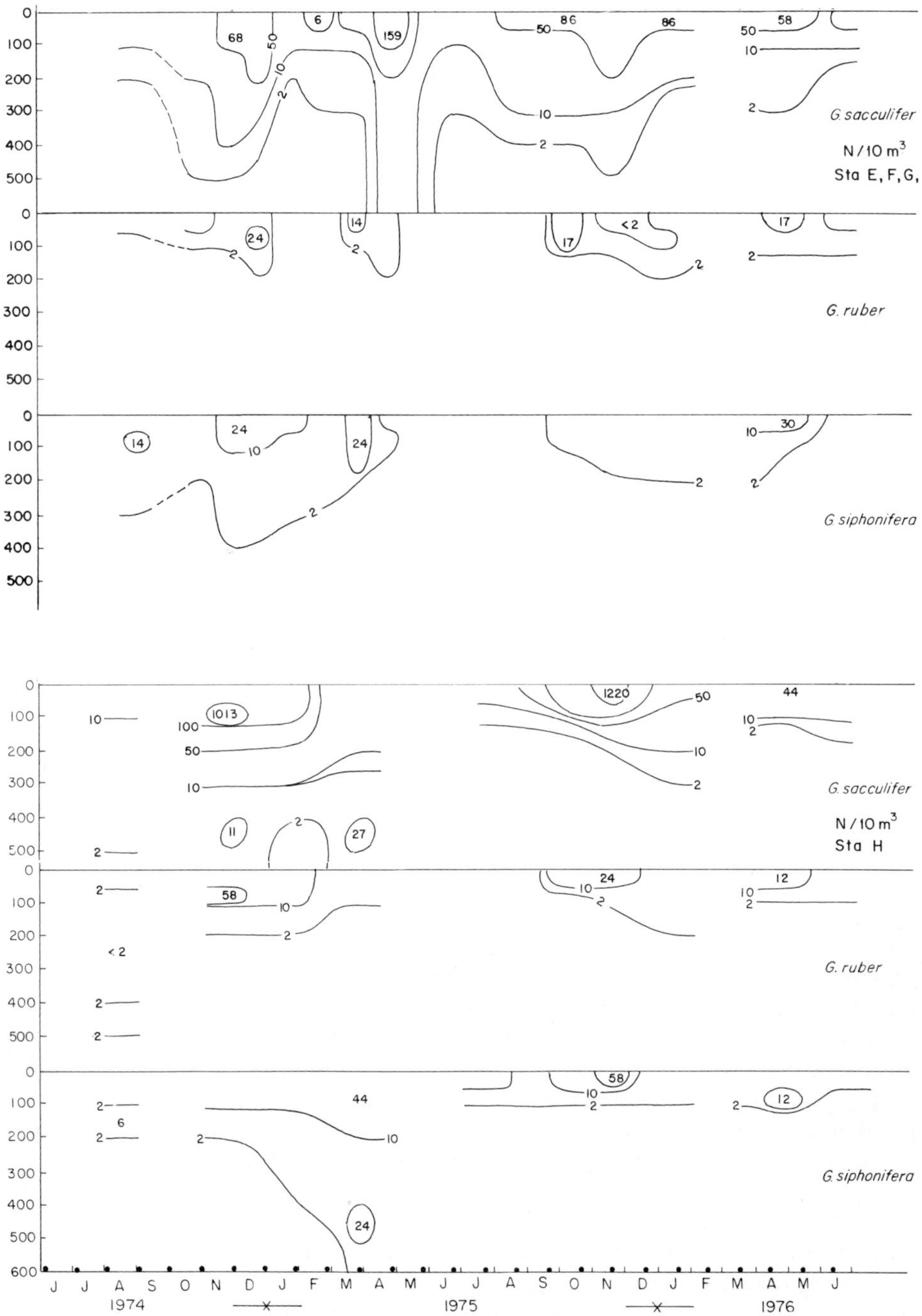

Fig. E.10. Depth distribution of dominant planktic foraminiferal species in the northern and southern sectors of the Gulf of Aqaba and in the northermost Red Sea during 1974–1976 (Almogi-Labin 1981)

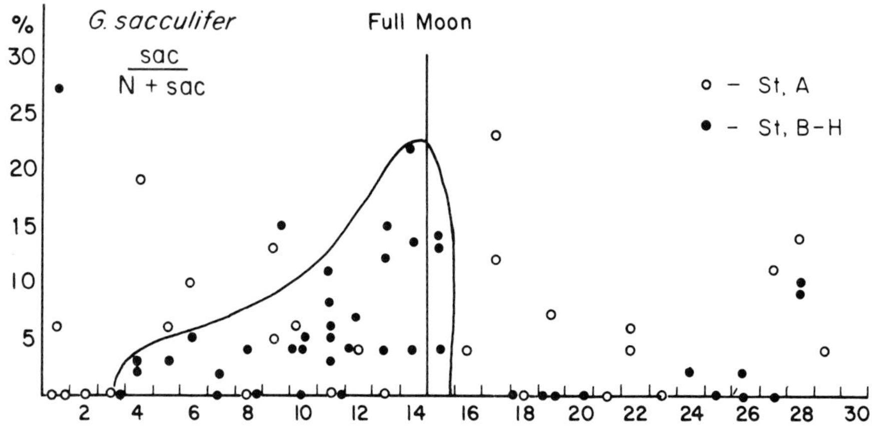

Fig. E.11. Ratio *G. sacculifer* with sac-like chamber to those without such a chamber in the upper 600 m at DCPE-Station A and at Stations B–H according to the day of the lunar month (Almogi-Labin 1981)

down to 200 and intermediate numbers down to 300 m. The northernmost Red Sea is similar in this respect to the southern sector.

G. ruber, although rarer and concentrated at shallower depths than *G. sacculifer,* shows the same trend, i.e., higher numbers restricted to the upper 50 m in the northern sector and to the upper 100 m in the southern sector of the Gulf and in the northernmost Red Sea. Peaks of *Globigerinella siphonifera* occur occasionally in winter in the upper 100 m in the northern sector of the Gulf and in the upper 200 m both in the southern sector and in the northernmost Red Sea (cf. Fig. E.7). In 65 μ plankton hauls *Turborotalita clarkei* is more abundant in samples deeper than 150 m. A **depth ranking by frequency** of *G. ruber–G. siphonifera–G. sacculifer–T. clarkei* is therefore apparent.

The **date of plankton sampling** is of great importance to the assessment of the relative frequency of individual species. Thus, 80% of the frequency peaks of *Globigerinoides sacculifer* (more than 100 specimens larger than 200 μ per 10 m^3) occur in the upper 600 m at all stations between new moon and full moon. The highest ratio between sac-bearing (gametogenic) and normal (largely nongametogenic) specimens of *G. sacculifer* is observed in the water column around full moon at all stations, except at Station A, where relatively high ratios occur also at other times of the month (Fig. E.11). One week before full moon the number of specimens increases significantly in samples deeper than 200 m.

A detailed quantitative and biometric study of *G. sacculifer* populations in the upper 10 m opposite the MBL Elat during 1979–1980 indicates that the highest percentage of specimens larger (up to 500 μ^{max}) than the average (about 200 μ) occurs 2–3 days before full moon and is drastically reduced within 2 days afterwards (Fig. E.12). The data for specific days of the synodic month are shown on a model for average size of specimens in the 10 m population in Fig. E.13, illustrating the monthly cycle of *G. sacculifer*.

Since both the size-frequency variation and the percentage of sac-bearing specimens, especially at depth, reflect **reproduction** processes, it seems that *G. sac-*

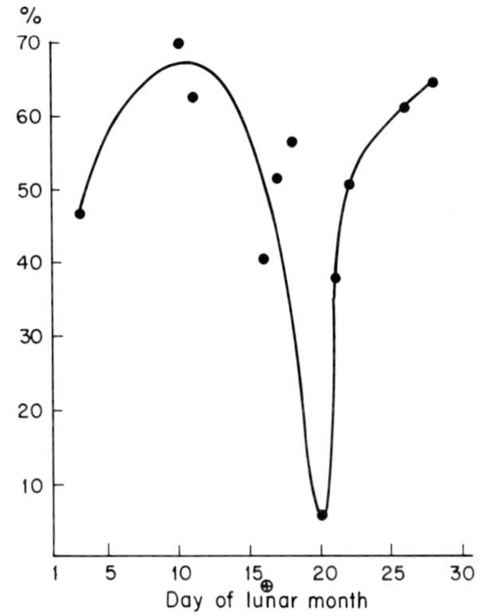

Fig. E.12. Frequency of specimens of *G. sacculifer* larger than the median in the upper 10 m in the northern Gulf of Aqaba according to the day of the lunar month (Avraham 1981)

culifer in the Gulf of Aqaba has a reproduction periodicity which is closely related to the moon cycle, like *Hastigerina pelagica* from the Atlantic Ocean. Laboratory experiments by J. Erez (unpubl.) indicate a life span (until gametogenesis) of 28 days for *G. sacculifer* from the Gulf of Aqaba.

The species composition of planktic foraminifera >63 μ in **sediment assemblages** between 50 and 600 m depth in the northern Gulf of Aqaba (see Fig. E.5) is shown in Fig. E.14. The strong dominance of *G. sacculifer* followed by *G. ruber* between 50 and 200 m on one hand, and that of *T. clarkei* between 200 and 600 m on the other, reflect dominance in the plankton hauls with mesh sizes of 200 μ and 65 μ respectively.

Judging from the abundance of aragonitic pteropods in sediments and from the high saturation values with regard to calcite (Chap. D), no selective dissolution of foraminiferal tests should be expected in the Gulf of Aqaba. This despite the fact that all species occurring are low-resistant to dissolution and the major dominant ones are among the least dissolution-resistant.

Indeed, the assemblage composition of specimens > 149 μ in surface sediment and core-top samples compared with plankton samples (200 μ net hauls), as

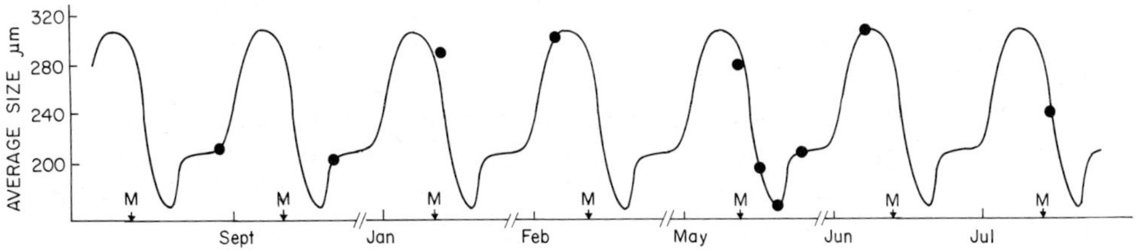

Fig. E.13. Average specimen size of *G. sacculifer* in upper 10 m in the northern Gulf of Aqaba in relation to lunar month-dependent scheme (Avraham 1981)

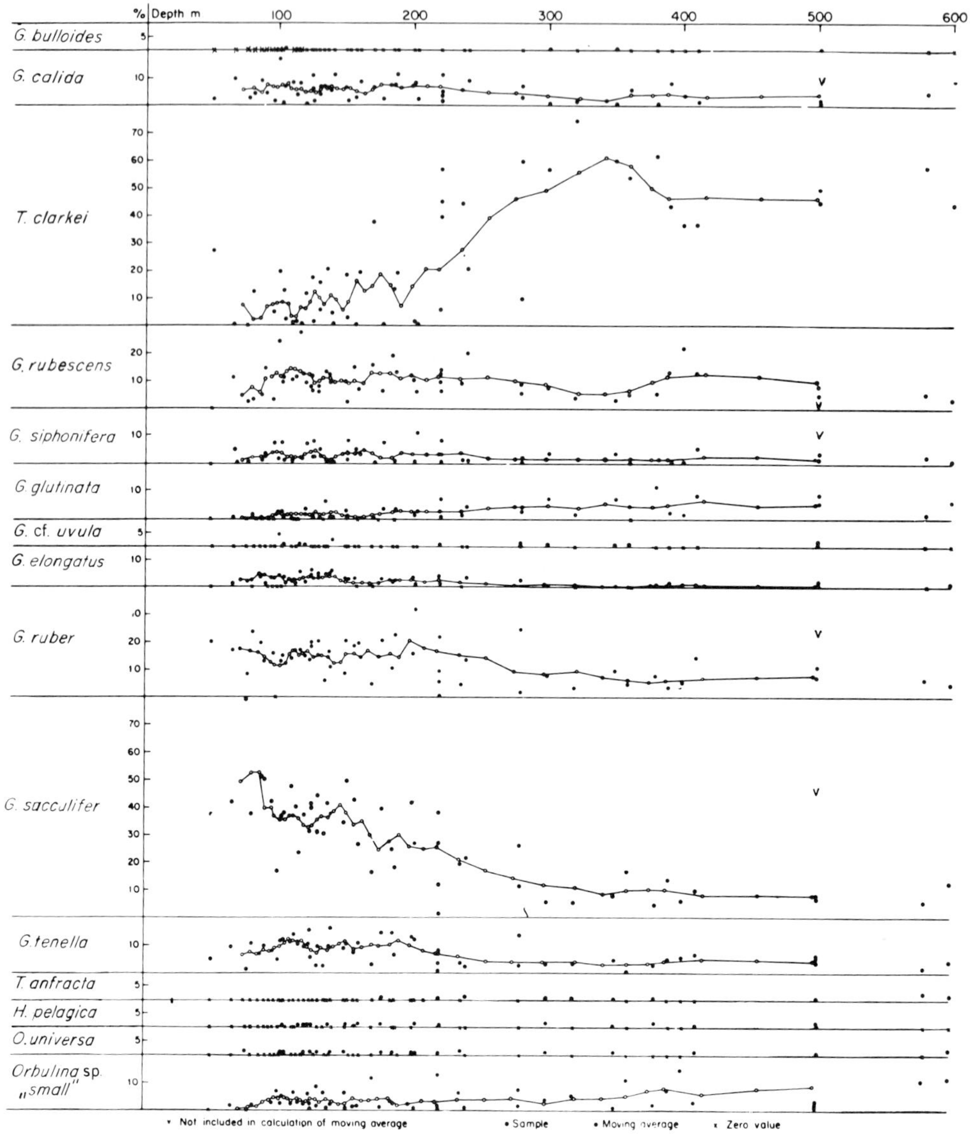

Fig. E.14. Percentage distribution and moving averages per five stations of planktic foraminiferal species >63 μ in surface sediment-samples, northern Gulf of Aqaba. Calculated from 100% planktic assemblage. *G. bulloides* includes *G. falconensis* (Reiss et al. 1974)

shown in Fig. E.15, indicates a close correspondence in percentage distribution, with the exception of *G. sacculifer* and *G. ruber*: the percentage of *G. sacculifer* is nearly half that of *G. ruber* in the sediment samples, as compared with the plankton populations. The reason for this discrepancy lies in the reproductive cycle of *G. sacculifer* which – at least in the Gulf of Aqaba – seems to be twice as long as that of *G. ruber*.

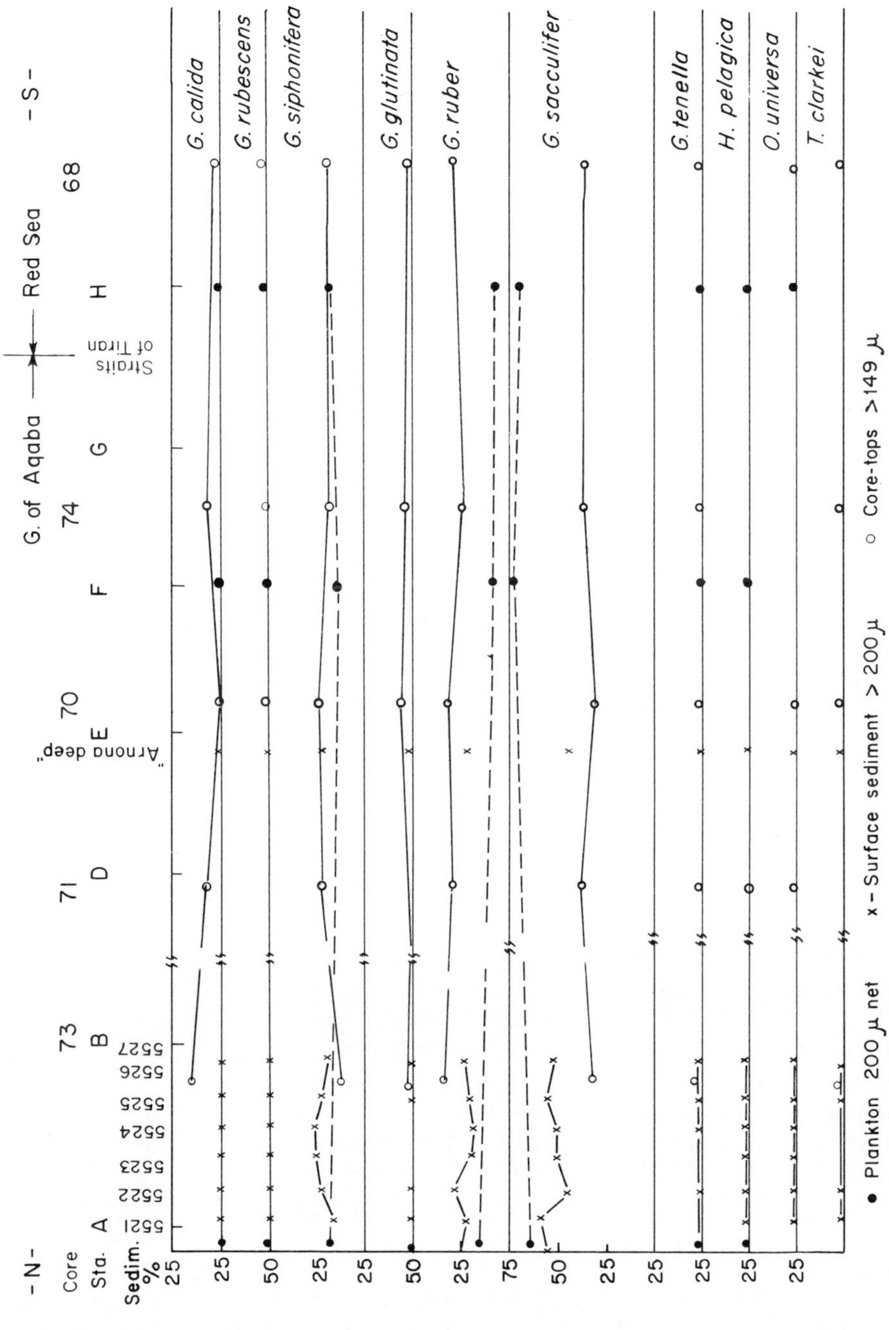

Fig. E. 15. Comparison of weighted average relative frequencies of planktic foraminiferal species in plankton hauls (> 200 µ), in surface sediments (> 200 µ) and in core-top samples (> 149 µ) (Almogi-Labin 1981)

An analysis by size fraction of the ratio between sac-bearing and "normal-form" *G. sacculifer* in surface sediments of the Gulf of Aqaba (Fig. E.16) indicates the highest numbers of sac-bearing, large specimens to be present at depths of more than about 150 m. This may be the approximate minimum depth of reproduction of *G. sacculifer*.

The **plankton/benthos ratio** of foraminiferid specimens >63 μ in sediment samples from the Gulf of Aqaba (Fig. E.17) shows a steady increase with depth. No planktic specimens are usually present in samples from less than 35 m. The P/B ratio reaches about 3% at 50 m, and is higher than 90% in core-top samples from more than 800 m depth (see Chap. H). The ratio is a function of the reduction in number of shallow water, largely light-dependent benthic species with high specimen frequencies and of the increase of planktic specimens in the deepening water column.

The plankton/benthos ratio in sediments also indicates that in the Gulf of Aqaba no significant transport of shallow water material occurs across the steep slopes and that sorting, winnowing, or bottom transport are not important factors (except in turbidites of the "deeps").

Interpretations in terms of ecological factors of the foraminiferal living and sediment assemblages in the Gulf of Aqaba and northern Red Sea have been attempted by Reiss et al. (1974), Reiss and Halicz (1976) and lately by Almogi-Labin (1981) (see also Chap. H and references). These interpretations involve empirical correlations of community structure with known ecological factors based mainly on (1) published information on distributional patterns of planktic foraminifera in the world oceans which has accumulated during the last two decades and which permits certain generalizations on the ecological setting of individual species (for references, see Bé 1977, Vincent and Berger 1981); (2) the synoptic sampling DCPE program in the Gulf of Aqaba and northernmost Red Sea; and (3) on (albeit still restricted) knowledge derived from in situ and laboratory experiments on planktic Foraminifera carried out in various places, including the MBL in Elat.

The component species of the assemblages occurring in the Gulf of Aqaba and northernmost Red Sea are known from the Indian, Pacific and Atlantic Oceans. With the exception of *Orbulina* sp. ("small") and *Globigerinita* cf. *G. uvula*, all species were recorded from the middle and southern Red Sea and – with the additional exception of *Tenuitella anfracta* – from the Mediterranean. Except for *Globoturborotalita tenella, Globoturborotalita rubescens, Turborotalita clarkei,* and *Tenuitella anfracta,* probably missed by large plankton mesh sizes, all species were reported from the Gulf of Aden.

All dominant species and most of those co-occurring in the Gulf of Aqaba and northernmost Red Sea are spinose, symbiont-bearing, epipelagic to shallow mesopelagic (see above). Conspicuously **absent** in the Gulf of Aqaba and in the northernmost Red Sea are tropical to subtropical nonspinose, finely perforated, carinated forms *(Globorotalia)* and nonspinose cancellate ("pitted") ones *(Neogloboquadrina* and *Pulleniatina)* which are, however, present in the middle and southern Red Sea and frequent in the Gulf of Aden.

Species of these genera lack symbionts and live below the mixed layer of the oceans, adult stages occurring predominantly below the upper thermocline (me-

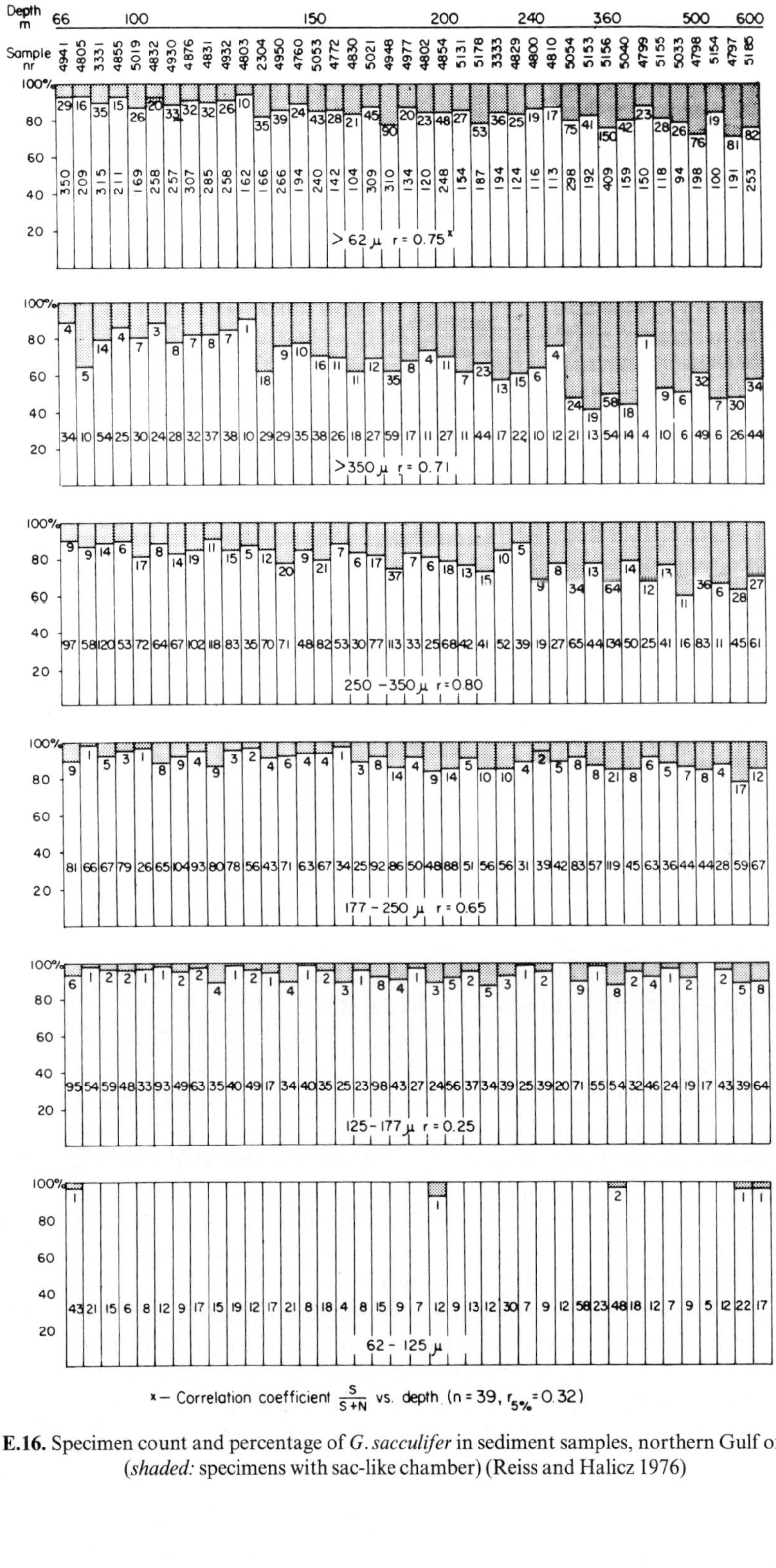

Fig. E.16. Specimen count and percentage of *G. sacculifer* in sediment samples, northern Gulf of Aqaba (*shaded:* specimens with sac-like chamber) (Reiss and Halicz 1976)

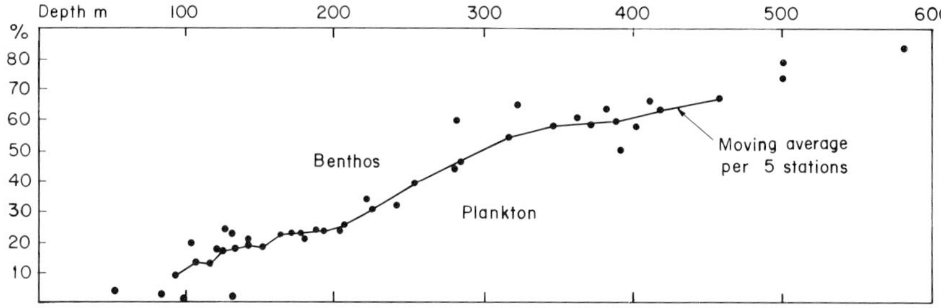

Fig. E.17. Foraminiferal plankton/benthos ratio, northern Gulf of Aqaba (Reiss et al. 1974)

so- to bathypelagic forms). Some species of *Globorotalia* and *Pulleniatina* seem to be omnivorous with specific multiple food requirements. Nothing is known on feeding habits of *Neogloboquadrina* or of *Sphaeroidinella.* Diel migration patterns recorded are low for *Pulleniatina* and intermediate to high for *Globorotalia* and *Neogloboquadrina.* It is also noteworthy that the initially spinose and finally cortex-forming *Sphaeroidinella,* whose adult stages are deep-water dwelling, is present in the southern Red Sea and Gulf of Aden, but absent in the Gulf of Aqaba and northernmost Red Sea.

The assemblage in the Gulf of Aden and southern Red Sea contains, in addition to the species occurring in the Gulf of Aqaba, *Boliella adamsi* (Banner and Blow), *Globigerinita humilis* (Brady), *G. iota* (Parker), *Globigerinoides conglobatus* (Brady), *Sphaeroidinella dehiscens* (Parker and Jones), *Neogloboquadrina conglomerata* (Schwager), *N. dutertrei* (d'Orbigny), *Globorotalia cultrata (menardii)* (d'Orbigny), *G. tumida* (Brady), and *Pulleniatina obliquiloculata* (Parker and Jones).

For the sake of brevity the assemblage in the Gulf of Aqaba and northernmost Red Sea will be referred to here as the "Aqaba assemblage" while that characteristic of the Gulf of Aden and main Red Sea will be called the "Aden assemblage".

The northern border of occurrence in the Red Sea of *Globorotalia cultrata* and *Pulleniatina obliquiloculata* is at approximately lat. 20° N and that of *Neogloboquadrina dutertrei* (*G. eggeri* of some authors) at approximately lat. 27° N. [The record of *Globorotalia cultrata* by Rosenberg-Herman (1965) from core VA14-127 at the entrance of the Gulf of Suez is questionable.] The boundary between the typical Aden and Aqaba assemblages lies therefore **within** the Red Sea.

Both the Aqaba and Aden assemblages are composed of species with a combined latitudinal range of about 5°–35° N and S and with a combined surface temperature range of 18°–30 °C. They are, therefore, of a general **tropical-subtropical** character. The gyral circulation of the oceans generates a distributional pattern that is like a moving ring, with subtropical faunas at its core and mixed faunas (tropical and – depending on the hemisphere – northern or southern) around its periphery. Thus, subtropical faunas are characteristic mainly of central water masses, being increasingly mixed with elements of the surrounding current systems at their borders. Boundary currents carry tropical faunas far from the equatorial regions and mid-latitude faunas into low latitudes. Isolation and mixing by currents, endemism, expatriation, and tropical submergence of foraminiferal species have been widely discussed in the literature.

The **Aqaba assemblage** is best compared, both by species composition and relative frequency, with those characteristic of **central water masses** of the oceans. Most relevant to this comparison are the North Atlantic Central Water Mass (Sargasso Sea), especially its southern part, e.g., between approximately lat. 22°–30° N and long. 40°–65° W, south of the "tropical convergence" and near the gyre center throughout the year, and its northern part, north of lat. 30° N, during summer only; the western North Pacific Central Water Mass, mainly the region of approximately lat. 30° N, long. 160° E; as well as the less well-defined Central Water of the Indian Ocean, mainly the area at approximately lat. 20° S and long. 60°–70° E (assemblages P5-P6 and S3 of Bé and Hutson 1977).

The **Aden assemblage** corresponds to the tropical-subtropical assemblages P7 and S5 of the Indian Ocean and can be compared closely to those of the southernmost **gyre-margins (equatorial current systems)** of the N. Atlantic and N. Pacific Oceans, as well as to some extent with the winter assemblage in the northwestern Sargasso Sea, north of lat. 30° N.

Thus, a gradual change from an assemblage characteristic of the open-ocean, tropical-equatorial zones to one characteristic of the central water masses of the ocean gyres can be discerned from the Gulf of Aden along the increasingly desert-enclosed stretch Red Sea–Gulf of Aqaba.

The known **temperature** ranges of the species characterizing the Aden assemblage do not explain either their absence in the Aqaba assemblage or their scarcity or complete absence in oceanic central water masses.

Absence of carinated planktic foraminifera has been correlated with absence of strong thermal gradients (Wonders 1980). Indeed, the thermal gradients within the upper 300 m weaken increasingly from the Gulf of Aden toward the northern Red Sea. On the other hand, keeled *Globorotalia* are nearly absent during the stratified summer stage of the northern Sargasso Sea and become abundant only during the isothermal winter-overturn, while in the Gulf of Aqaba no *Globorotalia* occur either during the stratified stage throughout most of the year or during the isothermal winter stage.

Salinity is certainly an important factor in distribution of planktic organisms and high salinity has been invoked to explain the changes of planktic associations in the Red Sea. However, salinity does not explain the changes from equatorial to central water assemblages in the open ocean; there, salinities are much lower than in the Red Sea and, moreover, the equatorial to central water salinity gradients are different from ocean to ocean (increase in the Atlantic and decrease in the Pacific). The strong dominance of *Globigerinoides sacculifer* in the faunas of the Gulf of Aqaba also proves that the dominance of *G. ruber* over *G. sacculifer* in subtropical regions of the ocean is **not** due, as widely claimed, to higher salinity, which is far below that of the G. of Aqaba. Laboratory experiments also prove that *G. sacculifer* still reproduces at salinities of about 48%. The absence of *G. sacculifer* in the Eastern Mediterranean (Levantine) Basin, explained by high – albeit lower than in the Gulf of Aqaba – salinities, is probably due to temperatures of less than 17 °C at depth of reproduction (approximately 150–170 m; see above as well as Chap. H).

Density of the water, as a function of temperature and salinity, has been regarded as a major limiting factor (through osmotic pressure effects) in the distribution of planktic foraminifera. Important evidence has been accumulated which indicates that at species level ranges of density are more restricted than either those of temperature or of depth.

Density limitations could well explain the depth stratification of deep-water species, as related to **diel** and especially to **reproductive cycle** migration patterns. The density ranges recorded for *Globorotalia cultrata, Pulleniatina obliquiloculata* and *Globigerinoides conglobatus* (present in the Aden assemblage and absent in the Aqaba assemblage) are about 24–26.2, 25–26, and 24–26 σt respecitvely. The isopycnal 26 in the Red Sea reaches the surface at approximately lat. 22° N during winter and lat. 25°–26° N in summer. Considering the recorded depth, temperature and diel migration ranges of the species mentioned, one is tempted to relate the northern boundary of their occurrence in the Red Sea to the configuration of isopycnals.

On the other hand, the density range of *Globorotalia cultrata* includes values available in the permanently stratified southern Sargasso Sea, where, however, this species is nearly absent. Furthermore, the density ranges recorded for *Globigerinoides ruber* and *G. sacculifer* (24–25.2 and 22–25.5 σt respectively) are far below the densities in the Gulf of Aqaba, where these species are abundant. Thus, while density is doubtless an important factor influencing planktic organisms, it does not fully explain by itself the foraminiferal presence-absence or abundance patterns in the Aden or Aqaba assemblages or for that matter in the open-ocean ones.

Oxygen content of the water is an important parameter affecting animals. Data on presence and relative abundance of *Globorotalia cultrata, Pulleniatina obliquiloculata,* and *Neogloboquadrina dutertrei* – all characteristic of current waters – may suggest a correlation with relatively high oxygen content. However, these species are absent from the Gulf of Aqaba and northern Red Sea, despite the high oxygen levels in these waters.

Diel migration patterns, specific depth distribution and seasonality of planktic foraminifera have been interpreted by various authors as a response to **food requirements** governed by the influence of **nutrient levels** and **light penetration** both on primary producers serving as food and on symbionts. The distinct and predictable seasonal succession of planktic Foraminifera in the N. Atlantic has been attributed to the direct dependence of the foraminifera on the **nutritional setting,** the various physicochemical parameters affecting the foraminiferal patterns only to the extent that they affect the abundance and particular suite of food phytoplankters at any given season.

The strong insolation throughout the year in subtropical regions facilitates the profusion of symbiontic algae with efficient nutrient assimilation. The similarity of the predominantly shallow water, spinose and symbiont-bearing foraminiferal Aqaba assemblage with those characteristic of central water masses is best attributed to the oligotrophic conditions characterizing them. The similarity of the Aden assemblage – comprising nonspinose, symbiont-lacking, intermediate to deep-water forms, having multiple food requirements – to those in equatorial current systems may be correlated with high fertility of the waters. It is also noteworthy that *Globigerina bulloides,* a species known from fertile upwelling regions, is abundant in the Aden assemblage, but extremely rare in the Aqaba assemblage.

Light penetration and primary production are intricately interconnected and related to climatic zones and to circulation patterns. Light penetration is deepest in oligotrophic waters, i.e., both in the central water masses and in the northern Red Sea–Gulf of Aqaba. It decreases rapidly with depth due to high biomass in productive regions of the sea like in current systems, upwelling areas or the monsoon-influenced Gulf of Aden and southern Red Sea. Deep light penetration is also enhanced by lack of significant amounts of abiogenic material in suspension both in central water masses, which are far from land, and in the Red Sea–Gulf of Aqaba, where transporting rivers from the surrounding land are absent.

Although little is known on **light-tolerance** levels of planktic Foraminifera, this may be a factor of great limiting importance, affecting geographical and bathymetric distribution of species, especially of those depending upon a **particular suite of food organisms.** Light-tolerance limitations may explain latitudinal and seasonal changes in depth ranges (tropical submergence), as well as the absence of certain intermediate to deep water, probably negative phototropic, foraminiferal species in oligotrophic waters with unusually deep light penetration. The required suite of food organisms – at present largely unknown – may not be available to these species **at depths to which they are forced by light levels.** It is also possible that foraminiferal species with restricted density ranges are prevented by light intensity from occupying water layers with suitable density, but at too shallow depths. Thus, nutritional setting and light levels may determine the absence of such species despite available density values, which are optimal to them. Possibly the specific food organisms required by deeper-dwelling Foraminifera may be sensitive by themselves to water density–light patterns (cf., e.g., deep-dwelling coccolithophores) and thus influence distribution patterns of these Foraminifera.

The importance of light to planktic Foraminifera is further emphasized by the fact that calcification rates in symbiont-bearing species are enhanced by light itself, even after photosynthesis of symbionts has been artifically (DCMU) inhibited.

The gradual northward disappearance in the Red Sea of the species characteristic of the Aden assemblage follows the gradient of increasing oligotrophy, deeper light penetration and higher density in the upper 300 m (cf. the change in coccolithophore assemblages, Chap. E.2).

The boundary between the Aden and Aqaba assemblages corresponds closely to the location of the major drop in nutrient and primary production levels and of the steepening density gradient in the upper waters of the Red Sea. The absence north of this boundary of "seed specimens" of the characteristic Aden species prevents their occurrence in the Gulf of Aqaba even during the winter season, when nutrient supply to the upper waters is enhanced.

The nutrient and temperature patterns in the Gulf of Aqaba and northern Red Sea affect the **seasonal, biogeographic,** and **depth distribution** of the (shallow water) assemblage of planktic foraminifera. *Globigerinoides ruber* is known to be most abundant in the least productive regions of the sea, particularly during warm seasons. The affinity of *G. sacculifer* and *Globigerinella siphonifera* to relatively more fertile waters has been noted by various authors. The optimum temperature range of *G. siphonifera* reaches, however, lower values than that of *G. sacculifer*.

In the Gulf of Aqaba and northernmost Red Sea, *G. ruber* is relatively most abundant during the warmest and least productive season (April–October). *G. sacculifer* dominates at the beginning of the winter-overturn (November–December) and at its end (April), when the water temperature is above the lowest value (the optimum temperature recorded for *G. sacculifer* in the Indian Ocean is about 25 ± 3 °C) and nutrient content is high. *G. siphonifera* is relatively frequent during the season of the lowest water temperature (January–March) and equally high nutrient content.

The north–south trend of species abundance is also correlatable with nutrient content and to some extent with temperature. Thus, *G. ruber* is percentage-wise more abundant (approximately 50%) in the northern Gulf of Aqaba during summer than in the southern sector of the Gulf or in the northernmost Red Sea (10% average), where *G. sacculifer* dominates. At the beginning of winter the latter reaches 75% of the assemblage in the northern Gulf of Aqaba and 90% in the northernmost Red Sea. *G. siphonifera* composes on the average 40% of the assemblage in the Gulf and in the Red Sea during the coolest and most fertile months.

Since *G. sacculifer* is generally the dominant species in the Gulf of Aqaba and northern Red Sea, the total standing crop varies with temperature, nutrient content and location, i.e., it is highest toward the south at the beginning and the end of the winter.

An exceptional peak of *G. sacculifer* in August 1975 at station D may be due to the arrival of nutrient-enriched northern Red Sea mid-waters, upwelling in May–June (see also Chap. D.4 and compare patterns of pteropod abundance, Chap. E.4).

The north–south trend in abundance variation with depth, especially of *G. sacculifer,* reflects the total frequency in the upper 100 m and, therefore, the general nutrient–temperature pattern. It is, however, noteworthy that in the Gulf of Aqaba–Red Sea *G. sacculifer* lives at greater depth than recorded from other regions.

The north–south trend of increasing standing crop of planktic Foraminifera in the Gulf of Aqaba and northernmost Red Sea also strongly resembles that from central water masses into equatorial current systems of the oceans: e.g., 50 specimens per 10 m^3 in the northern Gulf of Aqaba as compared with 3–50 specimens per 10 m^3 in the southern Sargasso Sea; and 450 specimens per 10 m^3 in the northernmost Red Sea as compared with 30–900 specimens per 10 m^3 in the equatorial western Atlantic. The lowest standing crop in the Gulf of Aqaba occurs not only during the warmest and least productive season, but also at the time of strongest insolation. The possibility is not excluded that photoinhibition of symbionts in *G. ruber*, the shallowest water, typically summer species, also plays a role.

It seems, therefore, that **nutritional regime, light penetration** and **temperature** are the major factors influencing the composition and seasonality of the planktic foraminiferal assemblages in the Gulf of Aqaba. There seems to be a good correlation between the standing crop of planktic Foraminifera (dominated by *Globigerinoides sacculifer*) and the abundance of copepods, at least as known from synoptic information derived at Station A during 1974–1975 (cf. Figs. D.45, D.49, D.52, and E.7). The same is true of abundance of *Globigerinella siphonifera* as compared with the major winter peaks (February–March) of diatoms and dinoflagellates. On the other hand, the complete food requirements of the different species of planktic Foraminifera are unknown as yet. Also, there are no long-range observations on the phytoplankton in the Gulf of Aqaba and no information at all on picoplankton, which may be important as a food source in oligotrophic waters. More knowledge in this respect would probably explain the long-range seasonal variation in abundance patterns of planktic Foraminifera observed during 1974–1976, especially such localized peaks as, e.g., at Station D in August 1975 (see Figs. E.6 and E.7).

References: Almogi-Labin (1981), Avraham (1981), Bé (1977), Bé and Hutson (1977), Bé et al. (1977), Cifelli and Smith (1974), Erez (1983), Hansen (1975), Reiss and Halicz (1976), Reiss et al. (1974, 1980), Vincent and Berger (1981), Wonders (1980).

E.4 Pteropoda

Living assemblages of **Pteropoda** in the Gulf of Aqaba and northernmost Red Sea were studied in 1,268 stratified, vertical haul samples, obtained at DCPE stations A–H (Figs. D.9 and E.5) with 65 μ, 200 μ, and 330 μ nets (diameter 58 cm), and with 333 μ Indian Ocean Nets (100 cm diameter) during 1974–1976. The data discussed here refer mostly to 1,068 hauls obtained with 200 μ nets. Since the

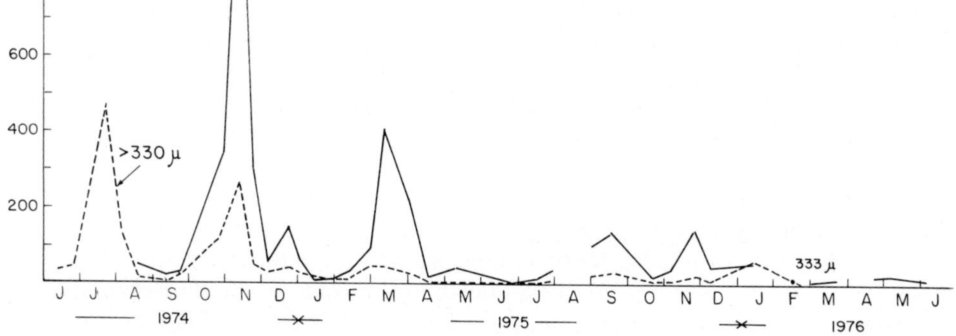

Fig. E.18. Standing crop of Pteropoda >200 μ and 330 μ in upper 600 m at DCPE-Station A during 1974–1976 (Almogi-Labin 1981)

differences between results derived from 200 μ and 330 μ nets are more significant in the case of pteropods than in that of Foraminifera (see Fig. E.18 and cf. Fig. E.17), 330 μ and 333 μ samples were used for quantitative studies in a few instances only.

Numerous surface sediment samples were examined qualitatively and 45 samples were studied quantitatively. These latter samples are Willemoes chandelier and grab samples obtained in the northern Gulf of Aqaba and examined also for planktic Foraminifera (Figs. D.9 and E.5). In addition, five Atlantis II core-top samples were studied qualitatively and quantitatively (Fig. E.5).

The quantitative study dealt exclusively with thecosomate species and subspecies.

A total of nine pteropod taxa were recognized in the Gulf of Aqaba and in the northernmost Red Sea, i.e.:

Euthecosomata

E *Cavolinia longirostris* (de Blainville 1821) f. *flexipes*, Van der Spoel (1971)
(E) *Cavolinia uncinata* (Rang 1829) f. *pulsatoides* Van der Spoel (1971)
M *Clio convexa* (Boas 1886)
E *Creseis acicula* (Rang 1828)
E *Creseis virgula virgula* (Rang 1828)
E *Diacria quadridentata* (de Blainville 1821) f. *crassa* Van der Spoel (1971)
M *Limacina inflata* (d'Orbigny 1836)
E *Limacina trochiformis* (d'Orbigny 1836)

Pseudothecosomata

M *Peraclis reticulata* (d'Orbigny 1836)

All species listed were illustrated and discussed by Almogi-Labin and Reiss (1977). However, the species recorded by the latter authors as *Cuvierina* cf. *C. columnella urceolaris* is a gastropod. The species recorded as *Clio polita* is a gymnosomate pteropod, while the specimens of *Limacina bulimoides* found in surface sediment samples are older Holocene, fossil ones (see Chap. H).

The species marked "E" are **epipelagic,** while those marked "M" are **mesopelagic** (brackets denote insufficient data). For information on distribution and ecology of the species mentioned, the reader is referred to the work of Van der Spoel (1967) and Bé and Gilmer (1977), as well as of Almogi-Labin (1981, 1982).

Of the nine species occurring in the Gulf of Aqaba, eight are **herbivores,** only *P. reticulata* being possibly carnivorous.

The assemblage comprises species widely distributed in **tropical-subtropical** regions of the ocean, as well as three subspecies endemic to the Gulf of Aqaba and Red Sea.

The total **standing crop** of pteropods in the upper 600 m water column **at all DCPE-stations** during 1974–1976 is shown in Fig. E.19 (see also Fig. D.49). Strong year to year variation is observable, particularly in the northern sector of the Gulf of Aqaba. There is no overall north–south trend in standing crop, like that observed in planktic Foraminifera. The high standing crops at Stations D and F in August 1975 are noteworthy (see also planktic Foraminifera, Chap. E.3 and cf. Chap. D.4).

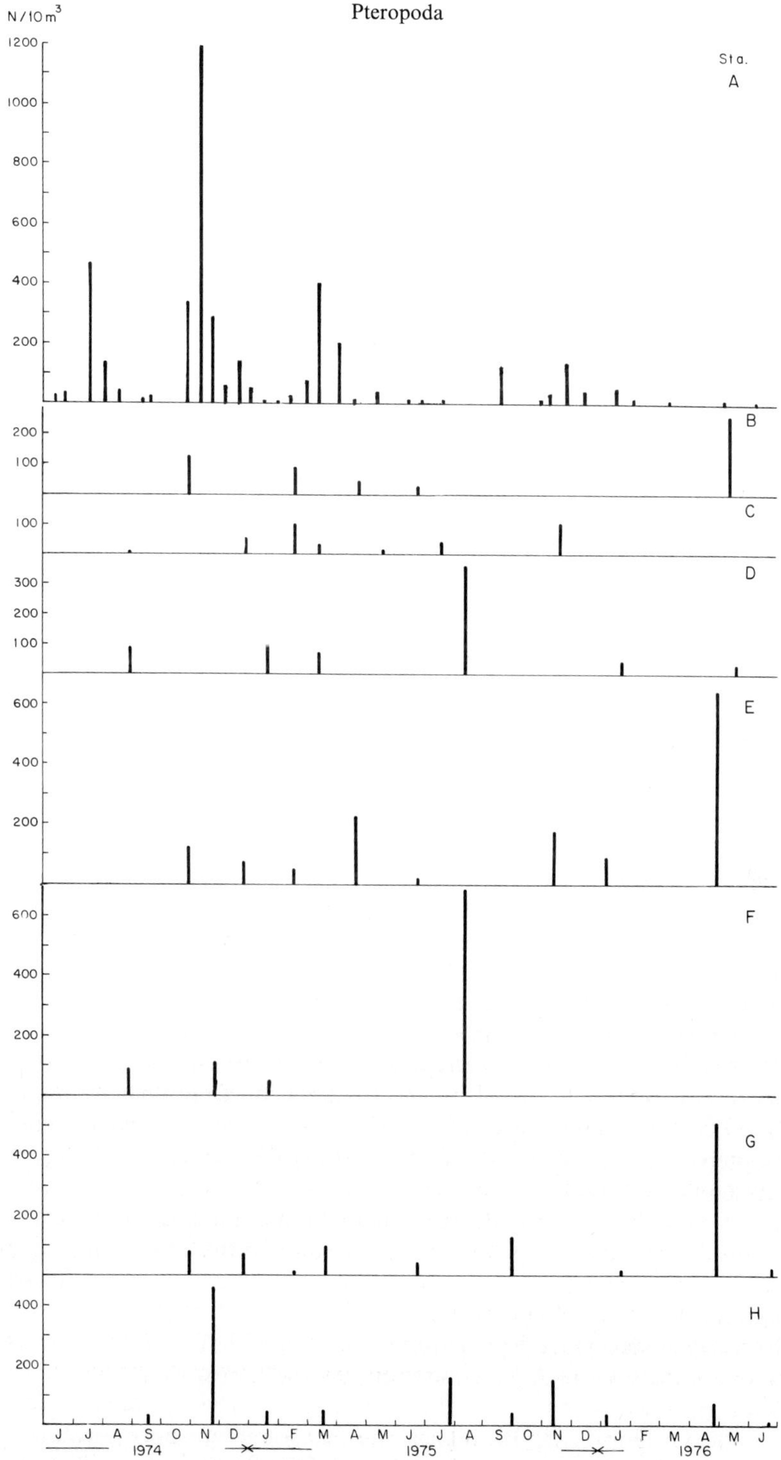

Fig. E.19. Standing crop of pteropods >200 μ in upper 600 m by station and sampling date, for DCPE-Stations A–H (Gulf of Aqaba and northernmost Red Sea) (Almogi-Labin 1981)

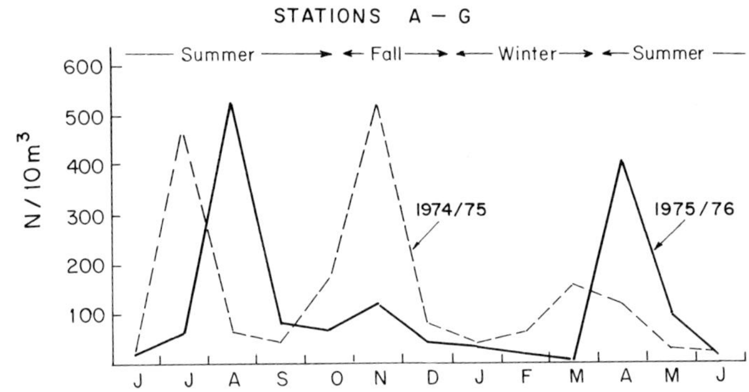

Fig. E.20. Monthly average total standing crop of pteropods >200 μ in upper 600 m integrated for DCPE-Stations A–G (Gulf of Aqaba), during 1974/1975 and 1975/1976 (Almogi-Labin 1981)

The **total seasonal standing crop** in the upper 600 m water column integrated for DCPE-Stations A–G within the Gulf of Aqaba for 1974–1975 and 1975–1976 is shown in Fig. E.20. Although long-range variation is again clearly observable, seasonal trends are well expressed. Especially noteworthy are the very low standing crops in June, September, and January–beginning of March.

The **seasonal abundance pattern of all species,** integrated for DCPE-Stations A–G within the Gulf of Aqaba is depicted in Fig. E.21. **Diversity** is very low as compared, e.g., with the southern Red Sea–NW Indian Ocean, while **dominance** is high, three species – *Limacina inflata, L. trochiformis* and *Creseis acicula* – making up 87% of the assemblage.

Despite the year to year variation, which includes also shifts of abundance peaks by about 1 month, a seasonal frequency pattern can be discerned. The lowest-diversity assemblages during summer are clearly dominated by July–August peaks of *L. inflata,* accompanied by *C. acicula, C. virgula* and (in 1975/1976) by frequent *Peraclis reticulata.* The highest diversity assemblage occurs during the transitional ("fall") season, (end October–December) when *L. inflata* is greatly reduced in numbers, *C. acicula, C. virgula, Diacria quadridentata,* and *Clio convexa* are more frequent and *L. trochiformis* is frequent to exceptionally abundant (as in 1974/1975). The winter (January–beginning of March) is a time of general decline in pteropod frequencies (see above), with species more equally distributed. Late March–May assemblages are characterized by disappearance of *L. trochiformis* and *P. reticulata* and by the rapid increase in abundance of *L. inflata.* The general decline in June leaves few and rare species in the assemblage, mainly *L. inflata, C. acicula,* and *C. virgula,* and is followed again by the low-diversity, *L. inflata*-dominated, July–August assemblage.

The **seasonal abundance depth pattern** of pteropods in the northern and southern sectors of the Gulf of Aqaba and in the northernmost Red Sea is depicted in Figs. E.22–E.24.

At station A (Fig. E.22), the highest numbers of specimens per volume water are concentrated in the upper 100 m. Intermediate numbers occur down to 150–200 m in October–November 1974. Generally, higher numbers of specimens at

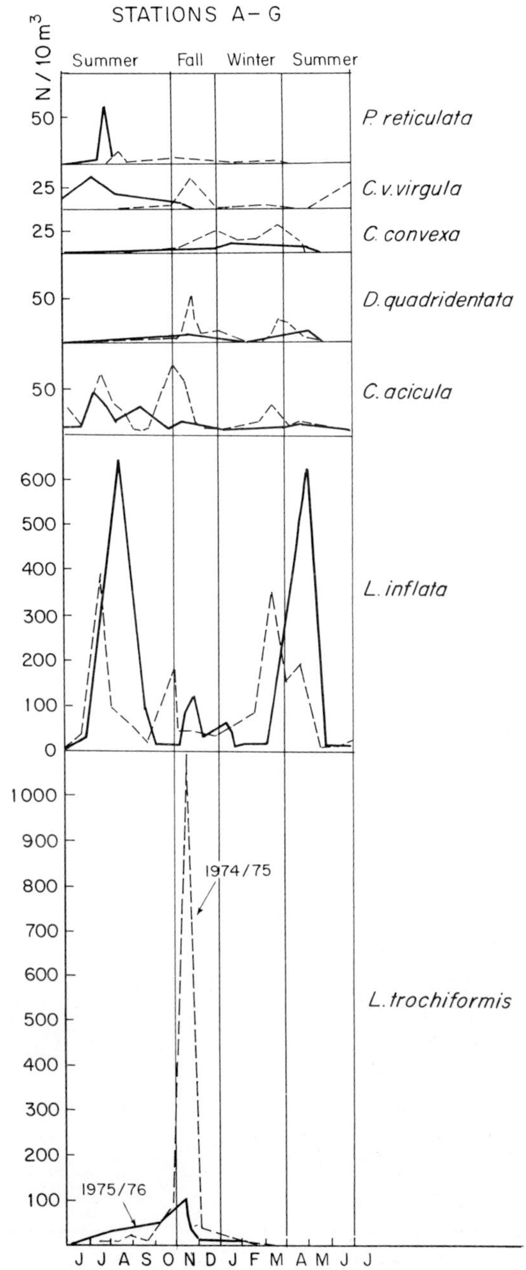

Fig. E.21. Monthly average standing crop of pteropods in the upper 600 m integrated for DCPE-Stations A–G (Gulf of Aqaba) during 1974/1975 and 1975/1976. Note change of scale between *L. inflata* and *L. trochiformis* and the remainder of the species (Almogi-Lobin 1981)

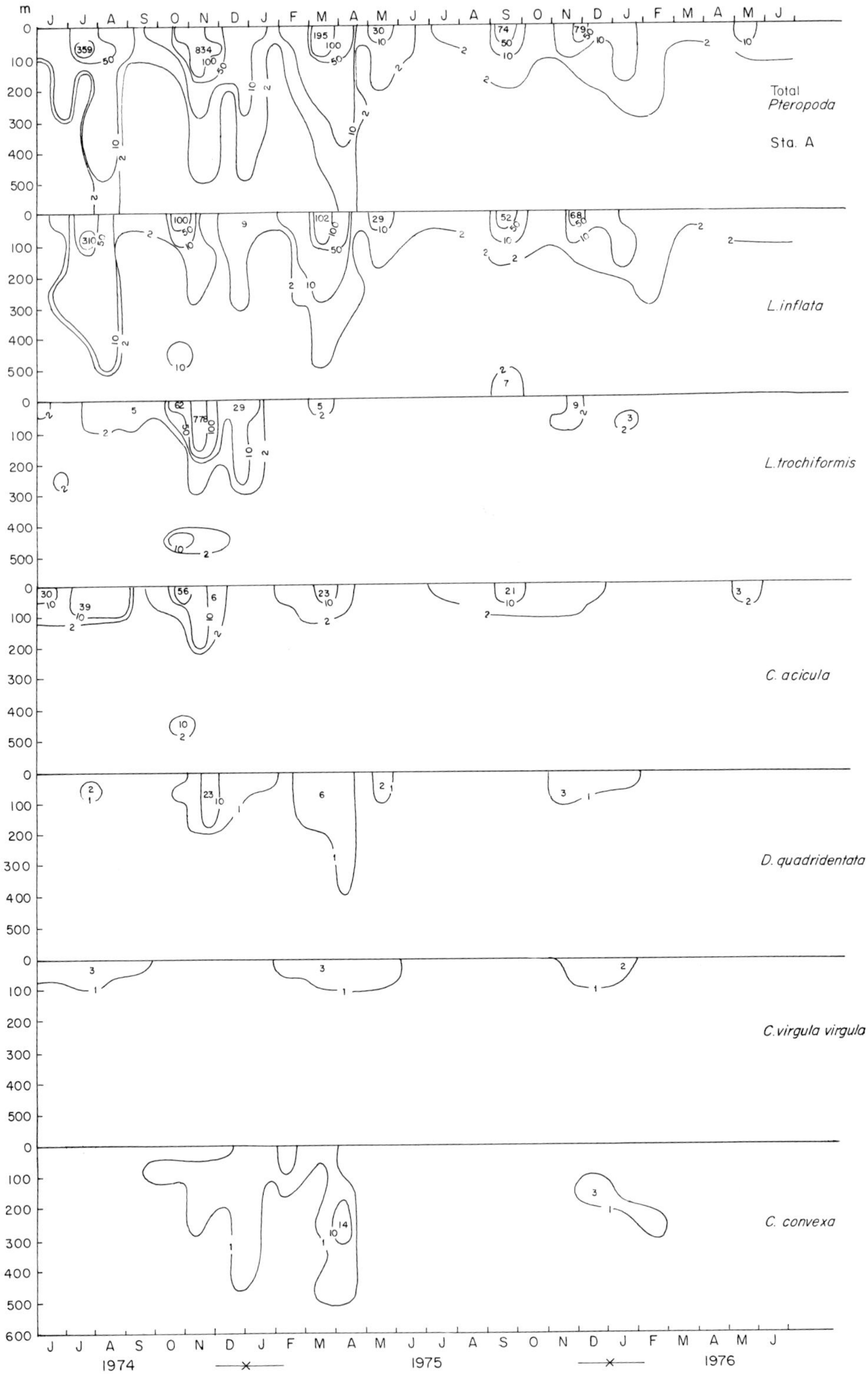
m
Total Pteropoda
Sta. A
L. inflata
L. trochiformis
C. acicula
D. quadridentata
C. virgula virgula
C. convexa
J J A S O N D J F M A M J J A S O N D J F M A M J
1974
1975
1976

Fig. E.22

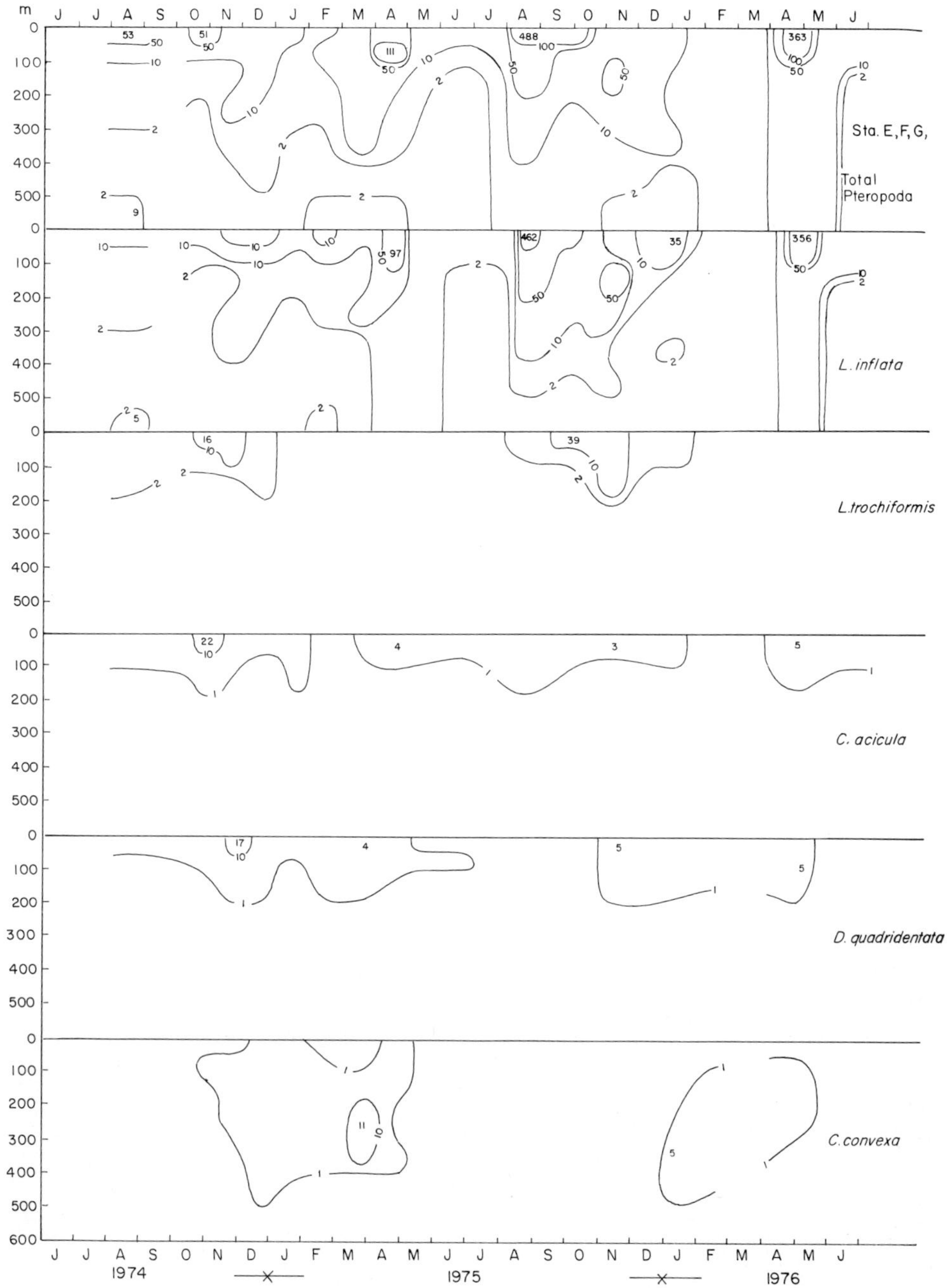

Fig. E.23. Depth distribution of standing crop of pteropods (total and per species) at DCPE-Stations E, F and G during 1974–1976 (Almogi-Labin 1981)

Fig. E.22. Depth distribution of standing crop of pteropods (total and per species) at DCPE-Station A during 1974–1976 (Almogi-Labin 1981)

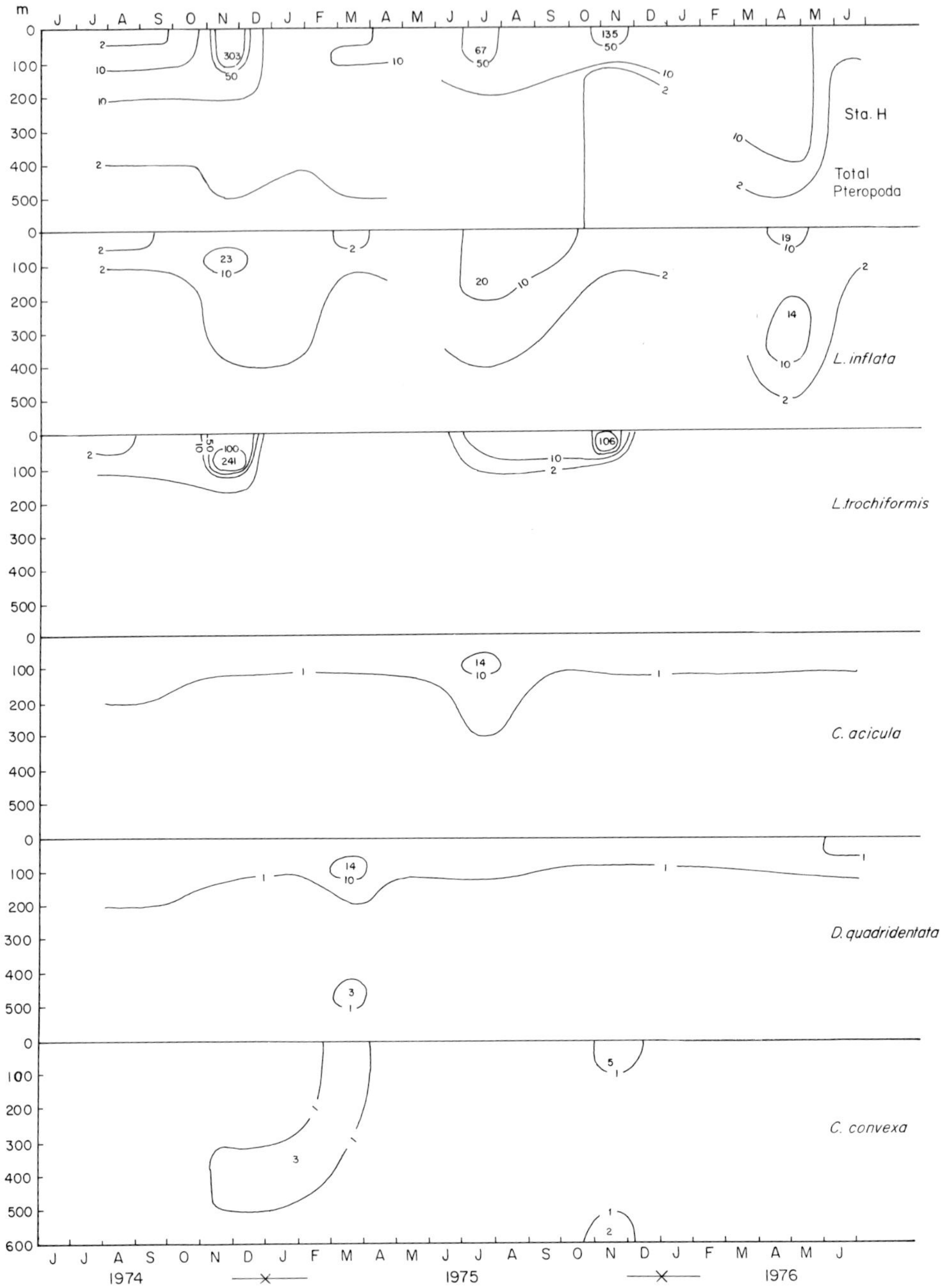

Fig. E.24. Depth distribution of standing crop of pteropods (total and per species) at DCPE-Station H (Red Sea) during 1974–1976 (Almogi-Labin 1981)

greater depths are correlatable with high standing crop in the upper 100 m water column. This is also seen from the considerable year to year variation between 1974/1975 and 1975/1976. Notwithstanding this variation, which includes also a shift of about 1 month in specific abundance peaks, the seasonal community structure is apparent, viz. a July–August assemblage of *Limacina inflata* and *Creseis acicula,* an October–November–December assemblage of *L. trochiformis, L. inflata, C. acicula,* and *Diacria quadridentata,* a March–May assemblage composed mainly of *L. inflata, C. acicula,* and *C. convexa.*

Stations E, F, and G (Fig. E.23) exhibit a year to year variation which is opposite that observed at station A, i.e., high frequencies of pteropods occur at E, F, G in 1975/1976 and low ones in 1974/1975. On the other hand, at stations E, F, and G frequencies of specimens down to greater depth are again correlatable with high standing crops which are mainly concentrated in the upper 100 m. The seasonal community structure is similar to that observed at Station A.

The general abundance pattern of pteropods at Red Sea Station H (Fig. E.24) is more similar to that of Station A in the northern Gulf of Aqaba than to its southern Stations E, F, G, viz. higher abundance in 1974/1975 than in 1975/1976. As within the Gulf of Aqaba, at Station H, the highest numbers of pteropods are found within the upper 100 m and intermediate numbers occur to greater – albeit shallower than within the Gulf – depth at times of highest standing crops.

The **preferred depth habitat** of the species is also expressed in their quantitative depth distribution. Thus, the mesopelagic *L. inflata* and *C. convexa* are concentrated in the upper 200 m, while the epipelagic *C. acicula* occurs mostly in the upper 100 m.

It is of importance to emphasize that in the case of pteropods the time of sampling is extremely important, because of **diel migration** of the different species. *L. inflata* is abundant at depths greater than 200 m during day time and at much shallower depths during the night. *L. trochiformis* is frequent in the upper 200 m in day time, but mostly concentrated in the upper 100 m during the night. *C. acicula* occurs during the day over the upper 100 m, being concentrated during the night in the upper 50 m. *D. quadridentata* occurs in day time between 50 and 200 m depth and is concentrated during the night in the upper 100 m. *C. convexa* occurs during the day between 200 and 600 m and throughout the upper 400 m during the night.

The distribution of total standing crop and of the most important five species of pteropods is integrated according to the summer, fall and winter seasons of 1974/1975 for the **whole Gulf of Aqaba and northernmost Red Sea** in Figs. E.25–E.28.

In fall (November–December) high standing crops (composed mainly of *L. trochiformis*) occur in the northernmost Gulf of Aqaba and in the northernmost Red Sea. Winter (January–March) standing crops are lower in general, with relatively high values in the northern sector of the Gulf of Aqaba (mainly *L. inflata*) and intermediate values to greater depths in the southern one *(L. inflata* and *C. convexa).* During the summer 1975 very high standing crops (mainly *L. inflata* and locally *C. acicula*) occur to great depth in the central and southern sectors of the Gulf of Aqaba, as opposed to low numbers of pteropods both in the northern sector of the Gulf and in the northernmost Red Sea.

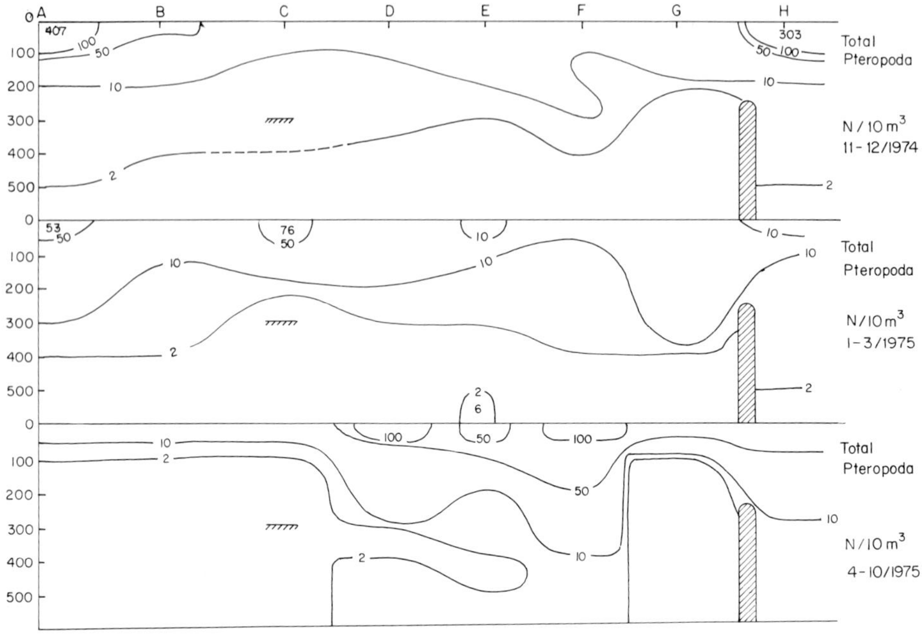

Fig. E.25. Average seasonal total standing crop of pteropods in the Gulf of Aqaba and northernmost Red Sea (Almogi-Labin 1981)

A comparison of the **percentage composition by species** of the **living pteropod assemblages** from plankton hauls (averaged over 2 yr) with that in **Recent sediments** (chandelier, grab and core-top samples) (Fig. E.29) shows a close correspondence, with the exception of the samples from the Arnona Deep, where the ratio *Creseis acicula*/*Limacina inflata* is higher than expected. This can be explained by the fact that the sediment samples originated from turbiditic sediments (compare chapter C) where the *Creseis*-specimens have been concentrated because of their elongated shape. In general, specimens of *Cavolinia* are relatively more abundant in sediments in the northern sector of the Gulf of Aqaba than in the plankton hauls from the same region. It seems most probable that *Cavolinia* escapes more easily from approaching plankton nets and is thus underrepresented in the living population. Masses of *Cavolinia* are often thrown on the shore by southern storms.

Interpretations in terms of ecological factors of the pteropod assemblages in the Gulf of Aqaba and in the northernmost Red Sea have been attempted by Almogi-Labin and Reiss (1977) and Almogi-Labin (1981), as well as – in the light of information obtained from deep-sea cores (see Chap. H) – by Almogi-Labin (1982), Reiss et al. (1980) and Winter et al. (1983).

Since published information on distributional patterns of pteropods, as related to measured ecological parameters, is much less comprehensive than in the case of, e.g., planktic Foraminifera, generalizations based on **empirical correlation** of pteropod community structures with known ecological factors are more diffi-

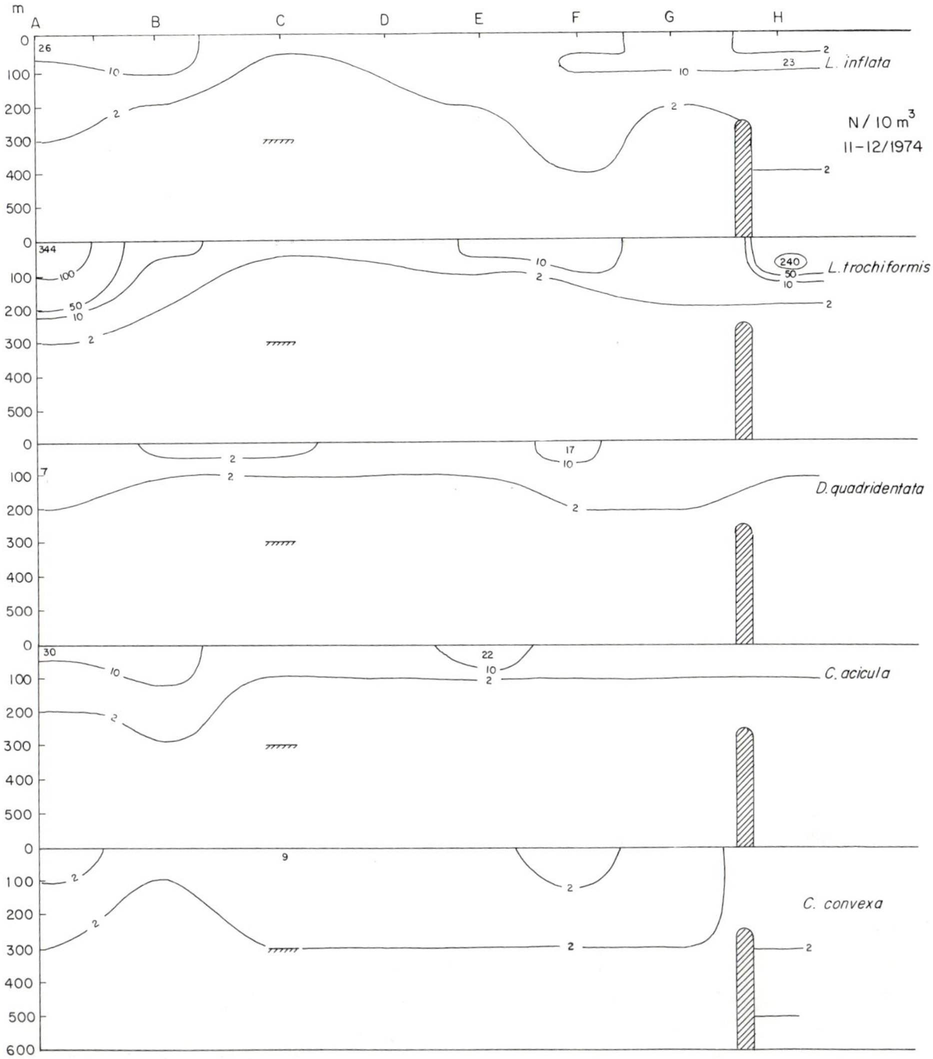

Fig. E.26. Average seasonal standing crop of pteropod species in the Gulf of Aqaba and northernmost Red Sea. Fall, 1974 (Almogi-Labin 1981)

cult. Furthermore, there is still very little knowledge on important aspects of pteropod life habits such as e.g., food requirements, reproduction, etc. Finally, three out of nine species and subspecies are endemic. On the other hand, the DCPE program, as well as the study of deep-sea cores in the Gulf of Aqaba and northernmost Red Sea (CATE) provided important information on the ecological and paleoecological significance of pteropods.

A most striking feature of the pteropod assemblages studied is the **absence of bathypelagic** and some mesopelagic species, which occur in the southern Red Sea, in the Gulf of Aden, and in the Indian Ocean. (As pointed out above, the record

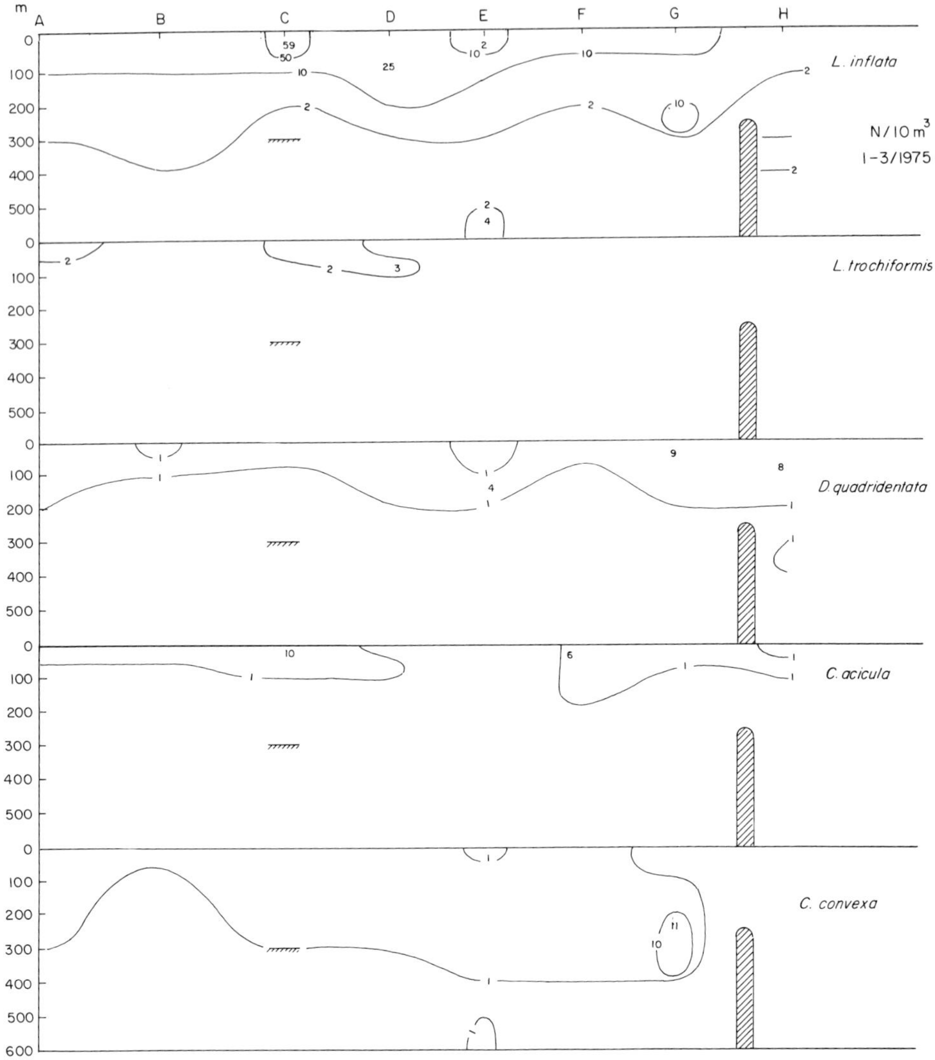

Fig. E.27. Average seasonal standing crop of pteropod species in the Gulf of Aqaba and northernmost Red Sea. Winter, 1975 (Almogi-Labin 1981)

of the bathypelagic *Clio polita* in the Gulf of Aqaba is erroneous.) This explains the low diversity of the Aqaba–N Red Sea assemblages: 9 species as compared with 14–17 species recorded from the Red Sea and with 20 species occurring in the NW Indian Ocean.

In this respect, pteropod assemblages in the Gulf of Aqaba and northernmost Red Sea resemble the planktic foraminiferid assemblages which also lack deep-water species, occurring in the southern Red Sea, Gulf of Aden and Indian Ocean.

Distributional patterns of pteropods were generally related by authors to **temperature** and **latitude.** According to this classification, the Aqaba–N Red Sea as-

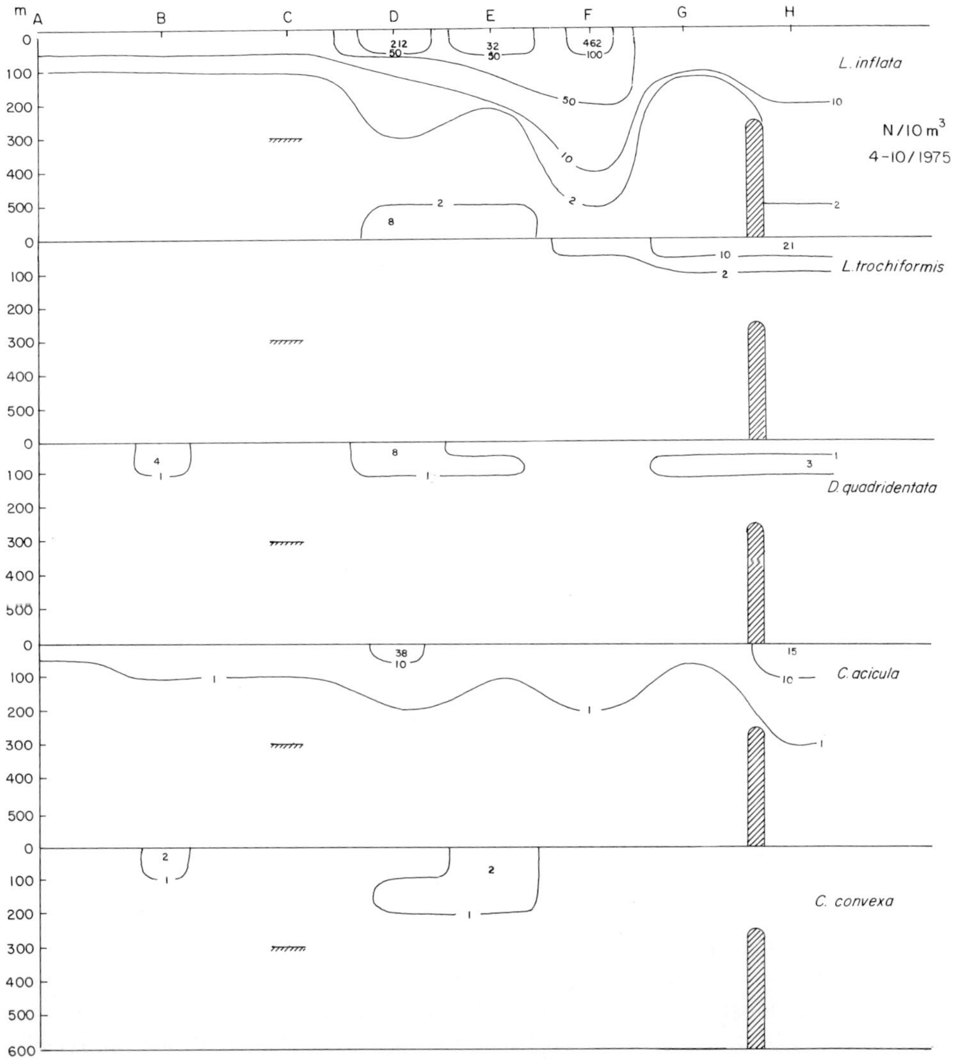

Fig. E.28. Average seasonal standing crop of pteropod species in the Gulf of Aqaba and northernmost Red Sea. Summer, 1975 (Almogi-Labin 1981)

semblages are composed of subtropical "cold-tolerant" species, i.e., *Limacina inflata* and *Peraclis reticulata;* subtropical "warm-tolerant" species, i.e., *L. trochiformis, Creseis acicula, C. virgula virgula;* of the tropical species *Clio convexa,* and of the endemic subspecies of tropical species *Cavolinia longirostris flexipes, C. uncinata pulsatoides* and *Diacria quadridentata crassa.*

As can be seen from Fig. E.30, all species present in the Gulf of Aqaba and northernmost Red Sea live there at temperatures well within the general range recorded from the world oceans. Although (e.g., in the North Atlantic Ocean) *L. trochiformis* occurs at slightly higher temperatures than *L. inflata,* the maximum

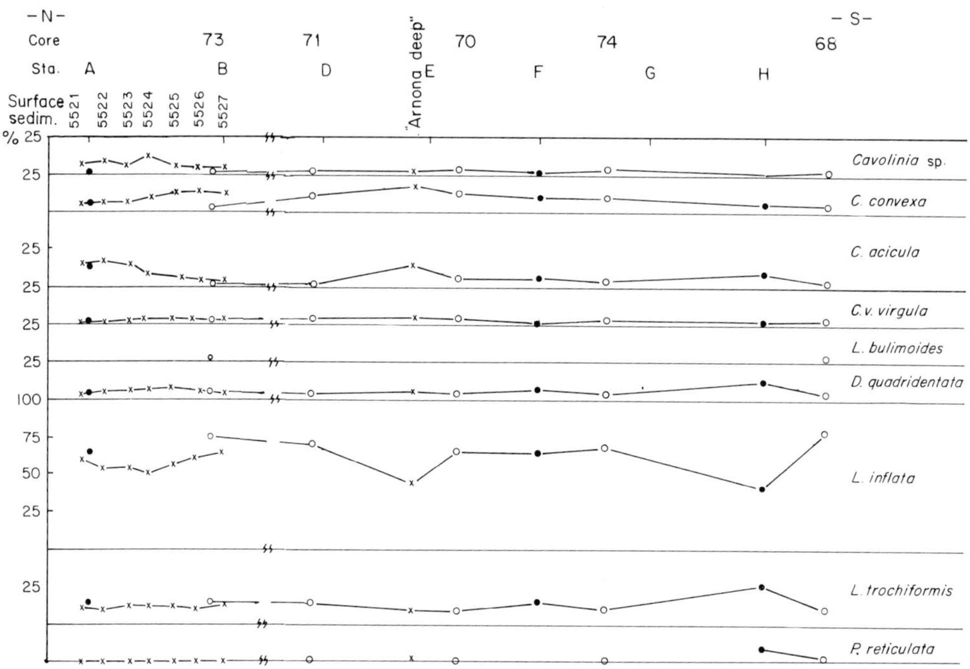

Fig. E.29. Comparison of weighted average relative frequencies of pteropod species in plankton hauls (>200 μ), surface sediments (>200 μ) and core-top samples (>149 μ) (Almogi-Labin 1981)

temperatures recorded for both species are above those in the Gulf of Aqaba and northernmost Red Sea. The abundance of *L. inflata* both at the end of winter and in summer and that of *L. trochiformis* in October–November (as well as the latter's absence between February and June) in this region can thus not be attributed to seasonal temperature changes. The seasonally fairly stable and high **salinity** in the Gulf of Aqaba and northernmost Red Sea also does not explain either the composition or the seasonality of the assemblages. Records from the world oceans refer mostly to water masses with salinities below 39‰ and the deep-sea core record from the Gulf of Aqaba shows that species which at present live at these salinities in the world oceans and are absent in the present-day Gulf of Aqaba, occur in the Aqaba cores in glacial intervals with calculated salinities well above 41‰ (see Chap. H).

A more adequate explanation is obtained by relating both the composition of pteropod assemblages and the seasonality of their species in the Gulf of Aqaba and northernmost Red Sea to other water-mass properties, viz. **nutritional regime, fertility of the sea** and **light penetration.**

The significance of the fertility gradient from the Gulf of Aqaba to the Gulf of Aden has been discussed above with regard to planktic Foraminifera and coccolithophorids and compared to that between central water masses and equatorial-boundary current regimes. As in the case of planktic Foraminifera, the absence of bathypelagic, hence negative phototropic, pteropod species is probably due to light penetration to very great depths (see Chap. D.8), where required food

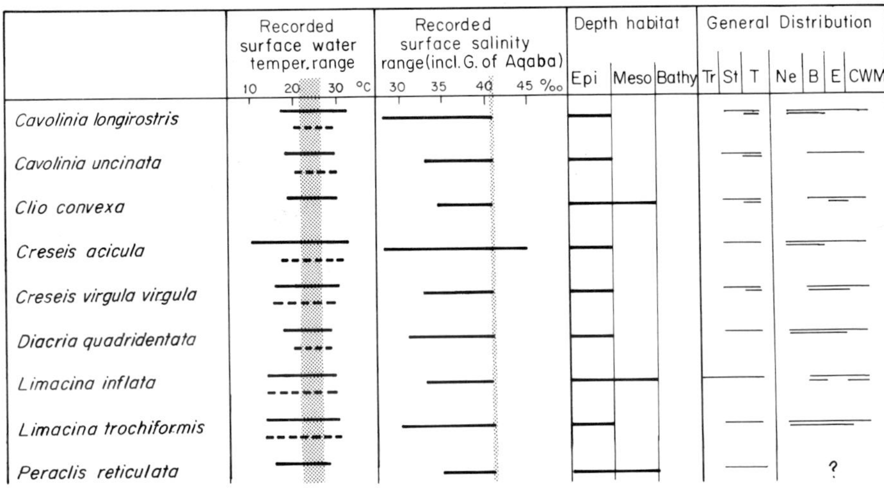

Fig. E.30. Recorded ecological ranges of pteropod species occuring in the Gulf of Aqaba and northernmost Red Sea. *Stippled:* present-day conditions in the Gulf of Aqaba. *Epi* epipelagic, *Meso* mesopelagic, *Tr* transitional zones, *St* subtropical, *T* tropical, *Ne* nerithic, *B* boundary currents, *E* equatorial, *CWM* central water masses (Almogi-Labin 1981)

organisms (among them animals) are not frequent enough and where oxidation of organic matter is strong in the oligotrophic and warm waters. The increase in fertility of the sea in the Gulf of Aden facilitates the frequent occurrence of deep-water pteropods.

L. inflata, although recorded as frequent in the Somali current, is mostly characteristic of gyre centers and is abundant in the Gulf of Aqaba and northernmost Red Sea during the generally oligotrophic season. (In this region it does not follow high chlorophyll a content, as reported, e.g., from the Indian Ocean by Sakhtivel, 1973.) *L. trochiformis,* which is characteristic of highly productive boundary currents and upwelling regions of the ocean, is abundant in the Gulf of Aqaba and northernmost Red Sea during the short fall period when productivity increases. The abundance peaks of the microphytoplankton- and detritus-feeding *L. inflata* may possibly be connected with the end of the winter diatom bloom (especially the minute *Thalassiosira*), with the summer–fall increase in coccolithophorids (Chap. E.2), and perhaps of other nanno- and picoplankters. *L. trochiformis* apparently feeds – because of the small dimensions of its ciliary system – mainly on nanno- and picoplankton. Indeed, its abundance in the Gulf of Aqaba and in the northernmost Red Sea is correlatable with abundance in coccolithophorids. The relative high frequency of *C. acicula,* mainly in fall in the Gulf of Aqaba and in summer in the northernmost Red Sea, may be connected with the prominent blooms of *Trichodesmium,* as in the Indian Ocean (Peter and Paulinose 1978). The abundance of the relatively large *C. convexa* seems to be connected with the blooms of diatoms and dinoflagellates in winter time. The exceptional abundance of pteropods, mainly *L. inflata* and *C. acicula,* at Stations D and F during August 1975 may be due to heightened productivity derived from the May–June mid-water upwelling in the northern Red Sea (see Chap. D.5); however, no information on primary production or phytoplankton is available from these stations.

The higher abundance of pteropods to greater depths within the Gulf of Aqaba as compared to the northernmost Red Sea may be due to the presence in the Red Sea of the prominent oxygen minimum below 300 m which hampers deep diel migration of mesopelagic species.

Greater abundance (and to greater depth) of pteropods at Station A during summer (Fig. E.22) reflects increase in productivity due to northern upwelling in the Gulf of Aqaba, which differs in intensity from year to year.

In absolute numbers of specimens per volume water, the pteropod assemblages in the Gulf of Aqaba and in the northernmost Red Sea resemble generally those in gyre centers of the ocean. Highest densities of specimens in the Gulf and in the N Red Sea are similar to intermediate values of productive regions of the ocean.

It seems, therefore, that both the composition and the seasonality of pteropod assemblages in the Gulf of Aqaba and northernmost Red Sea depend upon feeding habits of the species, the circulation-determined fertility of the waters and the seasonality of the phytoplankton. As will be seen below (Chap. H), the evidence obtained from deep-sea cores strongly supports the interpretation of present-day, generally gyre-center pteropod assemblages in the Gulf of Aqaba and northernmost Red Sea in terms of fertility of the sea. It appears that in a marginal sea with extreme conditions, like the desert-enclosed Gulf of Aqaba, even minor year to year climatic changes influencing circulation, fertility of the sea, etc. strongly affect the seasonal and local variation in total pteropod abundance.

However, as indicated by the comparison between plankton- and sediment assemblages, the relative species frequences in sediments are not affected by this variation, a fact of great importance to paleoceanographic considerations.

References: Almogi-Labin (1981, 1982), Almogi-Labin and Reiss (1977), Bé and Gilmor (1977), Peter and Paulinose (1978), Reiss et al. (1980), Sakhtivel (1973), Van der Spoel (1967), Winter et al. (1983).

E.5 Aqaba Calcareous Plankton: Significance and Problems

The study of the calcareous plankton in the water column and in surface sediments of the Gulf of Aqaba and northernmost Red Sea points to the great ecological importance of **nutritional regime** and **light** in determining both composition and abundance patterns of coccolithophorids, of planktic foraminifera and of pteropods. These factors appear also to be mainly responsible for the similarity of the oligotrophic Aqaba – N Red Sea calcareous floras and faunas with those characteristic of the equator-near margins of oceanic gyre centers. However, while in the open ocean "expatriate" eutrophic elements of the calcareous plankton are introduced from boundary and equatorial systems into gyre centers by current transport, such a "contamination" in the Gulf of Aqaba and northernmost Red Sea is prevented by distance from the open ocean and by circulation regime.

The composition of all three groups, coccolithophorids, planktic foraminifera, and pteropods follows the gradient of increasing oligotrophy and illumination from the Gulf of Aden to the Gulf of Aqaba. The seasonal abundance pattern of all three groups is modulated in the Gulf of Aqaba in accordance with the – circulation-determined – seasonal and year to year changes in fertility of the upper Red Sea waters, as well as with the seasonality of stratification, mixing and local upwelling within the Gulf itself.

Thus, the Gulf of Aqaba facilitates a more meaningful analysis and perhaps better understanding of **oceanic gyre center type** assemblages of calcareous plankton.

However, many problems remain still to be solved, most of them pertaining to the **pathways and biological mechanisms** whereby nutrients and light influence composition and seasonality of calcareous plankton. For example, **efficiency of nutrient uptake** and possible **phagotrophy** of various important coccolithophorid species must be studied; **feeding habits in nature** of planktic Foraminifera and Pteropoda should be known better; the **role of symbionts** in the nutritional regime of planktic Foraminifera must be elucidated. Nothing is known on the role of picoplankton in primary production levels and as food organisms for planktic Foraminifera and Pteropoda. The physiology of **negative phototropy** must be studied.

Calcification processes in planktic Foraminifera as related to their light-dependent symbionts or as directly enhanced by light levels are not yet fully understood and require intensive biochemical work. The same is true of influence of light on morphology of coccoliths.

On the other hand, despite these problems, considerations derived from the Gulf of Aqaba study of fertility of the sea and light as major ecological factors influencing calcareous plankton are, as will be seen below (Chap. H), of crucial importance in providing an internally consistent paleoceanographic interpretation, independently supported by other methods, of the Late Quaternary deep-sea record of the Gulf of Aqaba and northernmost Red Sea.

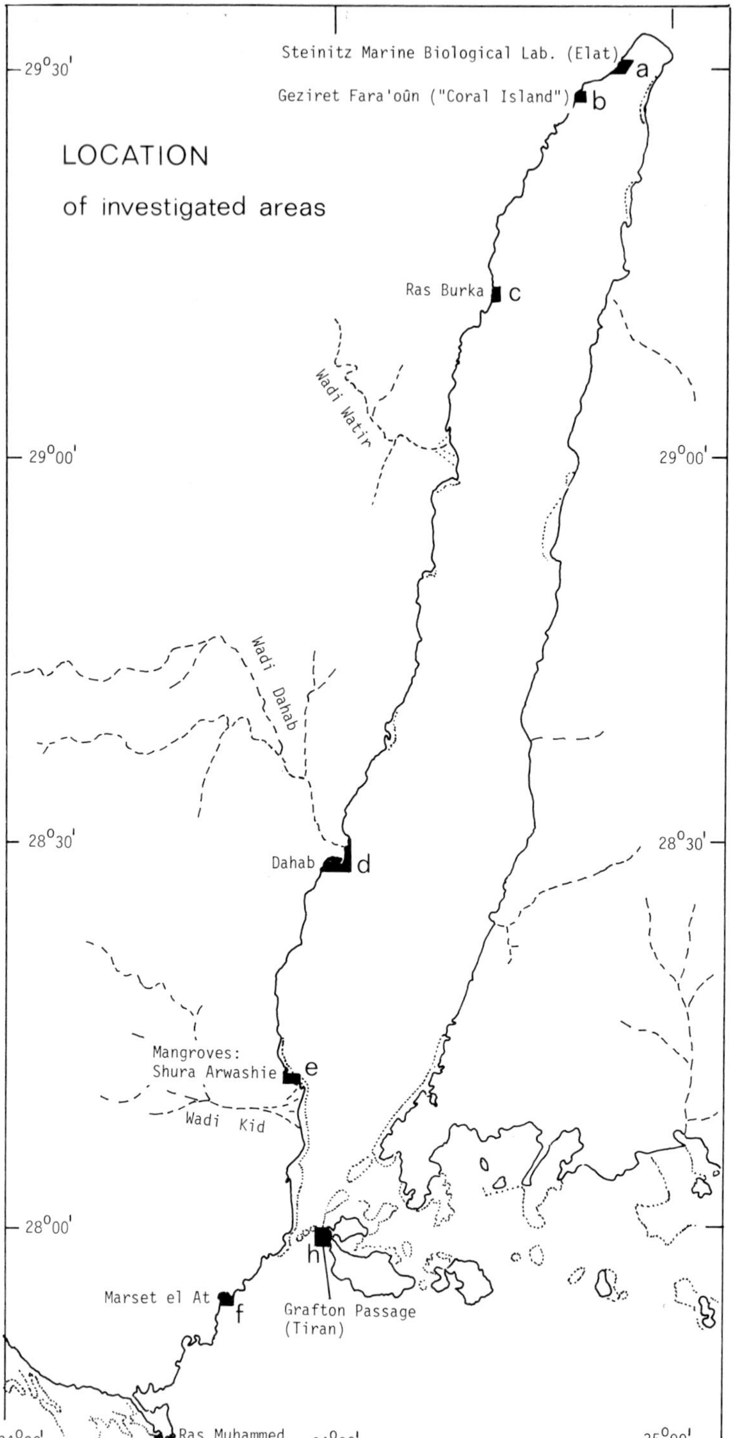

Fig. F.1. Location map of areas where the distribution of selected benthic organisms was explored

F. The Sea Bottom – a Mosaic of Substrates

F.1 Methods of Investigation

In contrast to pelagic environments, **the benthic realm** is characterized by two physicochemical subsystems below and above the bottom-water interface. Both subsystems work according to different mechanisms, but are closely interrelated. The organisms living on the bottom depend thus not only on the properties of their watery medium, but also on the nature of their substrate of sediment or rock. On the shelf at least, the substrates are changing over short distances, restricting the ecosystems to surface extensions in the order of some hundreds of meters to a few kilometers. These small-scale ecosystems cannot be surveyed by methods of remote, "blind" sampling or measuring from surface ships, as currently used to explore the much larger-scale planktic systems or the benthos of the deep sea. In order to explore the shallower benthic realm, **visual control** of the bottom must provide adequate accuracy to select samples or implant measuring devices at locations relevantly positioned with respect to ecosystem boundaries. Mapping under water by extensive **SCUBA diving** to 70 m assisted by **grab sampling** with the small research vessel **Arnona** and the use of the research **submersible GEO** down to 200 m depth revealed on selected parts of the Sinai "shelf" a mosaic of hard and soft substrates. The use of an underwater, sand-sucking **"vacuum cleaner"** device and sampling with the **submarine's artificial arm** provided ample evidence for early diagenetic processes progressing below soft surfaces of the sea bottom and participating in the evolution of the substrate. The benthic communities covering the bottom interact with the physicochemical processes and participate in various ways in the formation of their own substrate. As these larger organisms form the substrate for many benthic microorganisms, particularly for benthic Foraminifera, their distribution on the bottom also determines the distribution of their epizoans. The following descriptions of selected areas on the Sinai "shelf" (see Chap. C) try to exemplify the **complex relationships between physicochemical factors and benthic communities generating a mosaic of substrates.** For logistic reasons, the areas in front of the H. Steinitz Marine Biological Laboratory, S of Elat, and around "Coral Island" (Geziret Fara'oûn) in the northern part of the Gulf of Aqaba have been explored systematically, whereas in the regions selected for study farther south (Ras Burka, Dahab, Tiran, Marset al At, etc.) exploration was restricted to reconnaissance-diving, covering only part of

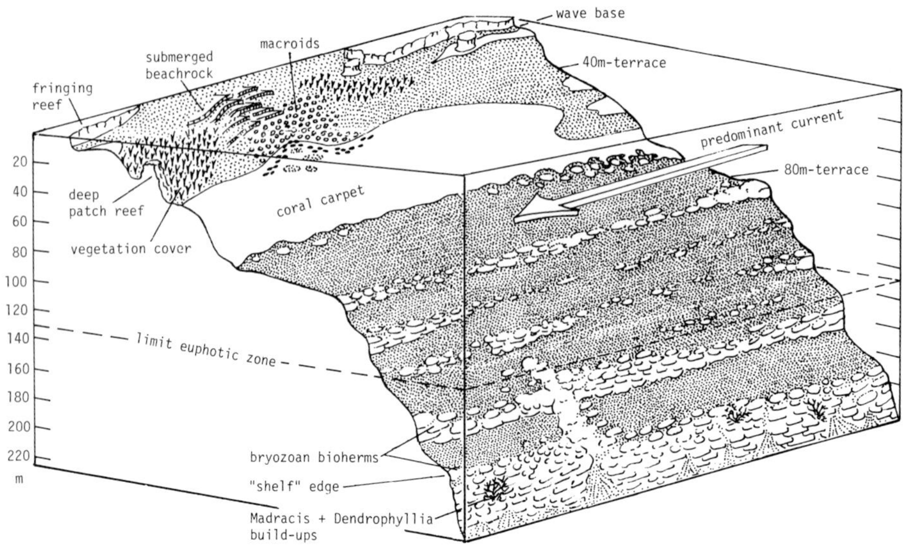

Fig. F.2. Stereograph of narrow "shelf" in front of the Steinitz Marine Biology Laboratory (MBL), 12 km south of Elat (Hottinger 1984a)

the depth range of the shelf (Fig. F.1). On the Arabian shelf south of Aqaba, fringing reefs are known in much detail from publications of Mergner and his collaborators, but these reef environments represent only the shallowest part of the benthic realm.

F.2 Selected Areas

F.2.1 The "Shelf" in Front of the H. Steinitz Marine Biology Laboratory

The H. Steinitz MBL has been built on the shore of a particularly narrow and steep section of the Sinai "shelf" favoring the study of the sea bottom according to its depth gradient (Fig. F.2). Numerous SCUBA dives, some of them reaching to the margin of the coral carpet at 70 m depth, and a dozen dives in the submersible GEO to the shelf edge permitted direct observation of the distribution of foraminiferal faunas and their collection under visual control.

In the northern part of this section of the shelf, small fringing reefs develop on a hardground 8–10 m deep, exposed by wave action below a large patch-reef called Moses Rock (Fig. F.3). Farther south, the subtidal bottom is characterized by strongly cemented, coarse conglomerates on which ephemeral larger algal thalli and pioneering *Porites* corals represent the major fixed benthic organisms. The wave base, experienced by the diver as a fading away of the rocking water movement, is marked on homogeneous mobile bottom by a change of dip in the slope and by the upper margin of the *Halophila* vegetation covers. According to the bottom's nature and exposure to the main wind direction, the depth of the

Fig. F.3 a–e. Hardground at 10 m depth constituting the basement of the fringing reefs south of Elat ("Moses Rock"). **a, b** Sampling the hardground under water with a pneumatic chisel. **c** Three generations of calcite cement harden the sediment: Surface of broken rock showing much reduced porosity (SEM, scale bar 20 µm). **d** Hardground surface with living diatom. Hardparts of epizoan organisms and cemented rock attacked by boring organisms (SEM, scale bar 20 µm). **e** Detail of corroded surface with corroded coccolith supporting bacteria (SEM, scale bar 1 µm)

wave base changes in front of the MBL between approximately 5 and 10 m. At the foot of the fringing reefs, coarse sand, produced in the reef, accumulates and is kept in motion by wave action and longshore currents.

Where the relief of the coastline exposes the subtidal bottom between 8 and 40 m depth to the strong northerly currents running along the shore, loose large pebbles of dark, crystalline rock accumulate during winter storms. During the calmer seasons they are overgrown by *Padina* and numerous larger Foraminifera, particularly *Amphisorus, Sorites, Borelis* and *Amphistegina. Heterostegina depressa* occurs less frequently than the other larger Foraminifera in this particular environment. Most Foraminifera live on the backside of the pebbles in the large interstitial spaces selected to collect several 1-kg-sized samples of living foraminiferans for biochemical analysis by the use of the underwater "vacuum cleaner" (Fig. F.4).

The inner border of the 40-m terrace (see Chap. C) is characterized by a deposit of large macroids growing around coral debris. Since the specific weight of the macroids is much lower than that of rock pebbles of similar size, only the macroids are moved periodically, whereas the pebble-covered slope remains stable. The belt of macroids extending along the 40-m isobath, along current-exposed stretches of the coast, marks thus the depth range of temporary high energy events capable of lifting the macroids and winnowing away the finer sediment particles.

The 40-m terrace and its gentle slope down to 65–70 m depth is covered by a dense carpet of soft and stony corals, dominated by *Alveopora* (Fig. F.5 a, b). Their coarse mineralized debris are cemented by *Acervulina inhaerens* and other encrusting Foraminifera and provide habitat and shelter for *Heterostegina depressa*. The finer sediment fractions are carried away by the current and accumulate farther south, where the terrace is overgrown by *Halophila* meadows, providing food for a dense burrowing infauna. In front of the lighthouse, a spectacular 12 m high, isolated patch reef emerges from the *Halophila* meadows in 40–45 m depth, offering shelter for swarms of young fish and a very diversified hard-bottom fauna (Fig. F.6 a, b).

Where the dip of the slope in 70 m changes abruptly, the coral carpet ends in a straight line along the isobath; where the dip changes gradually, oval, flattened build-ups submerged by shells of larger Foraminifera and small mollusks occur (Fig. F.7). The highest densities of *H. depressa* (relative to other larger Foraminifera) are found in this environment, where the number of adult microspheric forms relative to megalospheric ones and the number of dead juveniles is also largest.

The 80-m terrace is covered by sand inhabited by a burrowing infauna which is less dense than the one in shallower depth. Its activity, however, is intensive enough to destroy the stratification of the sediment. Unidentified brown algae with large, lobe-like, smooth and slippery thalli cover permanently the bottom to about 30% in spring and 2–5% in autumn. The thalli are fixed to coarse sand grains. They do not bear any epiphytes. Many thalli are transported with their support over the bottom with the slow currents, accumulating in large funnels produced by burrowing organisms (Fig. F.8). The transport is strictly parallel to the isobaths. Not a single thallus was observed below the margins of the terrace, fringed by a chain of small bioherms along the 90 or 100 m isobaths.

Fig. F.4 a–c. Pebble-covered slope (**a**) with algal cover in spring (**b**), bare in autumn (**c**), off MBL, Elat, 25–30 m deep

Fig. F.5 a,b. Coral carpet below 40 m depth (**a**) and detail (**b**) showing coral debris coated and cemented by the Foraminifer *Acervulina* (scale bar 2 cm)

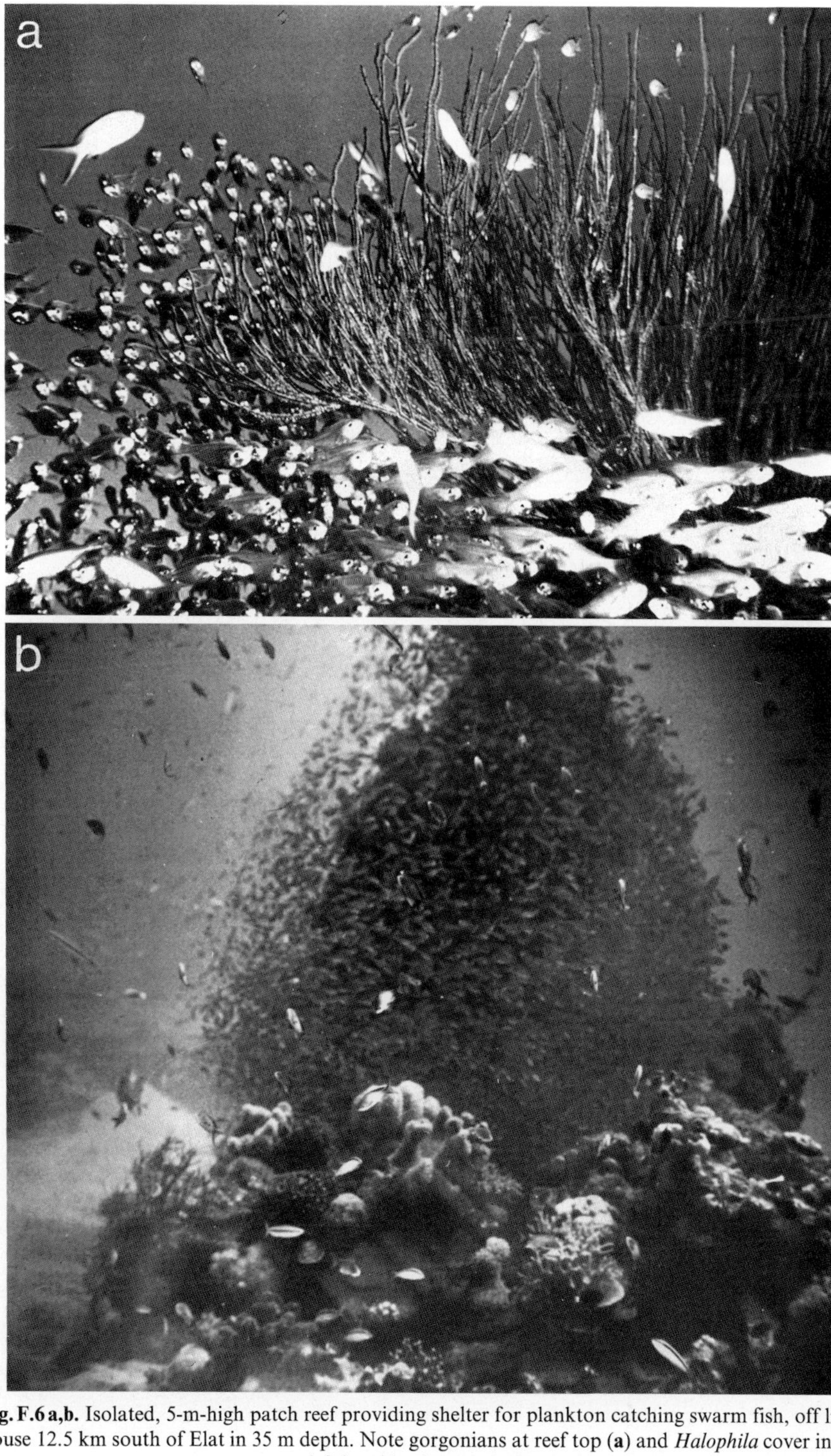

Fig. F.6 a,b. Isolated, 5-m-high patch reef providing shelter for plankton catching swarm fish, off light house 12.5 km south of Elat in 35 m depth. Note gorgonians at reef top (**a**) and *Halophila* cover in the left hand background of **b**

Fig. F.7 a, b

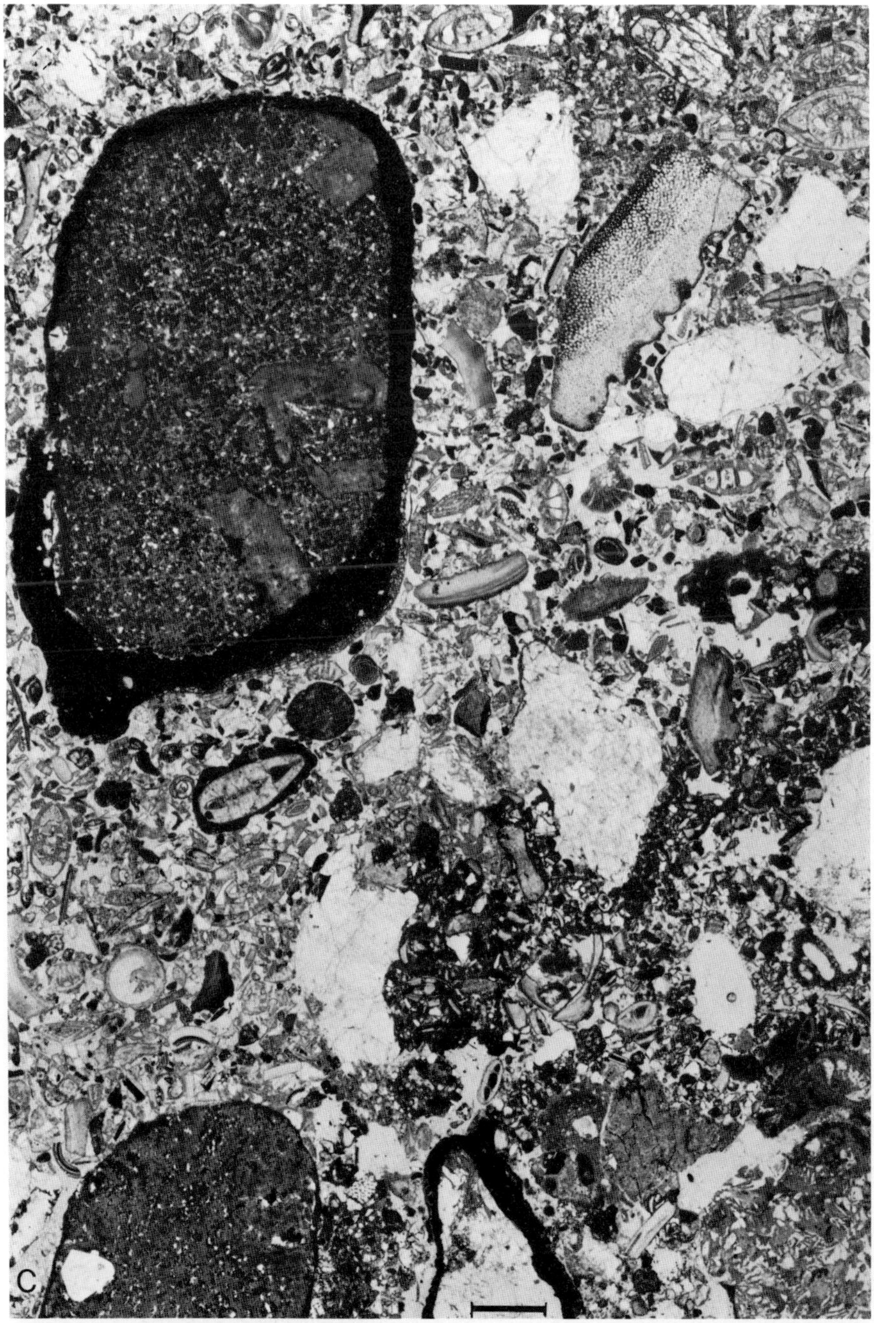

Fig. F.7 a–c. Lower limit of coral carpet at 70 m depth off MBL, Elat (**a**), detail of carpet margin and coarse sediment in front of margin (**b**) and vertical thin section (**c**) of artificially hardened sediment from core showing coated grains of crystalline rock or shells and larger Foraminifera. Note lack of stratification by burrowing. Scale bar 1 mm

Fig. F.8 a,b. Vegetation cover of 80-m-terrace off MBL, Elat (**a**) in spring constituted by a single species of unidentified, slippery brown algae attached to large sandgrains. Single thalli (**b**) 5–8 cm in diameter

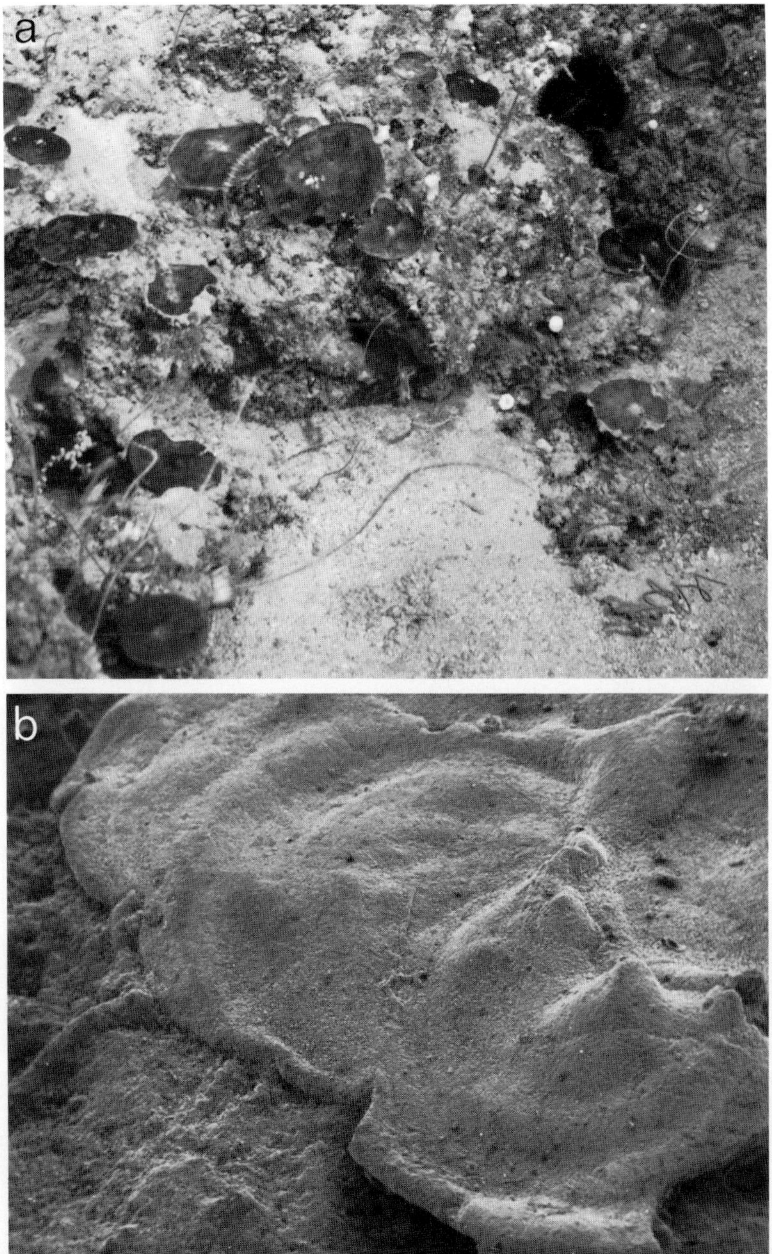

Fig. F.9 a,b. Bioherms at 130 m depth with the deepest symbiont-bearing coral species *Leptoseris fragilis* as seen from the submersible's window off MBL, Elat (**a**) and deepest corallinacean thallus collected (**b**), SEM, scale bar 0.1 mm

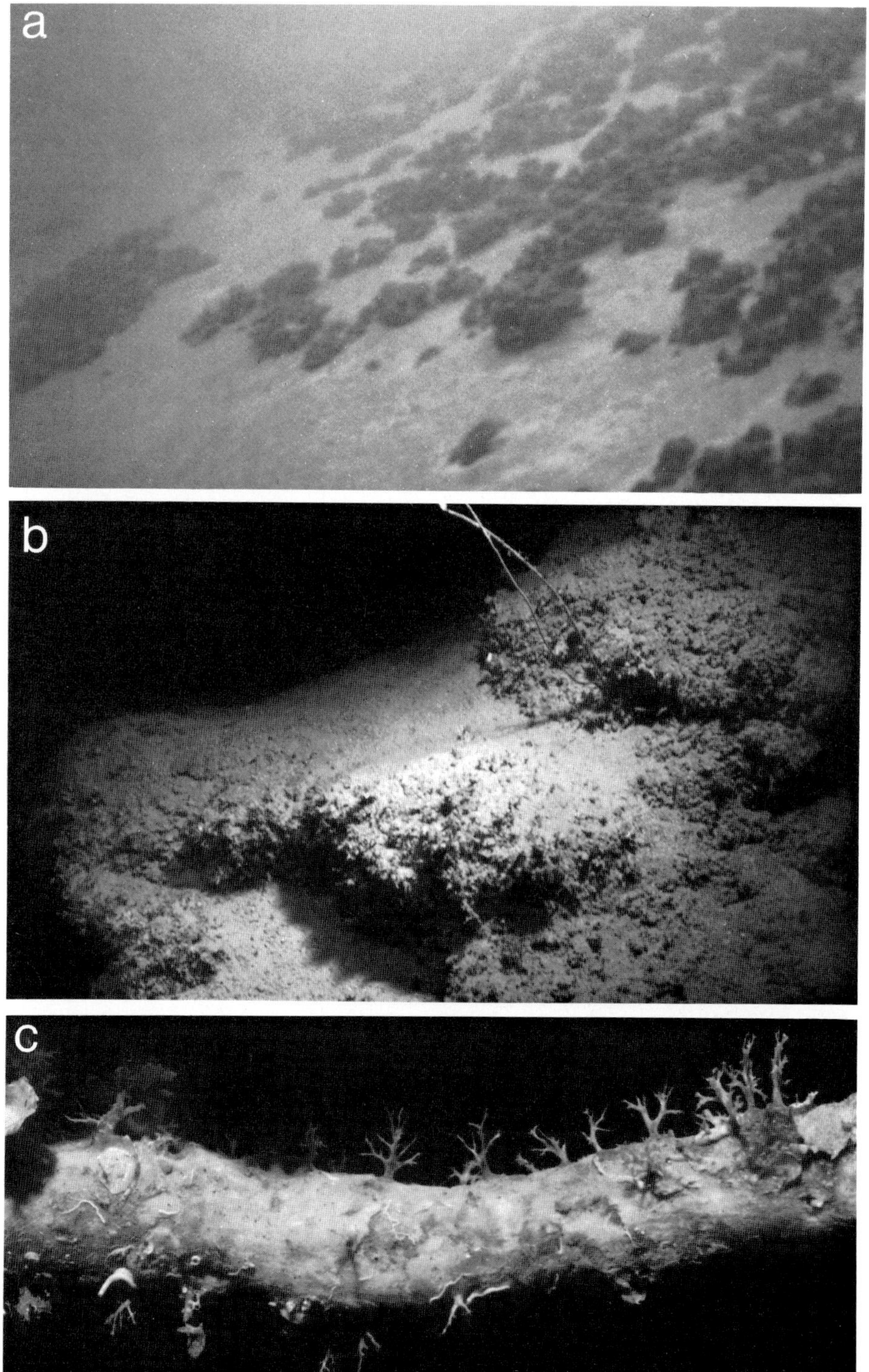

Fig. F.10 a–c

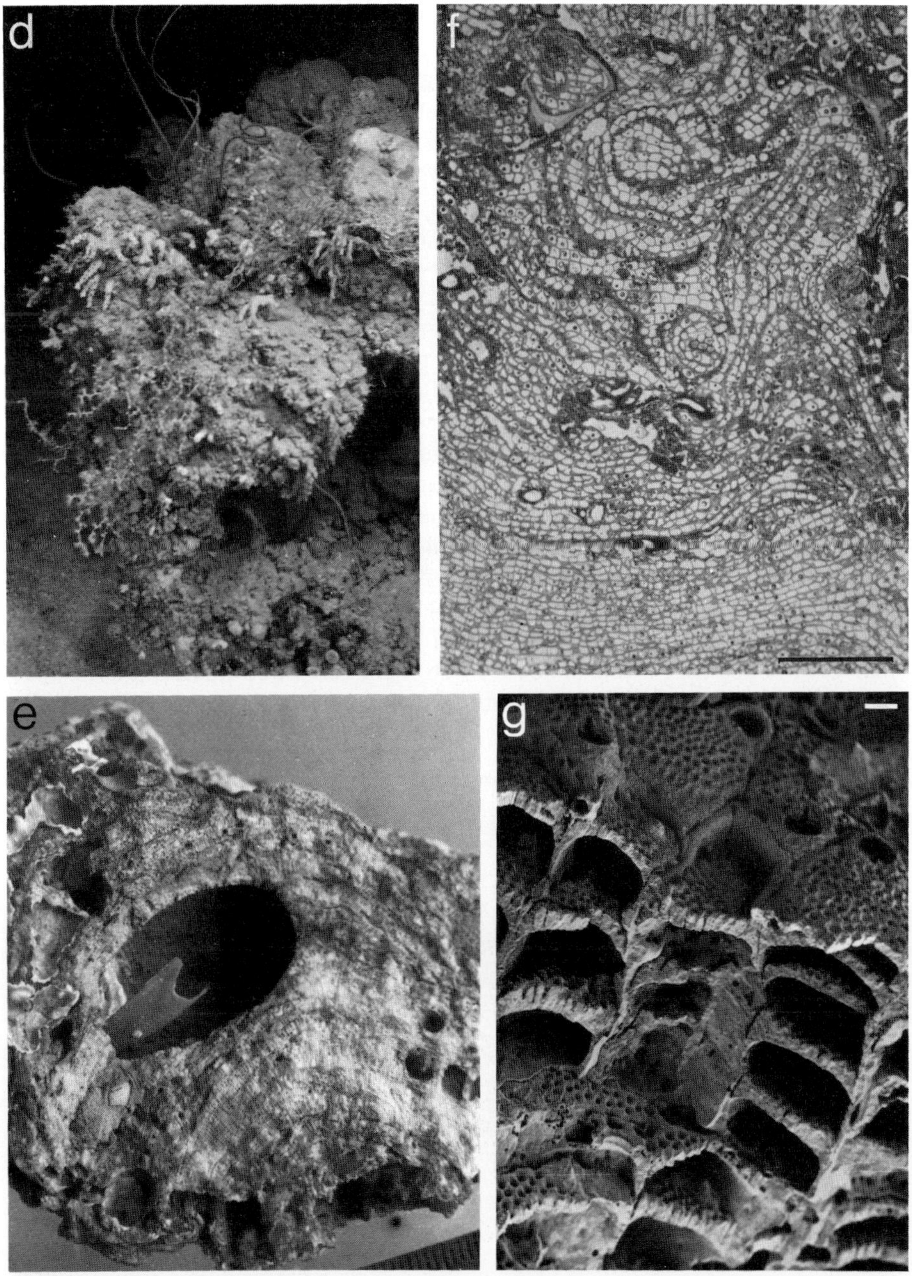

Fig. F.10 a–g. Bryozoan build-ups as seen from GEO submarine in 160 m off MBL, Elat, without artificial light (**a**), single build-up in the beam light of the submersible (**b**) and pioneering organisms, arborescent Foraminifera and encrusting Bryozoa, on sunken driftwood collected with the submarine in 180 m (**c**), scale bar 0.5 cm. **d** Ultimate epizoan organisms, ahermatypic corals and gorgonians on bryozoan build-up 1.5 m above sediment surface. **e** Bryozoan head with numerous boreholes in concentrically grown, massive bryozoan colony with thin section (**f**) oblique to surface of bryozoan colony (scale bar 1 cm). Detail of concentric colony (**g**), SEM, scale bar 0.1 mm

In contrast to the black sediments beneath the permanent *Halophila* cover in shallower depths, the sediments of the 80-m terrace are reddish-brown in colour and show no clear separation of an oxygenated surface layer. They reflect a change in the redox system affecting the bacterial activity below the bottom-water interface. Where the dip of the slope steepens again, at 90, 110, and 120 m depth, small bioherms emerge from the sand. They are constructed by worms, encrusting agglutinated Foraminifera, bryozoans, stony corals and gorgonians. Sponges are rare and small. The stony corals do not cover the surface as they do in the coral carpets at shallower than 70 m depth. The dominant species, *Leptoseris fragilis*, has symbionts and reaches 130 m depth, where 0.5% of the surface light intensity penetrates to the bottom (Fig. F.9; see Chap. D.8). Other organisms with symbionts, particularly the larger Foraminifera *Operculina ammonoides* and *Heterocyclina tuberculata* have their lower limit of distribution here, while *Amphistegina papillosa* may penetrate 10–20 m farther down. In the sediment samples, the deepest fresh debris of coralline algae were found in 135 m (Fig. F.9 b). The encrusting *Acervulina inhaerens* omnipresent in shallower depths, disappears abruptly at 130 m. This probably reflects the disappearance of its main food organism, benthic diatoms, near the limit of the euphotic zone.

Below 130 m depth, organic build-ups develop wherever the dip of the slope steepens. They are constructed by tubed worms and encrusting bryozoans, dendroid Foraminifera, "ahermatypic" small stony corals lacking symbionts and small gorgonians. The frame builders are so small in relation to the bioherms that their particular morphology has no impact on the general shape of the organic

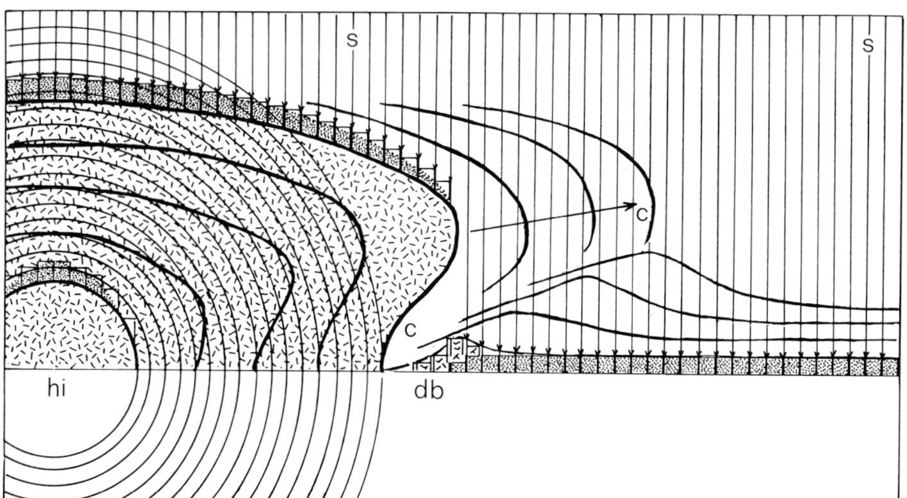

Fig. F.11. Schematic section through bryozoan build-up exposed to pelagic sedimentation. Growth rates of build-up depend on sedimentation rates, sedimentation rates on the inclination of the substrate modifying the isotropic (hemispherical) growth of build-up according to its size. *s* vertical rain of pelagic sediments; *shaded* sediment thickness per time unit; *hi* Initial hemisphere up to about 10 cm in diameter; *c* Deformation of successive contours of growing build-ups depending on diameter (age) of bioconstrucion; *db* biogenous debris produced by build-up and added to pelagic sediments. Not to scale (Hottinger 1984a)

Fig. F.12 a–d. *Madracis interjecta* reefs in 190 m depth off MBL, Elat, as seen from GEO submersible as 2.5 m high structure (**a**), detail with young *Madracis* colony on bryozoan build-up (**b**), fragments of *Madracis* colonies (**c**); dead parts covered by pioneering arborescent Foraminifera, about natural size. **d** Single *Madracis interjecta* calyx, soft parts partially removed by H_2O_2, SEM, scale bar 0.25 mm

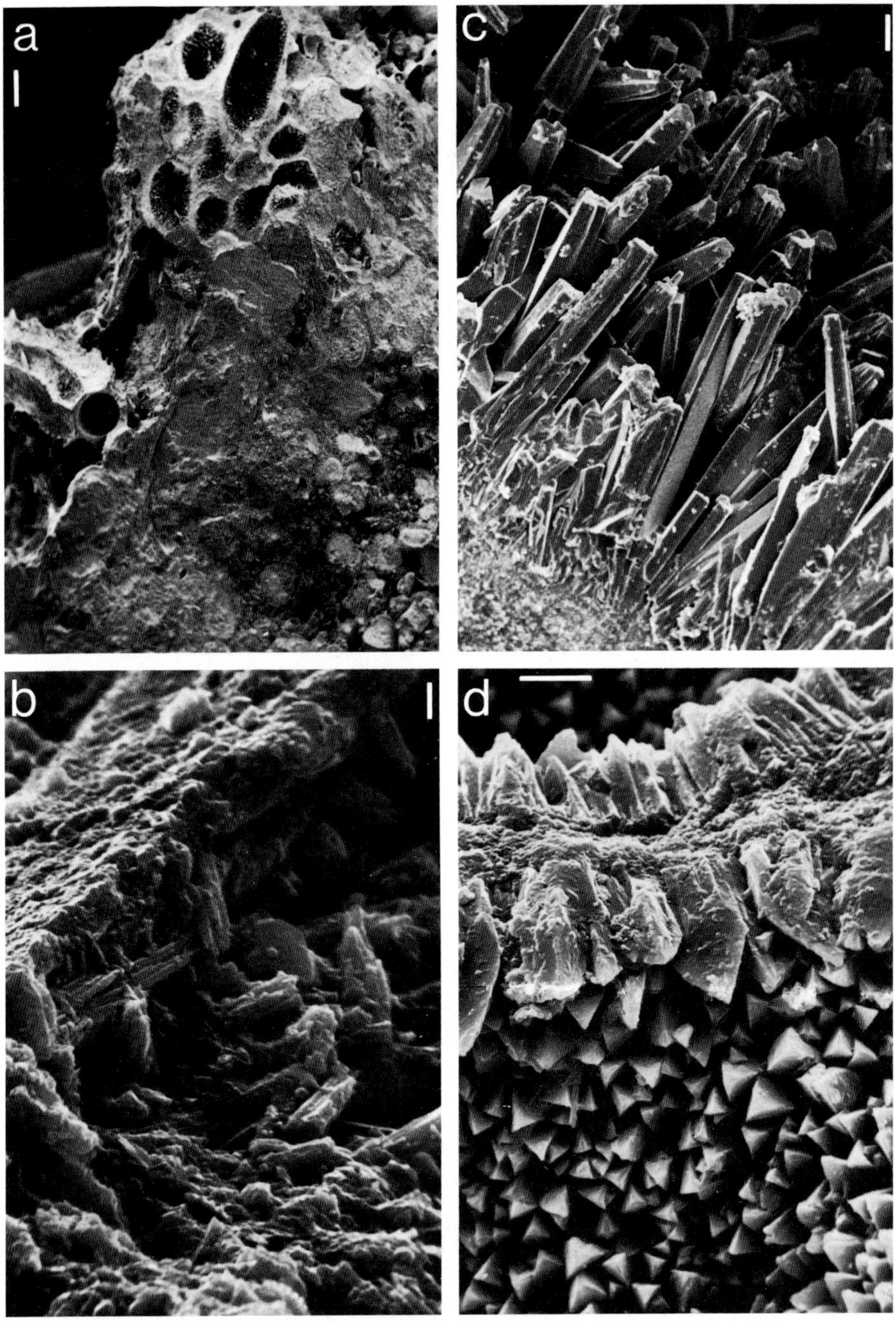

Fig. F.13 a–d. Cementation in deep hard-bottom and bryozoan build-ups from 130–190 m depth off MBL, Elat. **a** Hardened surface overgrown with serpulid worms and Bryozoa. Bore holes coated with aragonitic primary cement (**c**). Destruction of aragonite cements below long-exposed surfaces (**b**) and primary calcite cements coating calcitic bryozoan skeletons (**d**). SEM micrographs, scale bars (mm) **a** 0.25, **b, c** 0.01, **d** 0.005

Fig. F.14 a,b. Sediment (**a**) and insoluble residue (**b**) of fractions <0.112 mm. Note angularity of insoluble grains blown into the sea from nearby desert. Off MBL, Elat, 184 m. SEM, scale bar 0.1 mm

mounds. Small bioconstructions behave therefore like a single hemispherical colony with isotropic growth (Fig. F.10). Where the diameter of the bioconstruction reaches the size of a man's fist, pelagic sediments start to accumulate on top of the sphere and hamper vertical growth (Fig. F.11). Progressive flattening of the bioherms with growing size demonstrates, by the inverse relationship between growth and sedimentation rates, the role of substrate inclination in low energy environments deep enough to receive the sediment particles by vertical fallout from the water column. On subhorizontal terrace surfaces, the build-ups are much flattened, while densely grouped agglomerates of subspherical aggregates are constructed on steep slopes, demonstrating not only the difference of sedimentation rate on differently inclined bottoms, but also differences in growth rates, accelerated on steep substrates exposed to the slow currents carrying plankton.

In some exposed places, branching stony corals without symbionts, but with massive skeletons, *Madracis interjecta* and *Dendrophyllia* cf. *D. minuscula*, construct up to 2-m-high superstructures on bryozoan build-ups (Fig. F.12). Their debris accumulating at the foot of the bioherm are fixed mainly by encrusting bryozoans and worm tubes. The thickets of branching corals and the overhangs of the bryozoan bioherms provide sheltered space for numerous passive filter feeders, mainly gorgonians, and some active pumping organisms, in particular small brachiopods. The pteriid lamellibranch *Pterelectroma* and several small pectinids are the most important producers of coarse components of the surrounding sediment.

The bryozoan bioherms are cemented by one or two coats of low magnesium-calcite covering the calcitic skeletons of the bryozoans (Fig. F.13). The primary porosity is irregularly filled up to various degrees in contrast to the regular cementation of shallow hardgrounds. Despite the deep brownish colour of the cemented surfaces, no metal (particularly iron-) oxide concentrations can be demonstrated by energy-dispersive X-ray analysis under the scanning electron microscope.

Boring organisms of all sizes, from large *Lithophaga* to microscopic boring sponges, tend to destroy the bioherms and contribute to the production of larger debris, accumulating around the build-ups.

The mobile sediment on the shelf edge composed by shells of plankton organisms, shells of benthic animals produced on the spot, pellets, transported debris from hard substrate communities and very low percentages of angular quartz grains <0.112 mm (Fig. F.14) is bioturbated only on moderately inclined surfaces and shows a remarkable stability on steep slopes. Stirring the sediment with the submarine does not trigger larger slumps or turbiditic mechanisms in spite of the inclination of the slope reaching about 45°. Apparently, the lack of muddy fractions in the sediment and the low percentage of clay minerals is responsible for the stability of the sediments and the very low rate of particle transport along the depth gradient.

References: Cohen (1975), Fricke (1982a,b), Fricke and Hottinger (1983), Fricke and Schumacher (1984), Hottinger (1977a, 1983a, 1984a), Hottinger and Levinson (1978), Mergner (1971, 1980).

F.2.2 Geziret Fara'oûn ("Coral Island")

This area, located about 20 km south of Elat, was chosen for intensive field studies because it has a complex relief dominated by an island about 200 m long and by a separate submarine elevation 500 m off the shore (Fig. F.15). The topographic elevations modify the water energy regime in a particular depth zone, revealing its impact on substrate formation and benthic life independent of the depth gradient. A detailed nautical map with numerous soundings is available for this area. The map was completed and corrected by echo soundings from a large rubber boat positioned by sitometer in respect to topographic points on land. During numerous SCUBA dives, depths were measured by the diver's depth gauge, the distribution of living and dead larger Foraminifera was observed in situ and samples or photographs of the bottom were taken. Each dive gave access to three or four stations. Usually, samples of 150–250 g (dry sediment) were collected with a sweeping movement of the hand, covering 500–1,000 cm^2 bottom surface or by taking cores in 30-cm-long tubes of Plexiglas. Sampling in depths below 50 m was facilitated by the use of a grab from surface vessels. Density of sampling corresponded to about 10 m steps in depth. The diving activities concentrated on following the limits of conspicuous, mappable units of benthic communities, covering permanently the bottom, in particular *Halophila* meadows and coral carpets. Some limits were followed by underwater sledge driven by a rubber boat.

Subsurface processes were studied by sucking away superficial sand layers along profiles over several tens of meters with an underwater "vacuum cleaner." This sucking device consists of a 3 or 5 cm wide, flexible tube with a rigid, 1.5-m-long muzzle kept to the bottom by lead weights and at its end below the surface by a buoy. One meter from the muzzle's sucking mouth, compressed air from a spare diving bottle is introduced laterally in the rigid muzzle. The air bubbles expand during their rise and carry the water in the tube toward the surface, producing an effective stream easy to control by the amount of air admitted into the muzzle by an aluminum pistol valve like that used to inflate bicycle tires from containers of compressed air.

Nutrient concentrations in soft substrates were measured in samples of pore water extracted with a vacuum pump from about 20-cm-long sediment cores immediately after being taken by the diver in 5-cm-diameter Plexiglas tubes and closed in situ by rubber stoppers. Some dilution of the the pore water with open water during the sampling manipulations is inevitable. Nutrient measurements in the pore water represent, therefore, minimum values of concentration with high variability (Fig. F.21).

Water movement is characterized in this area by a long-shore current driven by the prevailing winds from the northeast. The current is divided by the Island of Geziret Fara'oûn into a near-shore branch, funneled into the passage between the island and the shore, and an off-shore branch bypassing the island on its eastern side. A particularly dense population of the fire coral *Millepora* colonizing large boulders and blocks (Fig. F.16) from the destroyed Black Tower on the island's northwestern shore pinpoint the location of the highest velocity of the near-shore current. The off-shore branch of the current loses much of its initial energy as observed in front of the MBL by broadening and splitting up during its crossing of the Wadi Taba bay, north of our test area. This must be the reason why no belt of macroids was found on the eastern slope of the island. The current is, however, strong enough to keep underwater elevations sufficiently free of fine sediment fractions to admit coral growth. The relative importance of current action and self-cleaning processes in the coral population may be assessed by the fact that no sediments are deposited on the 40-m-deep, current-exposed relief highs generated by dead coral carpets at the northeastern foot of Fara'oûn Island (Fig. F.17).

Wave action in this area has an impact on sediment movement to 5–8 m depth only, because the sea room in the northernmost part of the Gulf of Aqaba is much restricted. Where the island protects the mainland shore, the wave base is particularly shallow.

Sediment input from continental sources is restricted to the northern and southern borders of the test area. On the northern border, a small and most of the time inactive wadi has produced a delta of considerable dimensions, but the depositional rate of terrigenous sediments equals coral growth rates. Thus, a patchy distribution of coral heads disseminated in *Halophila* meadows is produced where most of the terrigenous sediment is trapped. Below 20 m no effect of terrigenous sedimentation can be observed in the closed cover of living corals, where the coarse sediments are exclusively biogenous.

Fig. F.15 a, b

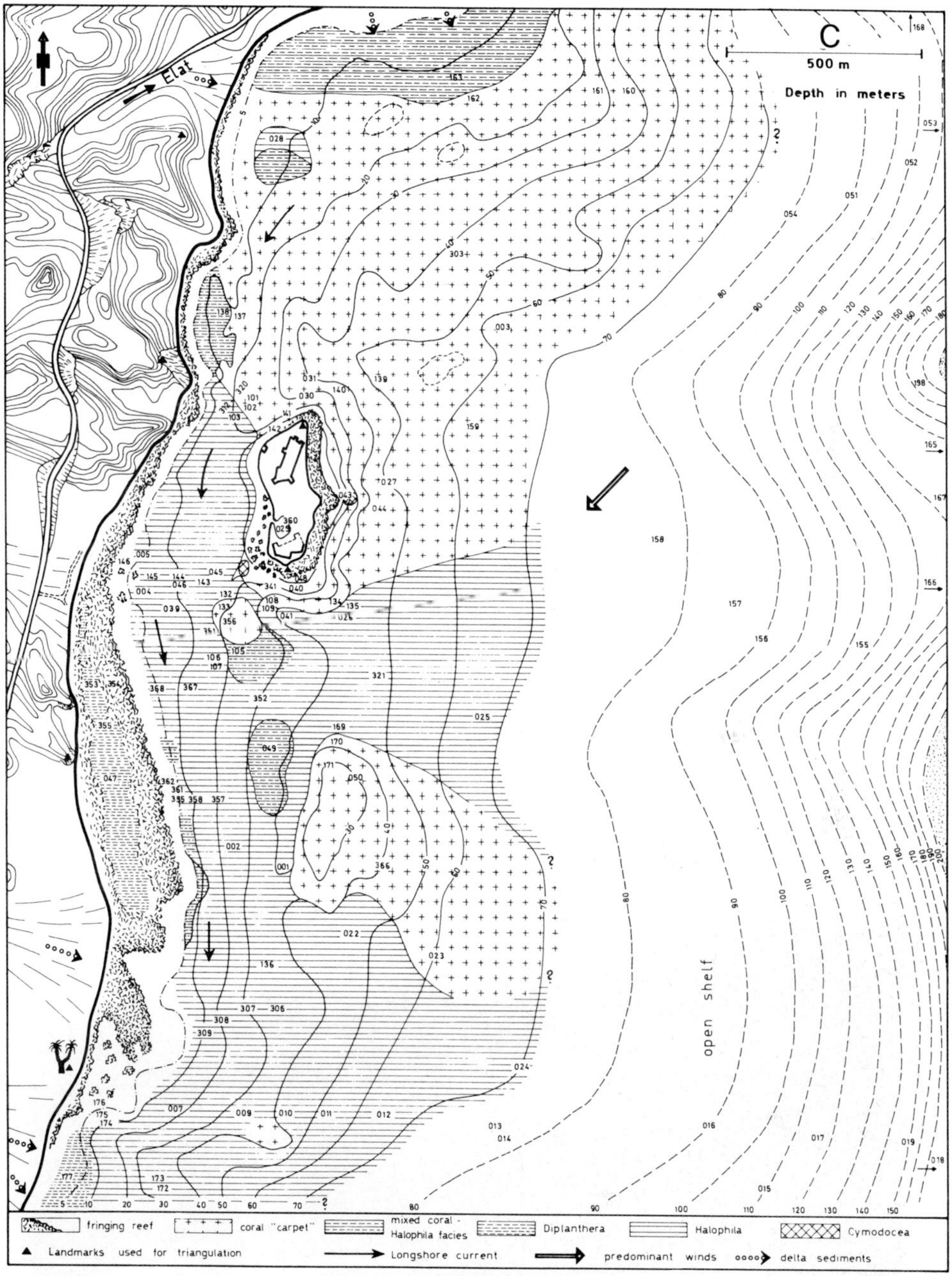

Fig. F.15 a–c. Distribution of substrates in Geziret Fara'oûn ("Coral Island") area south of Elat. **a, b** Mappable limits between solid substrate surrounded by bare streches of sand and soft substrates covered by *Halophila*. **c** Map of substrates with sampling sites. Note dependence of distribution pattern from exposition of sea floor topography in respect to main current direction from NNE and from sediment input by wadis. Between 70 and 200 m depth, the distribution pattern of hard and soft substrates is similar to the one explored by submarine off MBL, Elat (Hottinger 1977 a)

Fig. F.16 a,b. Reefs of the fire-coral *Millepora* below the "Black Tower" of Geziret Fara'oûn castle (**a**), site 142 on map in Fig. F.15c. Detail showing orientation of colonies perpendicular to longshore current (**b**)

Fig. F.17. Dead coral carpet overgrown with algae on 40-m terrace. Geziret Fara'oûn, site 140. Note pioneering coral heads at exposed sites

On the southern border of the test area, a very large delta has been built by a much larger wadi which is active at least every other year. Here the sedimentation rates forbid coral growth in spite of the favorable exposure of the shore line. Most of the terrigenous sediment is trapped in the dense vegetation cover down to 40 m depth. Sediment samples grabbed at the foot of the delta from 70 m depth and below show the same low percentages of terrigenous material as elsewhere.

Fringing reefs develop along the shores of the test area reaching very different sizes. In the shelter of the bays and of the island, the fringing reef platform is very narrow. Where the coastline is exposed to winds and waves, the reef platform widens, admitting shallow longshore channels trapping sediment. Where slopes are steep, as for instance on the eastern side of the island, the reef platform is narrower than on similarly exposed sections of the coast with less steep slopes.

The edges of fringing reef platforms are characterized by dense growth of the fire coral *Millepora* developing its fan perpendicular to the main, long-shore current direction. In the shelter of waves and currents, behind the island or outlying reef spurs, the fringing reef crest tends to be divided or split in a series of patch reefs. These are not relicts of destructive reef erosion, as claimed for the patch reefs on the Arabian side of the Gulf, but a product of active coral growth hampered by sedimentation of coarse particles in turbulent water, similar to spurs and grooves on large barrier reef fronts.

The frontal reef slope below wave base consists of naked coralline rock swept bare by currents and overgrown by disseminated small coral heads down to about 40 m. In places of reduced exposure the slopes are covered with thin layers of loose coral sand. Pottery fragments discovered on the most exposed northern flank of Geziret Fara'oûn, incrusting Foraminifera, red algae and occasional bryozoans, are held together by at least two generations of anorganic cements. Intensive carving by boring organisms tends to destruct the biogenic buildup and keeps the net **accretion rates** low (Fig. F.18). On parts of larger pottery fragments, the biogenous overgrowth reaches a thickness up to 8 cm, while other parts of the same fragment remain bare. Net accretion rates must therefore be very irregular over distances of decimeters, depending on the microrelief of the surface.

There are two varieties of pottery cemented into the reef slope: red fragments of large, crude amphoras, made with a potter's wheel, are possibly medieval Arab ware, cemented rather loosely onto the substrate, while black fragments of uneven thickness, made of very coarse clay without a potter's wheel, are cemented very tightly into hard rock; they have to be collected with their substrate. Some larger fragments of the black ware must belong to a pot or jar without base. Its rounded outline is subspherical, compressed with an opening about half as large as the total diameter. No remains of the neck have been found. Neither handle nor ornaments are visible. The surface is considerably eroded, but perforations by boring organisms have not been observed. The shape of the jar corresponds approximately to photo 86 in Amiran's (1969) corpus of ancient Israeli pottery, but is more compressed. According to Amiran, the poor quality of the clay worked on without a potter's wheel and particularly the method of firing, producing irregularly black and reddish colors in the pottery, are exclusively characteristic for Kirbet Kerah ware of the Early Bronze III period, corresponding approximately to the Old Kingdom in Egypt (2,650–2,300 BC). The cementation in at least two phases would thus be 4,500 years old and may correspond to the cementation of the 10 m hardground with a similar number of cement generations. The example shows the very low net accretion rate of the exposed reef slope where encrusting organisms are cemented, perforated by boring organisms and destroyed again, keeping in balance accretion and bioerosion for thousands of years.

Where the submarine relief is exposed to the main longshore current system, a **carpet of corals** covers the bottom. Between coral heads, coarse biogenous coral debris and loose shells are trapped. This sediment may accumulate locally in hollows or valleys, hampering coral growth. Where the accumulated layer of mobile sediment is thick enough (30–50 cm), infaunal life starts its burrowing activity. The shells accumulated in these depressions reflect the composition of the communities living a more or less cryptic life on the dead bases of the corals or on coarse coral debris. The low population density of coral fishes, particularly the many species of parrot fish (*Scarus* spp.) is largely responsible for the low contents of fine coral debris constituting in the fringing reefs the bulk of the sediment. The pelitic fraction of the sediment produced by the plankton in the water column and by the destruction of benthic skeletons or shells by microscopic endolithic organisms is winnowed away by the current. These sediments have therefore no anoxic layers below their surface, but their content in organic matter is still high enough to feed anaerobic bacteria when the oxygen supply is cut off: sediments of this type sealed in core tubes blacken in 3–5 h at room temperature.

Fig. F.18 a–c. Ancient pottery cemented into reef front above site 030 at northern tip of Geziret Fara'oûn Island (**a**). Thin section (**b**) through pottery fragment (*1*) and through its cemented coat with bore holes. Scale bar 1 mm. Detail of **b** in polarized light (**c**) showing three types of cement micritized (*1*), crystalline (*2*) and coating bore holes (*3*). Scale bar 0.1 mm

Fig. F.19 a, b

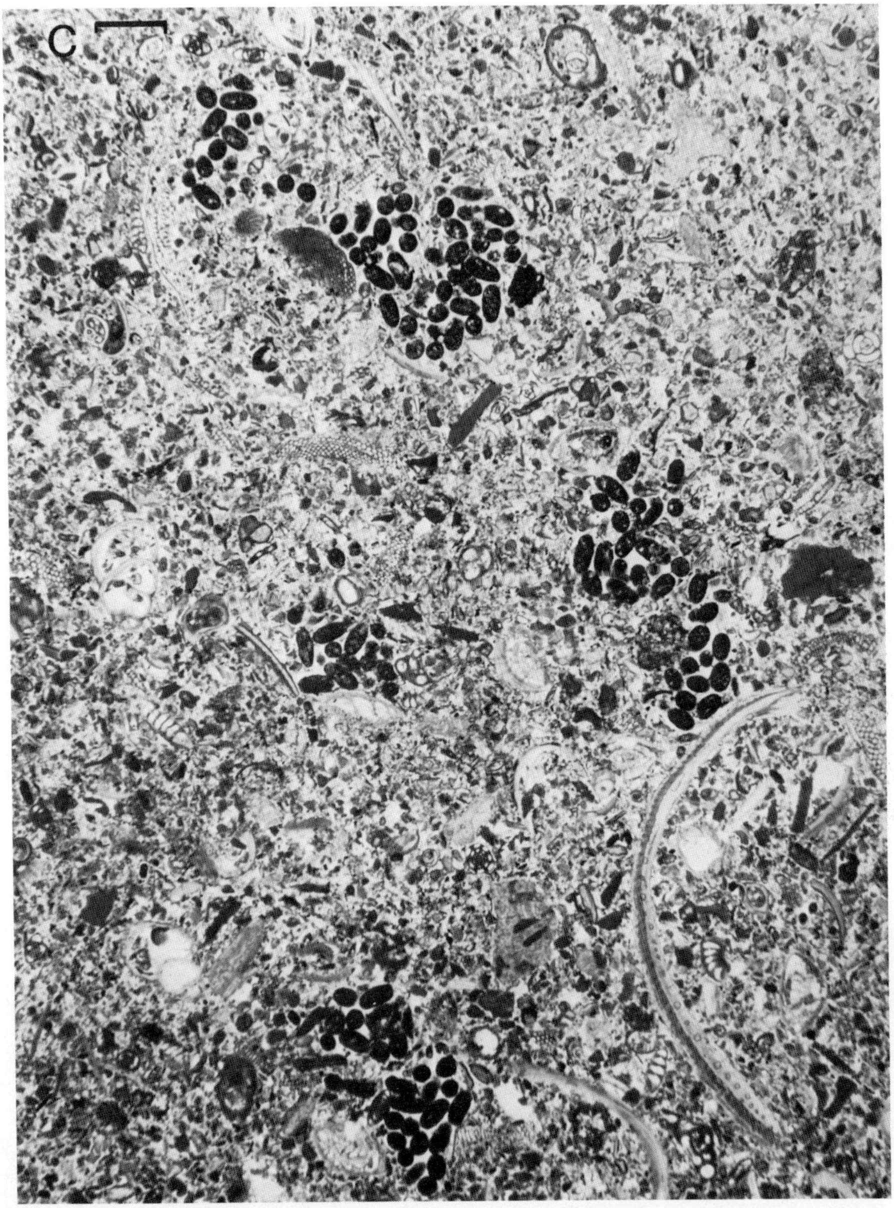

Fig. F.19 a–c. Burrowing mounds in *Halophila* meadow, 10 m, site 028 in Geziret Fara'oûn area (**a**) and detail (**b**) showing dead *Halophila* leaves on the bottom disappearing below the surface by burrowing activity of infauna. **c** Vertical thin section of corresponding sediment from artifically hardened core sample with fresh fecal pellets of burrowing animals and shells of epiphytic Foraminifera. Scale bar 1 mm

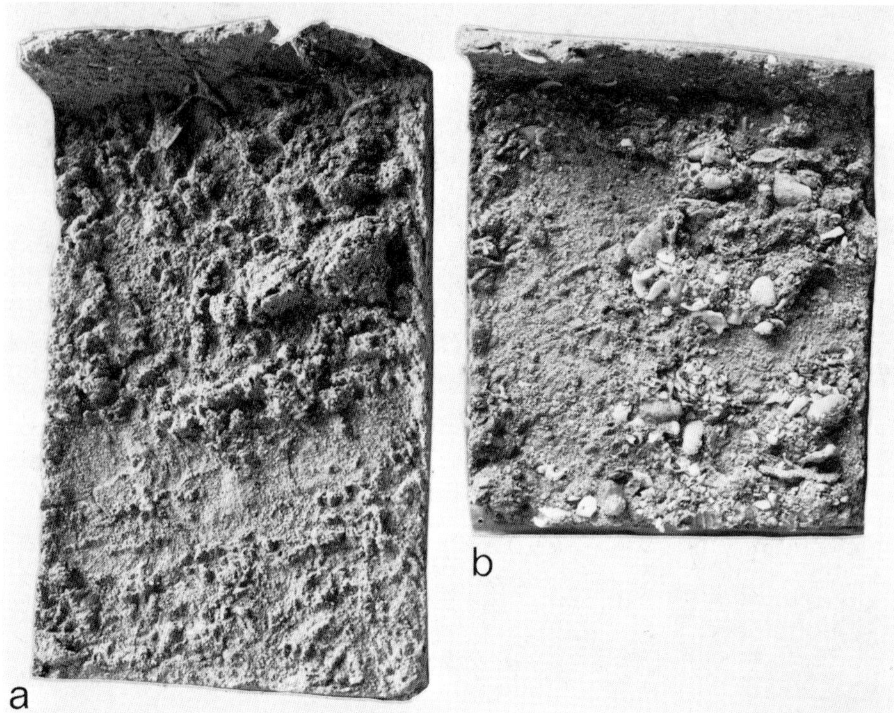

Fig. F.20 a,b. Relief casts from 20-cm-wide box cores showing absence of stratification by burrowing below *Halophila* cover (**a**) and below naked sand streches near coral heads (**b**). Location of samples see insert in Fig. F.25

Fig. F.21 a,b. Extraction of core from bare sand patch between corals, Geziret Fara'oûn, site 320 along line marking horizontal distances (**a**) and extraction of pore water from core with vacuum pump at MBL, Elat (**b**)

Where the water energy regime decreases locally by the widening of the funnel in the passage between the island and the mainland or in the shelter of relief highs, **sediment** accumulates, forming an unconsolidated, soft bottom. The finer fractions are trapped in the vegetation cover, fixed by organic slimes on the bottom and transported below the bottom surface by burrowing. The latter mechanism is very effective: after a particularly heavy thunderstorm in 1971, the wadis had carried large amounts of terrigenous sediments into the sea, covering the shallow shelf along the coast over many kilometers with a continuous mud coat easy to trace by its yellow color. Many corals were killed on the exposed side of coral heads, while on the protected side the corals cleaned themselves within hours. The mud coat on the bottom disappeared completely within 3 days, ploughed under the surface of the sediment by the burrowing activity of the infauna. This means that the entire surface of unconsolidated sediment is renewed at least weekly, ensuring a continuous flow of organic matter into the bottom to be consumed by the burrowing organisms. This **rate** of **burrowing** explains the high contents of plant remains and pellets below the vegetation cover in the unconsolidated sediments (Fig. F.19). The continuous burrowing activity is responsible for the lack of stratification in all shelf sediments of the Gulf (Fig. F.20). Larger burrows of the *Callianassa* type reach more than 1.5 m into the subsurface; smaller burrows are shallower. In generall, the infauna needs at least a 15–20 cm thick layer of unconsolidated sediment in order to develop an activity visible from the surface. Particular burrowing organisms, as for instance conoclypeid echinoids or sea-

Table F.1. Nutrient contents of bottom water (B) and pore water from cores (**C**) from different depths and substrate types (R. Buchmann, pers. comm.)

Sample no.	Depth (m)	NO_2 µg at l^{-1}	NO_3 µg at l^{-1}	PO_4 µg at l^{-1}
Vegetation cover				
367 B	12	0.15	Traces	0
367 **C**	12	0.23	1.47	3.88
357 B	20	0	0.20	0.10
357 **C**	20	0	0	7.16
352 **C**	30	0.15	2.45	2.12
348 **C**	31	0.06	0.12	0.24
No vegetation cover				
336 B	25	0	0	0
345 **C**	29	0.175	0.20	1.16
366 B	40	0	0.46	0
366 **C**	40	0.30	2.80	4.18
369 B	70	0	0.10	0.40
369 **C**	70	0	11.60	3.98
Surface water (date)				
23.5.1973		0	0.16	0
27.5.1973		0	0.22	0
Running sea water MBL, Elat				
28.5.1973		0	0.46	0.10

Retention capacity of sediments in 20 cm long core from site no. 352 washed with 350 ml KNO standart solution: 25%

stars, need much less sediment, but cannot be detected from the surface. In these layers, the mineralized nutrients gained from the consumed organic matter are dissolved in the pore water of the sediment where they reach considerable concentrations (Fig. F.21, Table F.1). As the free water immediately above the sediment surface in the vegetation cover has extremely low concentrations of nutrients (Chap. D), it is assumed that the plants constituting the vegetation cover take up their nutrients from the pore water by their superficial root system. Free benthic diatoms capable of retiring into the sediment may also profit from such a system of nutrient recycling.

Below 30 to 70 m depth, the sediment-water interface is covered by slimy brownish layers of microorganisms containing diatoms as inferred from their color (Fig. F.22a). These constitute a primary food source for bottom-dwelling organisms, particularly operculinids covering the sea floor in dense formations, and for burrowing organisms (Fig. F.22b). The latter are represented by **turritellas,** providing a model for the widespread Early Tertiary *Operculina-Turritella* marls. In the Gulf of Aqaba, the occurrence of *Turritella* cf. *columnaris* as identified by Por and Lerner-Seggev (1966) seems to be linked rather to fine-grained soft bottoms than to depth, as suggested by their distribution limited to 50 m in van Straaten's Rhone-Delta model (1960). In the grab samples of muddy sediments taken off Geziret Fara'oûn between 120 and 180 m depth, the *Turritella* shells constitute up to 30% of the coarsest biogenic fraction (Fig. F.22d). Many of them are juveniles. Adults often bear pioneering epizoans, indicating short residence times of empty shells at the sediment surface. Where numerous fragments of sessile Foraminifera, in particular *Miniacina* sp., indicate hard bottoms near the sampling site, *Turritella* disappears from the residues. In the euphotic zone, turritellas are rarely seen at the surface of the sediment. Dead adult shells emerging from the subsurface by the burrowing activity of the infauna are used by the sipunculid *Aspidosiphon* living in symbiosis with the carnivorous ahermatypic caryophyllid coral *Heterocyathus* growing on the shell. *Turritella* shells may thus be transported by secondary inhabitants of their shell into environments where the gastropod does not live.

The **vegetation cover** developing on unconsolidated sediments consists mainly of *Halophila stipulacea* and is perennial. Its ecological role for food supply and shelter is similar to that of vegetation covers on land. In very shallow water with comparatively high deposition rates of fine-grained sediments, *H. stipulacea* is replaced by the sea grass *Zostera* (*Zosterella*) sp. A unique, isolated patch of *Cymodocea* existed in 1971–73 at the most exposed spot on the east side of Fara'oûn Island in 30 m depth. The *Halophila* cover is densest between 15 and 25 m depth.

Fig. F.22a–d. Soft bottom and its communities. **a** Algal film (*AF*) in spring covering bottom and harboring Foraminifera over a few mm thick oxygenated sediment (*OX*) stirred up by metallic scale. Geziret Fara'oûn, 40 m. **b** Fraction > 1 mm of washed sediment dominated by *Operculina ammonoides* (*O*), with occasional *Heterostegina depressa* (*H*), some *Sorites variabilis* (*S*), *Dentalium* (*D*), *Fibularia* (*F*), gastropods, lamellibranchs and Bryozoa encrusting shell fragments. Off MBL, Elat, 70 m. **c** Soft bottom in autumn with smaller Foraminifera and single *Turritella* shell on surface. Wadi Taba, 45 m. **d** Fraction > 1 mm of washed sediment representing *Turritella* community sensu Por and Lerner-Seggev (1966), Geziret Fara'oûn, site 198, 175 m. Short-lived pioneers on *T.* cf. *columnaris* shells indicate short residence time of shells at sediment surface. Scale bars 1 cm

Fig. F.22 a–d

Farther down the density of the plants diminishes gradually. The deepest specimens were recovered by grab from 65–70 m depth. The biological decomposition of products of the vegetation cover in the sediments uses up the available oxygen in the interstitial water. Surface layers in contact with open water are oxydized one to several millimeters deep, as reflected by their bright color (Fig. F.22d). Below, the sediment becomes black by its reducing nature. Pyritization of foraminiferal shells is quite common, but always restricted to single specimens. There is an inverse relationship between the thickness of the oxygenated layer and the grain size of the sediment.

The pH was measured in situ at 10–12 m depth with an Ingold pH sensor packed into a watertight handle protecting the connections with the cable to the rubber boat. Measurements were made in fully oxygenated, naked sands and in reducing black sediments below the vegetation cover nearby. No significant differences in 10-cm-deep pore water were observed (Fig. F.23). The buffering capacity of the aragonitic grains in the sediment is apparently sufficient to keep the pH rather constant in the two different redox systems. The overall distribution

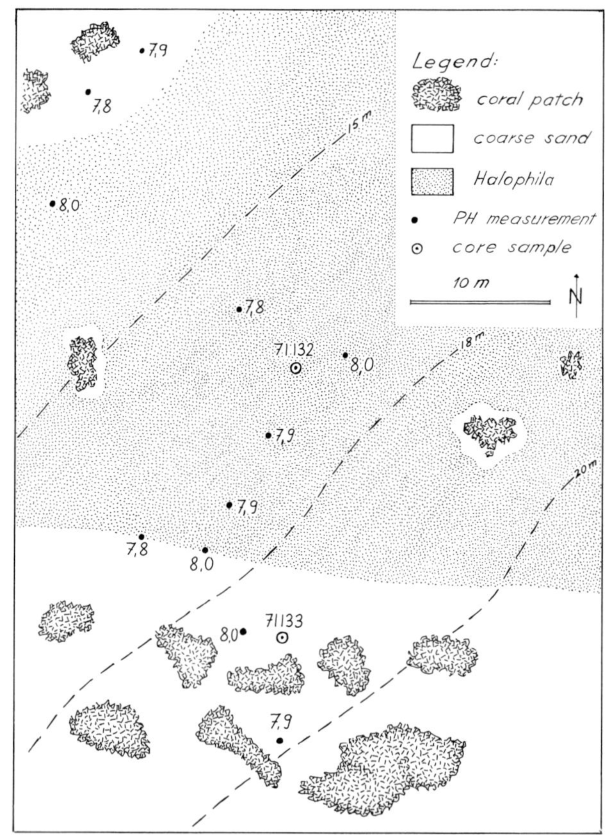

Fig. F.23. In situ pH measurements in pore water of sediment 5 cm below surface and location of core sample stations 132 and 133 (compare Fig. F.15c). No significant difference of pH values in different types of substrate due to the buffering effect of aragonitic grains in the sediment

pattern of solid and unconsolidated substrate obviously depends on the sedimentation rates determined by the water energy regime, modulated by the submarine bottom relief (Fig. F.15).

Solid substrates support coral covers, **soft substrate** vegetation covers where the moving sand mobilized by wave action does not prevent any permanent colonization of the bottom. During calm periods, the vegetation cover traps finer fractions of the sediment which are immobilized in the bottom by burrowing, whereas in the coral cover only coarse grains remain at their primary deposition site. Sedimentation rates are, however, not only determined by transport, but also by bioproduction in situ, involving biological processes in many ways. A closer look at the limits of the two main ecosystems in the photic zone determined by sedimentation rate reveals the generating mechanisms to be even more complex.

Around isolated coral heads in *Halophila* meadows and along limits of continuous vegetations covers, larger or narrower stretches of bare sand modify the submarine landscape in detail. The bare sand stretches lack any signs of burrowing activity. These small scale bottom patterns (Fig. F.23) cannot be explained by different sedimentation rates, or by differences in porewater chemistry. In the Caribbean, similar bare stretches of sand are explained as effects of grazing upon the plants by sea urchins or coral fishes. The grazing would be limited to an area corresponding to their extreme flight distance from the coral heads providing hiding places for the grazers. This explanation certainly does not apply to *Halophila stipulacea* meadows in the Gulf of Aqaba, where such grazers were neither observed, nor are the plants damaged. Observations in the Adriatic Sea, where local seagrass distribution can be correlated with minimum sediment depth over regularly broken subhorizontal limestone slabs outcropping nearby, led to investigations of the subsurface of the sandy bottoms near Elat. The "vacuum cleaner" helped to dispose of the unconsolidated sediment along transects perpendicular to the ecosystem boundaries. Below the sand irregular lumps of hard sandstone appeared. These consist of the same fully oxygenated coral sand as in the unconsolidated layer above, cemented by aragonite needles, forming a first generation of grain-coating cement (Fig. F.24). The lumps form a very irregular surface of cloudy masses below a layer of unconsolidated sediment usually more than 30 cm thick under the vegetation cover, less than 30 cm below naked sand. Thus *Halophila* vegetation covers depend on sufficient depth of unconsolidated sediments, admitting a burrowing fauna which facilitates nutrient recycling. The thickness of the soft sediment layer depends on an antagonistic system where sedimentation rate and rate of progress of **early diagenesis** in the subsurface are in equilibrium below the limit between the two ecosystems. When the sedimentation rate falls, the *Halophila* cover will retreat; when the rate rises, the cover will advance for distances depending on the inclination of the front of crystallization in the subsurface in respect to the sea bottom. Where the inclination of the diagenetic front is very small between the island and the larger underwater elevation farther south, migrations of vegetation cover boundaries of several meters per year were observed (Fig. F.25).

In the area of mixed *Halophila*-coral head facies on the delta at the northern border or our test area, no bare sand stretches form around the coral heads. In this area, the input of small amounts of terrigenous material keeps sedimentation

Fig. F.24 a–d. Early diagenesis by cementation of soft bottoms, Geziret Fara'oûn, near site 132–133. **a** Uncovering cemented layers below *Halophila* by sucking off the sand. **b** Detail of hardened surface in situ, ca. 20 cm below sediment-water interface. **c** Grains coated with aragonitic cement (SEM, scale bar 0.1 mm). **d** Aragonitic needles of cement (SEM, scale bar 0.01 mm)

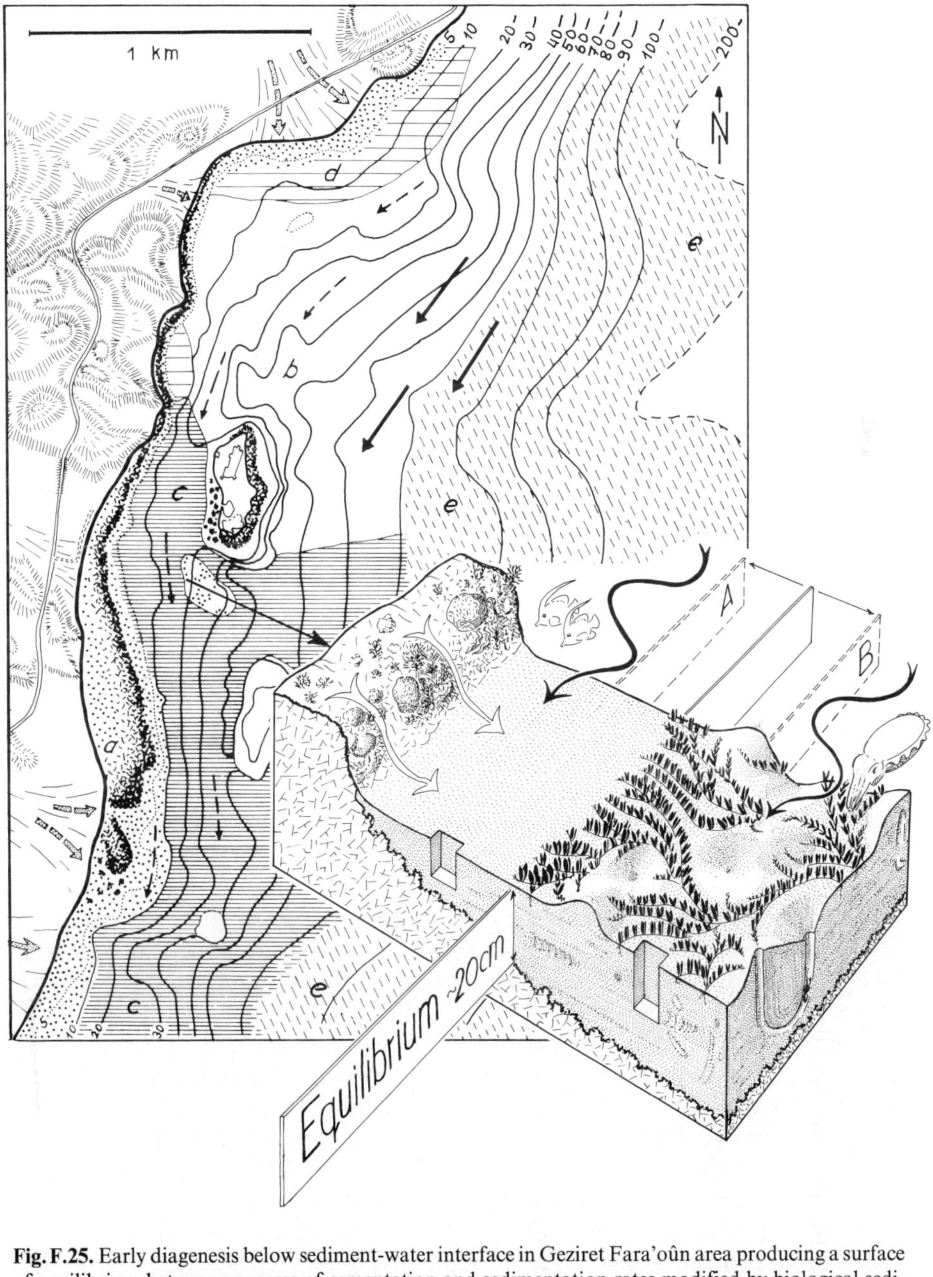

Fig. F.25. Early diagenesis below sediment-water interface in Geziret Fara'oûn area producing a surface of equilibrium between progress of cementation and sedimentation rates modified by biological sediment production on hard substrate (*white arrows*) and sediment production, trapping and burrowing in or below vegetation cover. 20 cm depth contour of equilibrium surface reflected by margin of vegetation cover. *Black arrows* currents. Substrates on map: *a* reef platform covered with *Diplanthera*; *b* coral carpet; *c Halophila* meadows; *d* mixed *Halophila*-coral head facies; *e* open shelf substrates (Hottinger 1984a)

rate ahead of progressing diagenesis without hampering coral growth. Soundings with an iron bar showed the soft sediment beside coral heads to be 40–50 cm deep, where *Halophila* is in direct contact with the coral build-ups. This is also proof that there are no biochemical repellent mechanisms separating the two kinds of permanent bottom covers, as they are known to condition the competition for space between slow- and fast-growing corals in a reef.

Coral growth is involved in the spatial distribution of ecosystems by modifying the local water-energy regime and thus the sedimentation rates. This biological factor introduces an independent element of disequilibrium into the physicochemical system reacting by displacement of the equilibrium boundary. A purely biological factor can thus produce nearby facies changes in the sediment without any alteration of the overall physical conditions. Facies changes in fossil sediments must therefore be interpreted with reasonable caution.

Precipitation of aragonite needs active transport of soluted carbonate by movement of interstitial water. The presence of accelerated porewater flux at the diagenetic front is documented by the full oxygenation of the hardened lumps below reducing grey and black sediments covered by vegetation. The porewater movement might be activated by the pumping generated by pressure differences at the sediment-water interface by waves. The effects of wave-driven porewater pumps were measured in littoral and sublittoral sands of the Mediterranean and on the Atlantic shelf of North Carolina. Whether this mechanism is wholly or partially responsible for the diagenetic processes in the subsurface of the shelf is not yet clear, the more so because the diagenetic process takes place to at least 200 m in depth, where wave-induced pressure changes are unlikely to have any effect.

The **distribution** of **larger Foraminifera** was registered by observation under water and by counting the shells in the sediment samples. Living and dead shells were not distinguished, in order to obtain representative measures independent of seasonal cycles and of patchiness in living populations. Larger Foraminifera were chosen because their size is sufficient to observe them directly under water (with the exception of *Borelis*), because of the importance of their Recent and fossil shells as sediment components and because of their close dependence upon facies types, as known from fossil representatives. Distribution studies of smaller Foraminifera at species level in and below the photic zone from different substrates are currently under way, but will be discussed elsewhere.

In our test area, the limits of distribution of larger Foraminifera coincide with isobaths of the underwater landscape and with limits between types of substrate, mapped as the boundaries between permanent vegetation and coral covers. Generally, the distribution patterns observed (Fig. F.26) indicate: (1) the distribution to be specifically distinct; (2) the larger Foraminifera to depend simultaneously on depth and substrate or on depth alone; and (3) the effectiveness of distribution patterns of sediment assemblages reflecting living populations, in spite of sediment mixture by burrowing and sediment transport by longshore currents. In order to be generalized, these observations must be tested in other areas with different oceanographic and local conditions.

References: Amiran (1969), Cohen (1975), Hottinger (1977a, b, 1978b, 1980, 1982b, 1984a), Por and Lerner-Seggev (1966), Riedl et al. (1972), Scheidegger (1972), Zweig-Strykowski (1973).

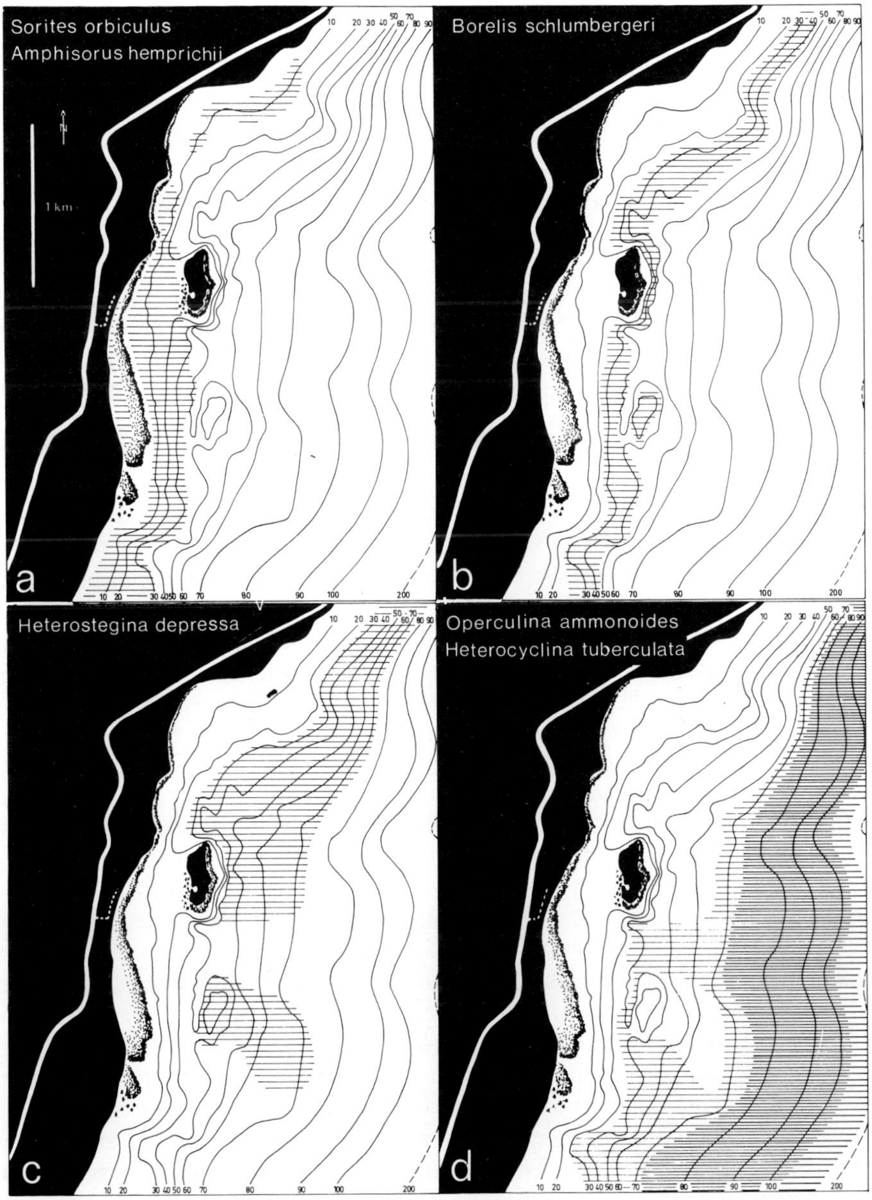

Fig. F.26 a–d. Distribution of larger Foraminifera in Gaziret Fara'oûn area as mapped under water. **a** *Amphisours* and *Sorites*. Note limits dependent on substrate and depth (30 m isobath). **b** *Borelis*. Note limits following the 15 and 45 m isobath irrespective of type of substrate. **c** *Heterostegina* dependent on depth and substrate. **d** *Operculina* dependent on depth and substrate complementary to the distribution of *Heterostegina* and largely overlapping with *Heterocyclina* on open shelf (Hottinger 1977a)

F.2.3 Ras Burka

Located 60 km south of Elat, the much-exposed cape Ras Burka presents advantages for the study of the effects of water energy on benthic life to 40 m depth. The sketch map (Fig. F.27) is based on the aerial photograph published in the *Red Sea Diver's Guide*, combined with observations gathered during one extensive reconnaissance dive in the company of J. Erez, in the middle of May, 1973 and with the data he published later (Erez and Gill 1977).

The bottom topography under water seems to continue the relief pattern on land, forming a kind of step running from the cape in southeastern direction into depth with a comparatively feeble inclination. Depth-dependent vegetation belts are, therefore, broadly developed in parallel, exposed or sheltered positions. Since the searoom of the waves is much larger in the southern sector than in the northernmost part of the Gulf, the wave base is correspondingly deeper, reaching 15 m depth in slightly agitated conditions of the sea. The longshore current from the North keeps topographic highs clean of the finer sediment fractions.

In spite of the exposed position of the shore line, far from any wadi mouth, the sediment input from land is important: north of Ras Burka, small grabens parallel to the coast have preserved Cretaceous sandstones outcropping along the shore. Salt, moisture and temperature changes during the day accelerate their degradation. The northerly wind steadily carries the Cretaceous sands along the coast, accumulates sand dunes in the shelter of the rocky promontory, and generates spectacular sandfalls on the latter's lee-side. Most of this sand is carried farther out into the sea, suppressing coral growth on the lee-side of the cape.

The transported terrigenous particles dilute the biogenous sediment produced in situ. Progressive dilution indicates the sedimentation rate to rise with depth within the *Halophila* lawn in this area. Highest dilution was found in the deepest sample at 40 m, while on deltas dilution diminishes with depth. This shows that the wind transports the sand through the air directely into the open sea, while the transport of sediments in the water is parallel to the coast and rather unimportant in this area.

The vegetation cover forms on both sides of the cape a shallow belt of *Zostera* sea grass followed at the most exposed spot by a large patch of *Cymodocea*. The latter is solidly fixed in the substrate. Its long, tenacious stems tolerate abrasion and rapid fluctuations of sediment levels. The basal parts of their stems are often overgrown by dense populations of hydrozoans, *Miniacina* and *Acervulina*, while their leaves remain without overgrowth, with the exception of thin coats of coralline algae of *Fosliella* type. Local distribution, morphology and overgrowth thus indicate *Cymodocea* to be a good indicator for high-energy turbulent waters.

Below wave base coral heads appear. They are associated first with *Halimeda* and form below 20 m more continuous covers, where large *Acropora* table colonies are frequent. In the shelter of the topographic high, *Halophila* meadows grow beyond 40 m depth, revealing the sudden drop of the water energy.

The distribution of larger Foraminifera closely reflects the different levels of water energy and its effect on substrate. In the most exposed parts of the area signalled by *Cymodocea, Calcarina calcar* appears in the sediment, while *Sorites*

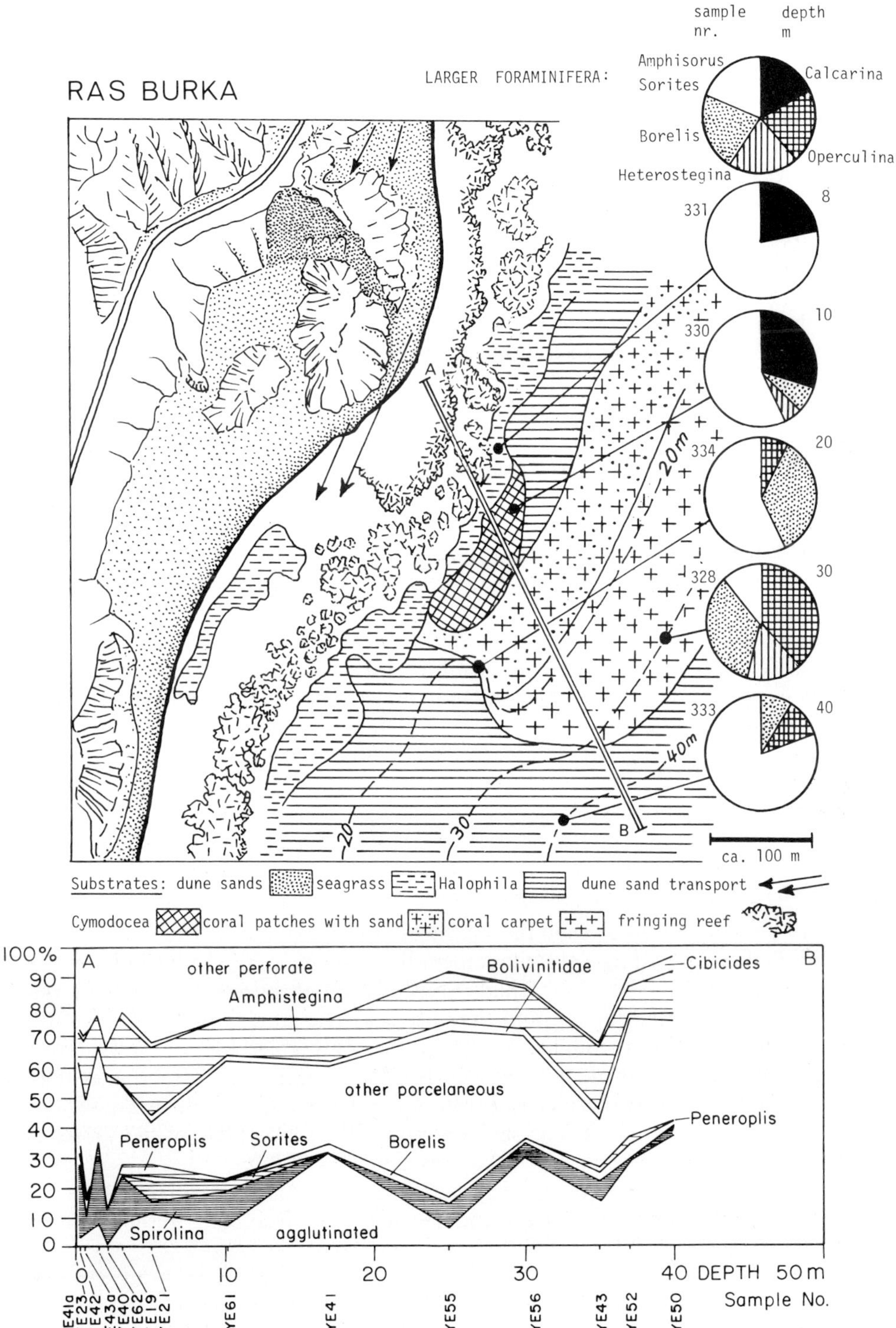

Fig. F.27. Distribution of larger Foraminifera at Ras Burka in selected samples from different types of substrate and of all benthic Foraminifera along traverse *A–B* (*B* from Erez 1972)

dominates clearly over *Amphisorus*. Soritids which are permanently fixed to their plant substrate, in particular to the stems of *Zostera* and *Cymodocea*, adjust their shell to the shape of their substrate. In the samples from *Halophila* meadows the soritids are uncommonly abundant by comparison to the frequency of *Operculina*. At the same time, small, epiphytic *Acervulina* and *Textularia* are extremely numerous. This is most probably due to a seasonal effect.

Multivariate analysis of the biogenic sediment constituents in general was carried out in this area. The samples were collected in a linear traverse over the exposed side of Ras Burka to 40 m depth. The analysis shows a clear classification of the samples into three depth zones corresponding to a near-shore, an upper-slope and a lower-slope section of the shallow shelf. Association analysis using presence or absence scores provided better results than modal ordination analysis based on metric modal data. Bioclasts other than shells of Foraminifera did not contribute to distinguish different environments. This is due to the difference in identification level: while the Foraminifera were determined at generic level, the other biogenic constituents were grouped into very large units, corresponding to class levels. There is, however, little doubt that, e.g., associations of smaller gastropods which are rather frequent in many sediment types, would very well permit the establishment of similar significant zonations if identified at generic or specific level. Multivariate statistics as used in this area are, however, a quantitative argument, strictly independent of other methods correlating distributional patterns, to verify that biogenic sediment constituents do in fact reflect different biota in spite of postmortem disturbances even in areas of comparatively high water energy.

References: Cohen (1975), Erez (1972), Erez and Gill (1977).

F.2.4 Dahab

This test area is located on the southwestern edge of the ancient delta cone of Wadi Dahab, providing a flat fundament for many types of shallow sea bottom extending in the shelter of the fringing reef. The latter reflects the extreme exposure of the site, built far out into the open sea. The large seaway, the constant northerly winds and the higher fertility of the southern section of the Gulf (see Chap. D) obviously enhance coral growth. The fringing reefs are broader than elsewhere along the western coast north of the straits and develop a distinct system of spurs and grooves. The reefs determine the topography of the shore, as well as the overall distribution of the substrates (Fig. F.28).

On the outer slope of the fringing reefs, *Acervulina* macroids occur in belts below 40 m. In contrast to the large and irregular macroids found in the northern

Fig. F.28 a,b. Dahab reefs and Kura Bay: Distribution of substrates (**a**) and location of intertidal *Chama cornucopiae* colonies (**b**). Elements of *Chama* colony: *a* Algal thalli as protection against the sun during low tide. *b* Spiral valve fixed to anchor and keeping aperture more or less parallel to bottom surface during growth. *c* Dead juvenile valves. *e* Hollow anchor formed by early cementation of intertidal, mobile sediment (*d*) (Hottinger 1984a)

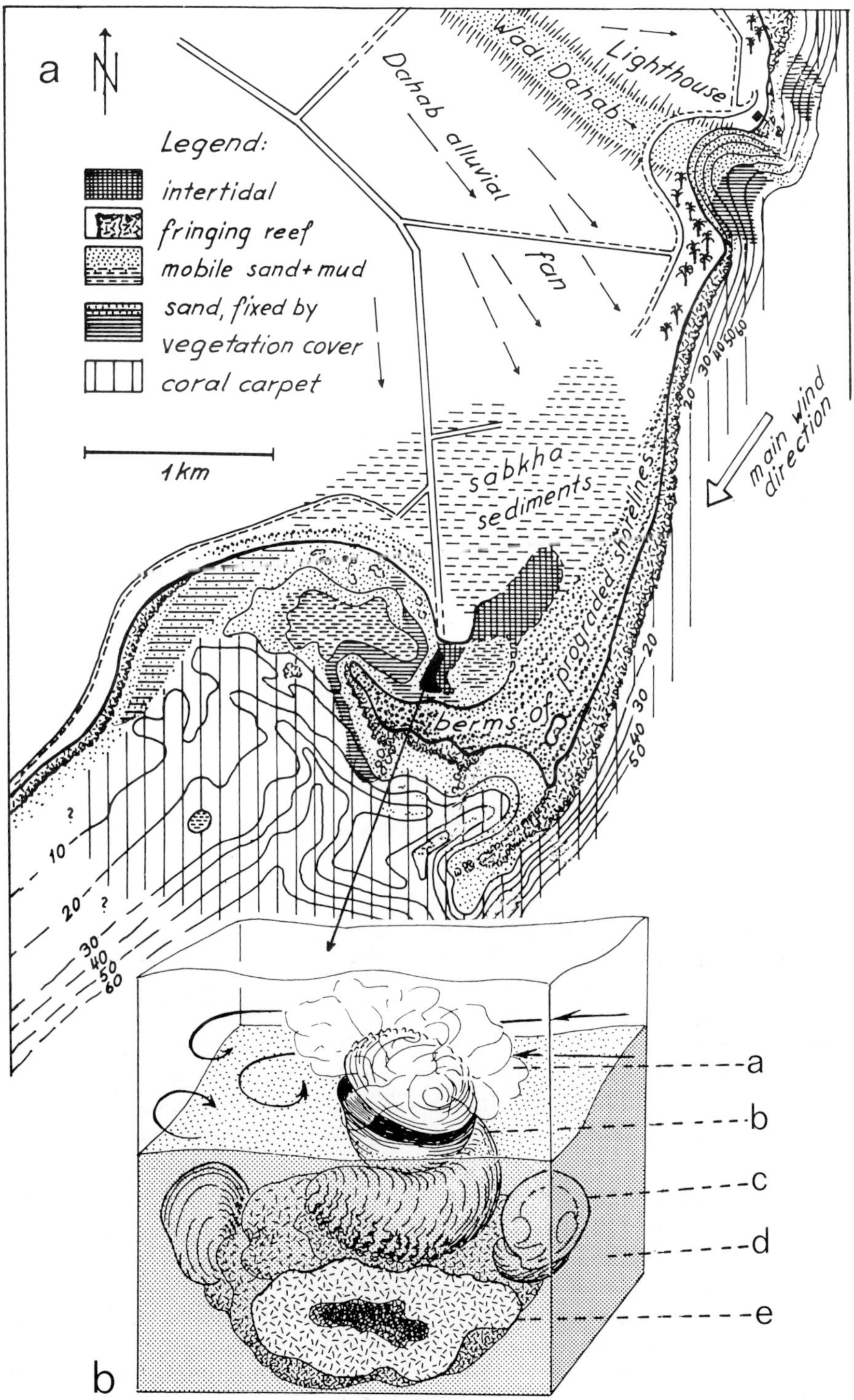
a
N
Legend:
intertidal
fringing reef
mobile sand+mud
sand, fixed by
vegetation cover
coral carpet
1km
Wadi Dahab
Lighthouse
Dahab alluvial
fan
sabkha
sediments
berms of prograded shorelines
main wind
direction
b
a
b
c
d
e

Fig. F.28 a, b

part of the Gulf in similar position and depth, the Dahab macroids are almost perfectly spherical and rarely larger than 5 cm in diameter. A single specimen of *Acervulina inhaerens* usually coats approximately one half of the sphere. As many large-sized Foraminifera have 1-yr growth cycles, we presume that at least once a year the water energy reaches the necessary level to lift the macroids in 40 m depth, while farther north such energy levels might be reached at longer intervals only.

Terrestrial sediment input is restricted to small amounts of clastics carried by occasional flash floods of Wadi Dahab over the large fan into the narrow embayment of Dahab lighthouse, where they disappear into depth, as through a trap door.

In front of the wadi mouth a slope of about 45° inclination is covered by pebbles and coarse sand. Repeated attempts to get the sediments moving downslope by digging under water failed. The sediment settles at once on the slope, a few meters below the site of triggering. In spite of the stability of the sediment, no infaunal life nor plant cover was observed. Apparently sudden, episodic floods of the wadi wipe out each attempt to colonize the slopes.

The terrigenous sediment accumulating in the sheltered el-Kura bay is lifted from the ancient cone by the winds and transported into the inner bay through the air. The surf separates the different fractions and sorts out the mica platelets which accumulate on the sandy beach. These glittering micas have given the name Dahab ("the golden") to the whole region.

The biogenic constituents of the sediment are produced mainly in the reefs, carried along the shore and deposited in the shelter of the large reef-spur, pointing southwestward and protecting el-Kura bay. The combined terrigenous and marine sediment production lags much behind coral growth. These sediments are finally trapped in the inner el-Kura bay by an ancient reef barrier overgrown by living corals. The barrier keeps the sediment away from the outer bay, where a luxuriant coral carpet develops. At station 191 an almost circular patch of mud – keeping the coral larvae from settling – has developed gradually into a 15-m-deep and 3–4-m-wide chimney during the growth of the coral carpet, starting off from the 40-m terrace (Fig. F.29). This is a clue to the minimum 15 m thickness of the coral carpet on its original substrate.

Dahab is the only place along the western shore of the Gulf known to have an **intertidal sand flat.** The latter forms a broad barrier separating a semi-closed, shallow lagoon from the inner Dahab bay and supports on the bay-ward side scattered patches of corkscrewed, heavy *Chama cornucopiae* shells (Fig. F.30). Protected by a "sun shade" of algae against desiccation and heat during low tide, the shells are deeply anchored in the sand against the tidal current by a foot of diagenetically cemented sediment. They may provide geology with a valuable actualistic model for the interpretation of Cretaceous *Requienia* or other Rudist intertidal environments, where foraminiferal faunas are poor, consisting only of shells grown in neighboring subtidal environments and transported by the tidal current. The *Chama* colonies demonstrate how a sessile organism profits from a diagenetic process to penetrate an environment lacking stable substrates in order to use – as the only first-comer at the surface – the rich intertidal food source available otherwise for burrowing organisms only. In the North Sea, where for

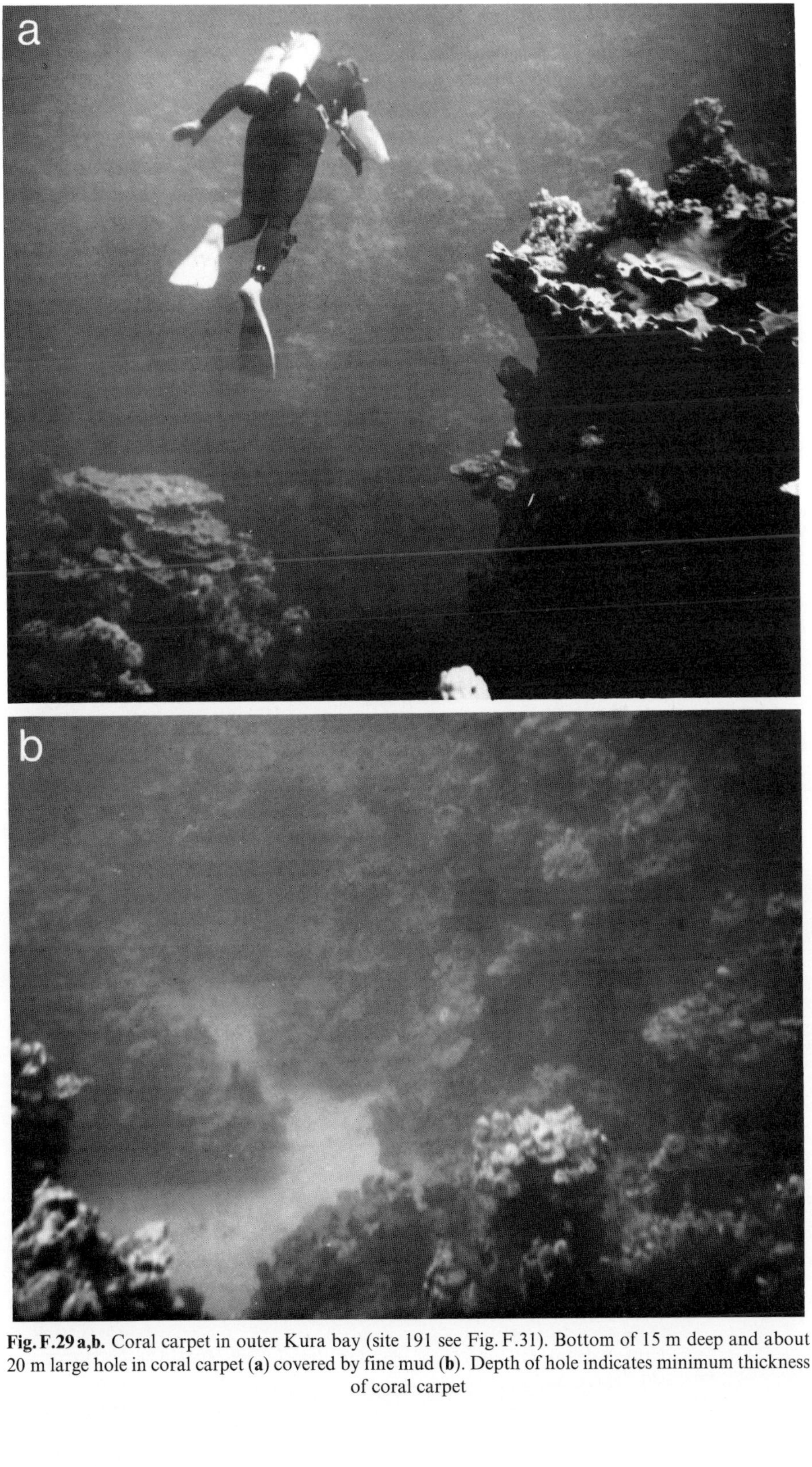

Fig. F.29 a,b. Coral carpet in outer Kura bay (site 191 see Fig. F.31). Bottom of 15 m deep and about 20 m large hole in coral carpet (**a**) covered by fine mud (**b**). Depth of hole indicates minimum thickness of coral carpet

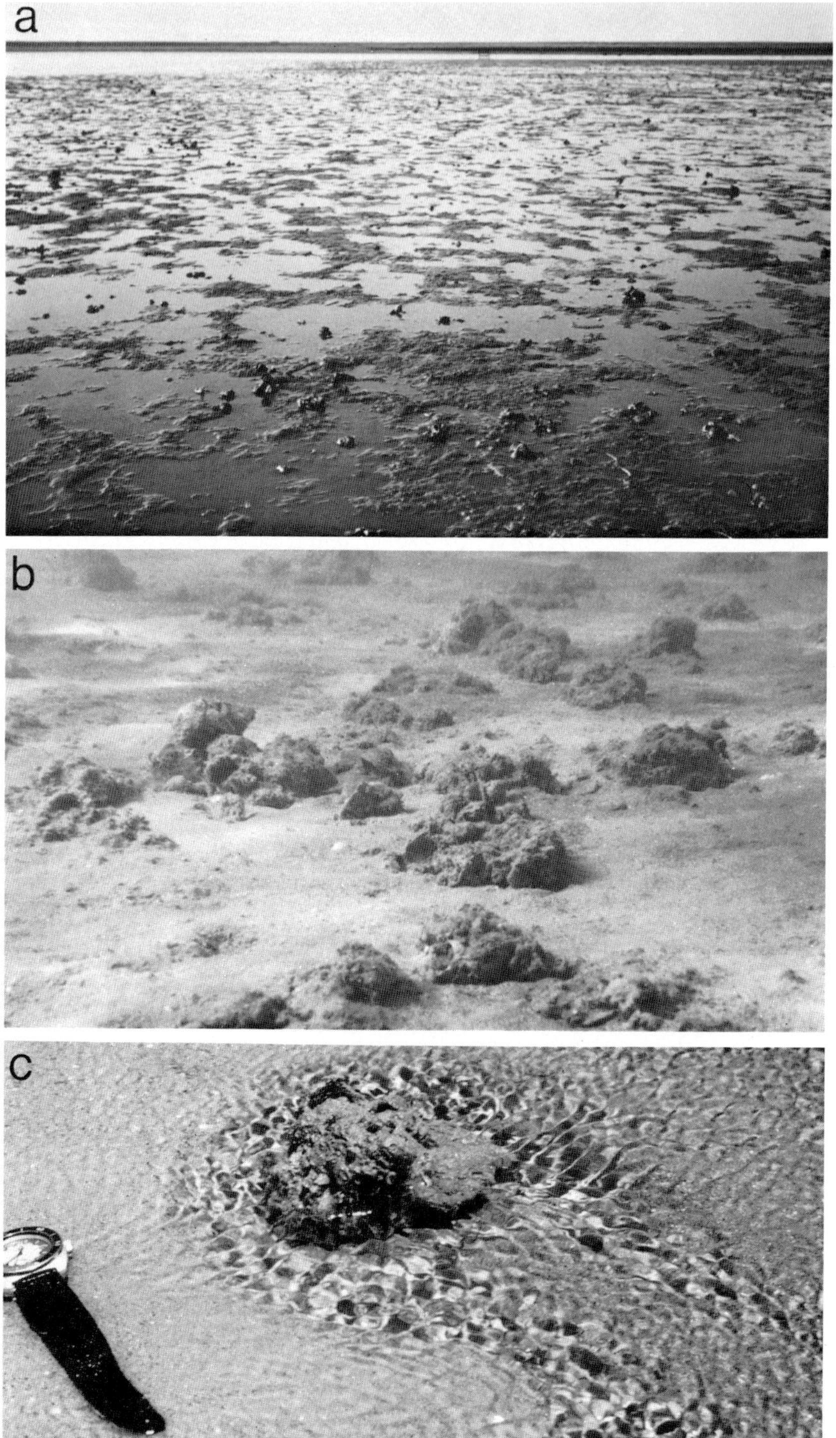

Fig. F.30 a–c

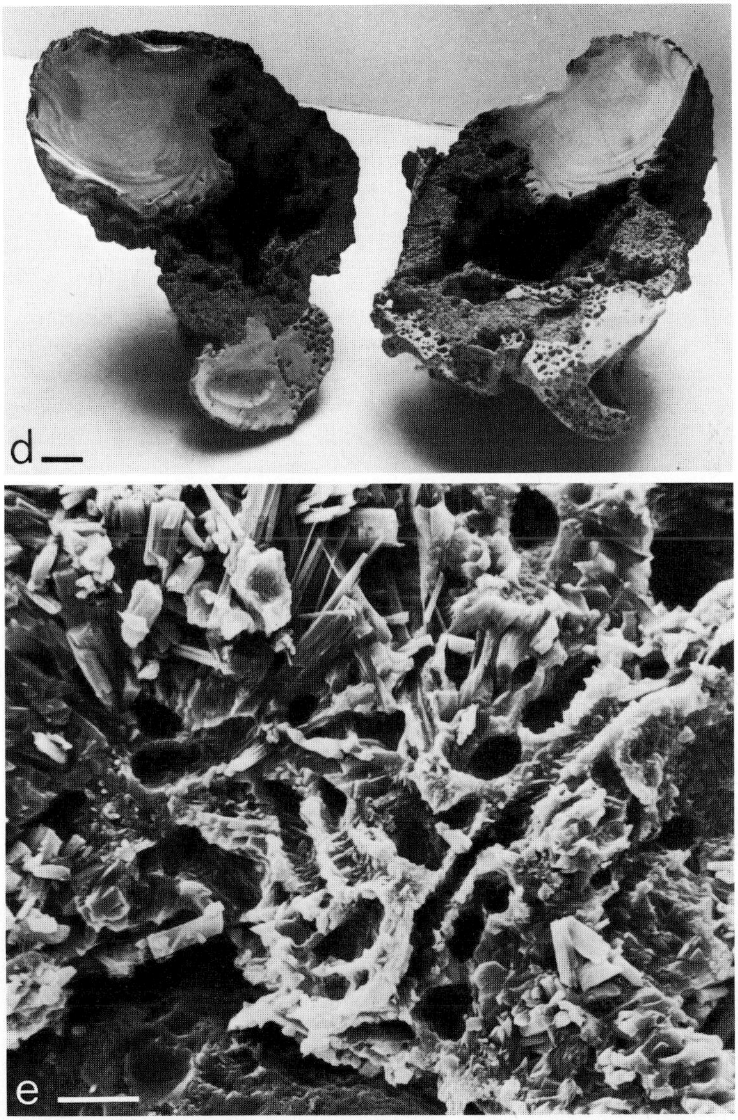

Fig. F.30 a–e. *Chama cornucopiae* colonies in Kura bay as a model for some rudist biota (location see Fig. F.28). **a** Tidal flat during low tide. **b** *Chama* colonies under water during high tide. **c** *Chama* colony uncovered by tidal current. **d** Sectioned *Chama* colony with hollow anchor formed by early diagenetic process; scale bar 1 cm. **e** Outer layer of *Chama* shell perforated by boring organisms and aragonitic cement coating the bore holes; SEM, scale bar 0.01 mm

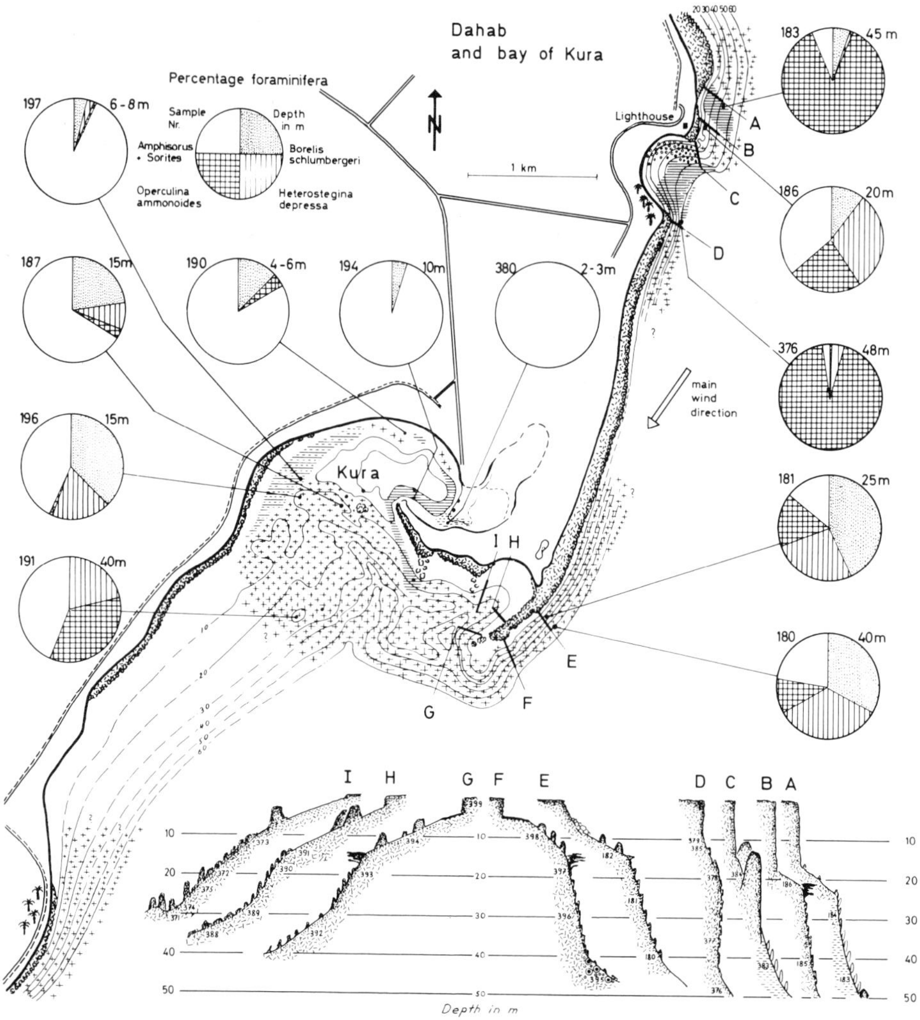

Fig. F.31. Frequency and distribution of larger Foraminifera on exposed Dahab reef slopes and in sheltered Kura bay. Note influence of type of substrate on percentage distribution. Substrates see legend Fig. F.15c (Hottinger 1977a)

reasons of lower temperatures this diagenetic process does not take place, *Mytilus edulis* occupies an analogous position in the intertidal food chain, by fixing itself in the mud substrate with byssus strings attached to dead shells.

The distribution of the larger Foraminifera in Dahab (Fig. F.31) is in principle the same as in areas with a lower energy regime, reflecting comparable substrates by identical associations. The frequency of the foraminiferal shells in the sediment is, however, often lower than usual because of dilution by other biogenous constituents produced by the strong activity of dense swarms of parrot fish living in the luxuriant reefs. On the outer slope of the fringing reef, north of the lighthouse

Fig. F.32. *Halophila*-covered, stable sediment on steep outer reef slope. Dahab, site 183, 45 m

(section I in Fig. F.31), indentations in the steep slope harbor soft bottoms inclined to 20°–30°, covered by *Halophila* and extending over a few hundred square meters only (Fig. F.32). The foraminiferal fauna clearly reacts to this small-scale facies change, while the permanent vegetation cover demonstrates the stability of the sediments on the slope, despite the intensive, potentially destabilizing burrowing activity and general high water energy conditions.

References: Cohen (1975), Gavish et al. (1978), Gvirtzman and Buchbinder (1978), Gvirtzman et al. (1977), Hottinger (1972, 1977a, 1983a, 1984a, b).

F.2.5 Mangroves

About 20 km north of the Straits of Tiran, the wreck of the Maria Schroeder on the reef edge spectacularly marks the world's northernmost mangrove swamps at Shura Arwashie. The *Avicennia* community grows on the reef platform, in front of the mouth of an inactive wadi, around a closed pool with *Caulerpa* and a semi-closed lagoon with *Halophila* and *Laurencia* covered bottoms (Fig. F.33).

On the aerial roots of *Avicennia*, no Foraminifera were found. The density of the epiphytic foraminiferal populations living in June was highest on *Laurencia papilosa* and *Halodule uninervis* while *Halophila stipulacea* supported smaller, and

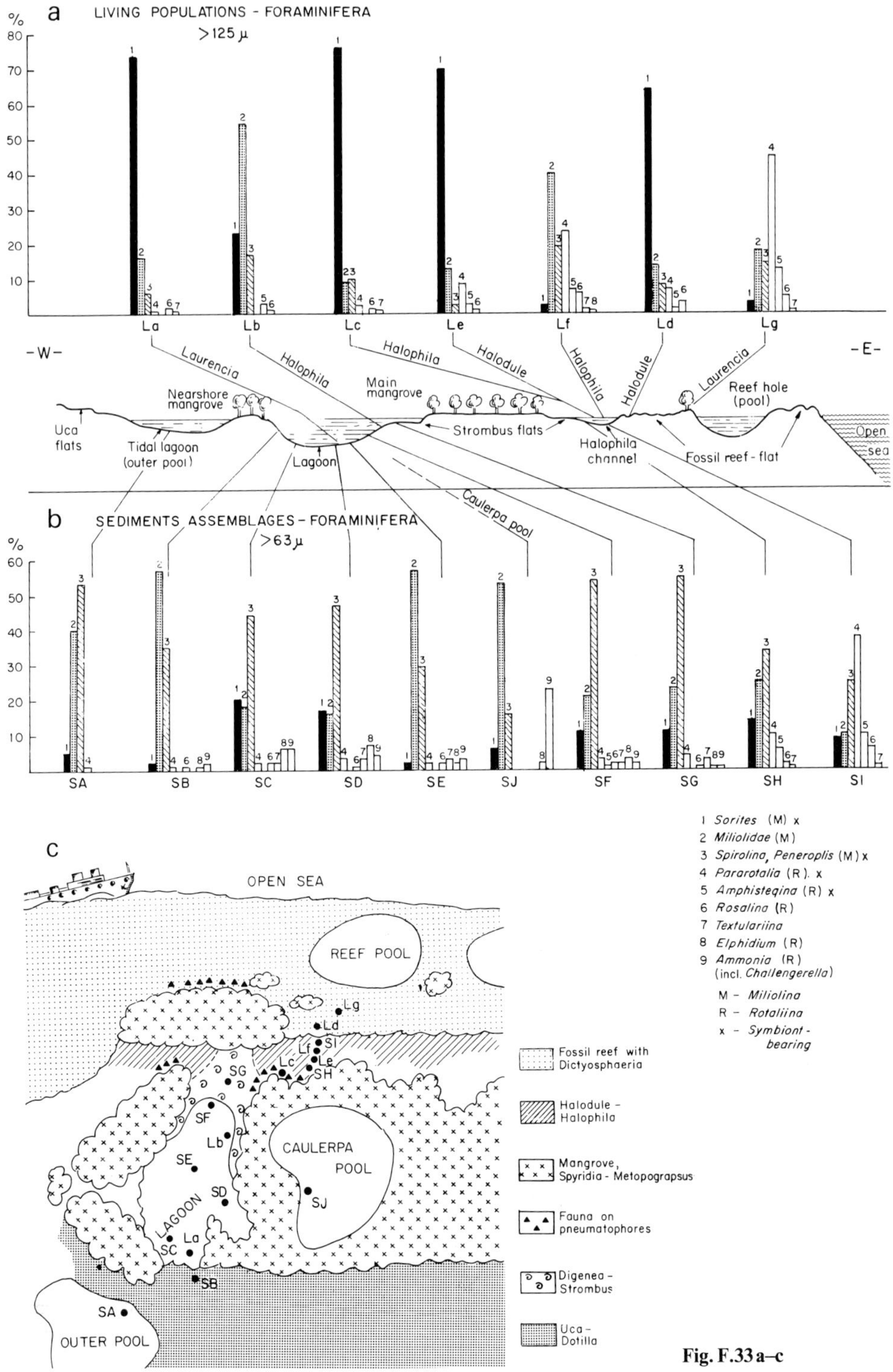
a
LIVING POPULATIONS - FORAMINIFERA
>125 μ
La
Lb
Lc
Le
Lf
Ld
Lg
-W-
-E-
Laurencia
Halophila
Halophila
Halodule
Halophila
Halodule
Laurencia
Nearshore mangrove
Main mangrove
Reef hole (pool)
Uca flats
Tidal lagoon (outer pool)
Lagoon
Strombus flats
Halophila channel
Fossil reef - flat
Open sea
b
SEDIMENTS ASSEMBLAGES - FORAMINIFERA
>63 μ
Caulerpa pool
SA
SB
SC
SD
SE
SJ
SF
SG
SH
SI
c
OPEN SEA
REEF POOL
CAULERPA POOL
LAGOON
OUTER POOL
1 Sorites (M) x
2 Miliolidae (M)
3 Spirolina, Peneroplis (M) x
4 Pararotalia (R). x
5 Amphistegina (R) x
6 Rosalina (R)
7 Textulariina
8 Elphidium (R)
9 Ammonia (R) (incl. Challengerella)
M - Miliolina
R - Rotaliina
x - Symbiont-bearing
Fossil reef with Dictyosphaeria
Halodule - Halophila
Mangrove, Spyridia - Metopograpsus
Fauna on pneumatophores
Digenea - Strombus
Uca - Dotilla

Fig. F.33 a–c

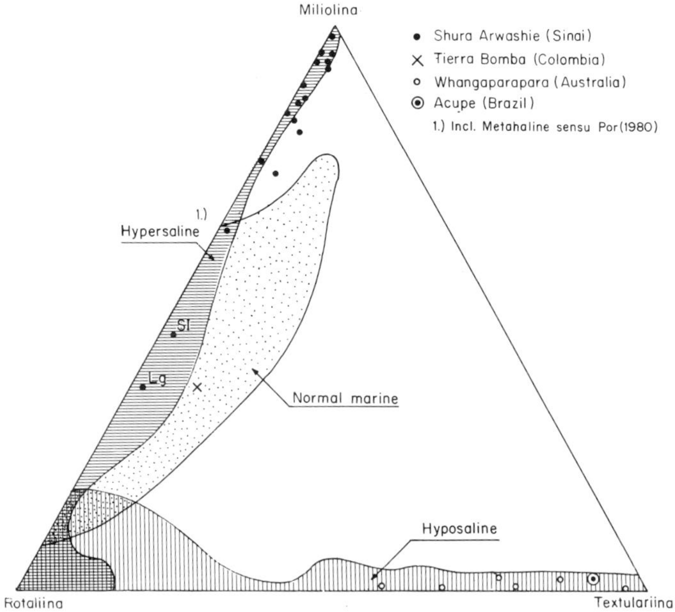

d

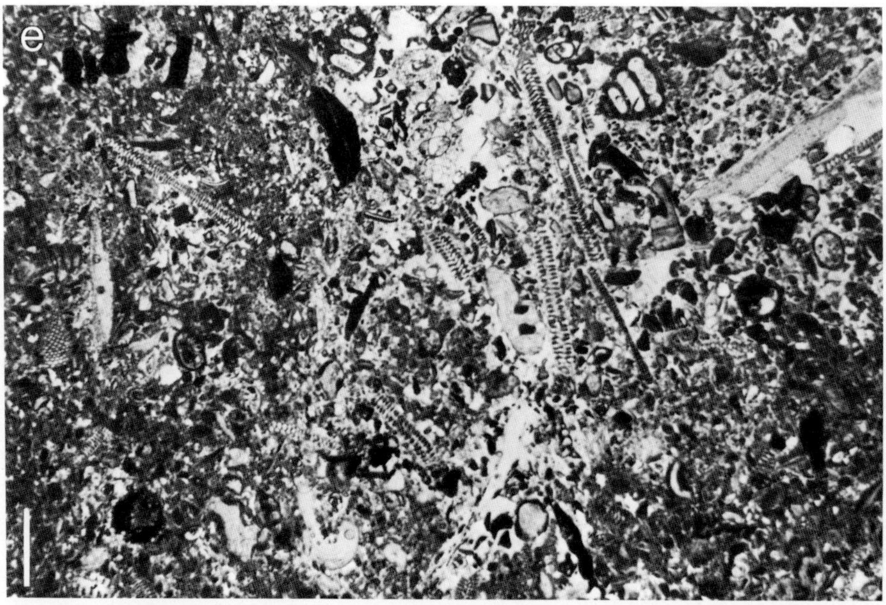

Fig. F.33 a–e. Mangroves at Shura Arwashie: Distribution of living (**a**) and dead (**b**) foraminiferal assemblages on different types of substrate (**c**). **d** Plot of foraminiferal assemblages from mangroves in different parts of the world according to nature of the shell wall and with reference to salinity (**a–d** from Halicz et al. 1984). **e** Vertical thin section of artificially hardened sediment from Nabek mangal showing usual shallow-water characteristics, i.e., *Amphisorus* and *Sorites* shells. Scale bar 1 mm

Halophila ovalis the smallest populations. The epiphytic populations were dominated by *Sorites orbiculus* and *S. "orbitolitoides"*, dead assemblages in the sediment by miliolids and *Peneroplis-Spirolina* (Fig. F.33 a). Evidently living *Pararotalia* (= *Calcarina calcar*), amphisteginids and *Rosalina* occurred mainly in the sediment samples form the "*Halophila* (tidal) channel", where water energy is comparatively high. *Elphidium* and *Ammonia* were absent from the epiphytic populations. *Ammonia* and related genera were, however, frequent in the sediment. Agglutinated Foraminifera are a very minor component of all assemblages recovered in the Sinai mangroves (including those of Ras Muhammed).

The specific composition of the foraminiferal assemblages in the Sinai mangals corresponds thus to the assemblages observed in the shallowest lagoons on reef platforms elsewhere in the Gulf of Aqaba, where the absence of *Amphisorus* and the abundance of *Peneroplis-Spirolina* mark high salinities and *Pararotalia* high water energy, mainly tidal currents. A plot of the assemblages from mangroves in various parts of the world and from Sinai (Fig. F.33 b) indicates the brackish-water mangroves (forming the majority of the world's mangals) to be associated with hyposaline marsh-assemblages of spiral agglutinated Foraminifera, normal saline (35%) mangroves to be characterized by normal marine assemblages and the highly saline (up to 47%) Sinai mangal to support meta- to hypersaline assemblages. There is no typical mangrove association of Foraminifera indicating some direct relationship between mangrove and foraminiferal communities.

References: Cohen (1975), Gvirtzman and Buchbinder (1978), Halicz et al. (1984), Por et al. (1977).

F.2.6 Marset el At

This area is one of the smaller embayments in the steep rocky coastline of the Gulf located 15 km S of the straits (Fig. F.34). It represents a common outlet of two small, ephemeral wadis. The inflow of terrigenous sediments is additionally hampered by touristic installations along the shore, particularly by concrete walls supporting the promenade. Between 1977 and 1981 no large amounts of terrigenous sediments were observed to enter the embayment. Sediment and substrate distribution in this bay represent, therefore, an undisturbed reflection of the interplay between coral growth, topography, sediment production and transport according to water energy levels. However, in 1982, H. Fricke (pers. comm.) observed a heavy rainstorm destroying part of the touristic installations. Large quantities of sediment were carried into the bay, killing most of the corals on the shallow patch reefs mapped on the enclosed sketch as observed prior to this event.

The outer part of the bay is fringed by reefs, which have their largest development at the southeastern edge of the bay most exposed to the prevailing winds. On the sheltered side, columnar patch reefs develop in front of the main reef slope. The northeastern section of the bay is characterized by patch reefs forming long ridges first and smaller patches of coral mounds toward the interior of the bay. The patches are kept separate by coarse coral sand distributed by the longshore

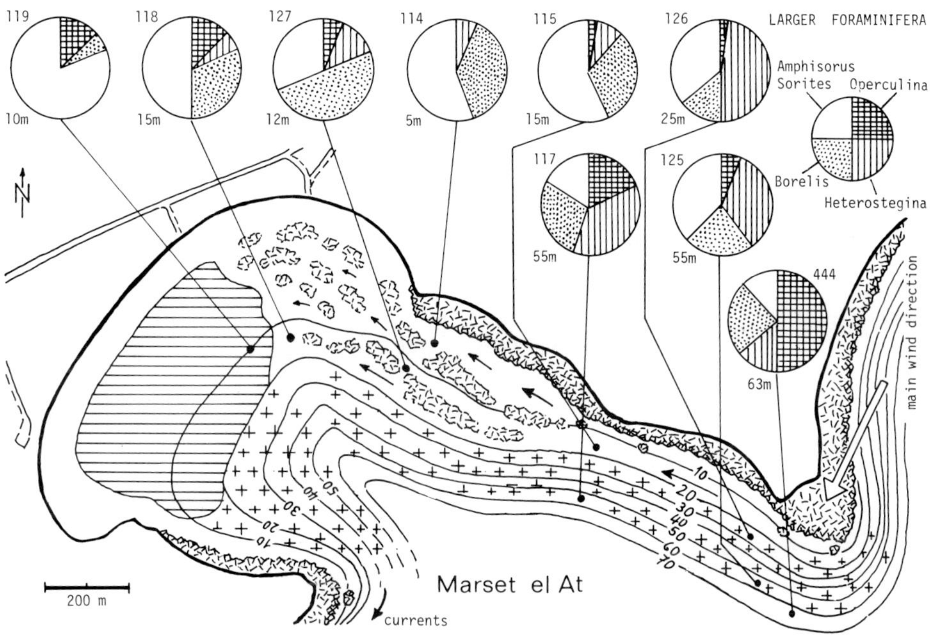

Fig. F.34. Frequency and distribution of larger Foraminifera at Marset el At. Substrates see legend Fig. F.15c. Note effects of gradient in water energy diminishing from E to W

current. In places, diagenetically cemented deeper layers are excavated, showing the net sedimentation rate to be very low.

The southwestern section of the bottom in the bay is covered by *Halophila* meadows colonizing the sediment accumulation in the zone of lowest water energy. The sediments do not penetrate the center of the embayment below 20 m depth. Most of it is trapped by coral heads growing faster than the sediment can accumulate. Toward greater depths these heads develop into a more or less closed coral carpet. The very simple substrate distribution pattern is reflected in the usual way by the foraminiferal associations dominated by *Heterostegina depressa* on hard substrate and soritids on *Halophila*. The notably higher fertility of the southern sector of the Gulf seems not to have any effect on the foraminiferal faunas.

References: Hottinger (1977a).

F.2.7 Ras Muhammed

The striking underwater beauty of this area has tempted many scientists to investigate the most diverse aspects of this region. In the present study emphasis is laid on the fact that almost the entire depth range of the shelf is exposed on quasi-vertical walls. The steepness of the bottom supporting the benthic communities is reflected by the narrowness of the fringing reefs, if compared to the opposite

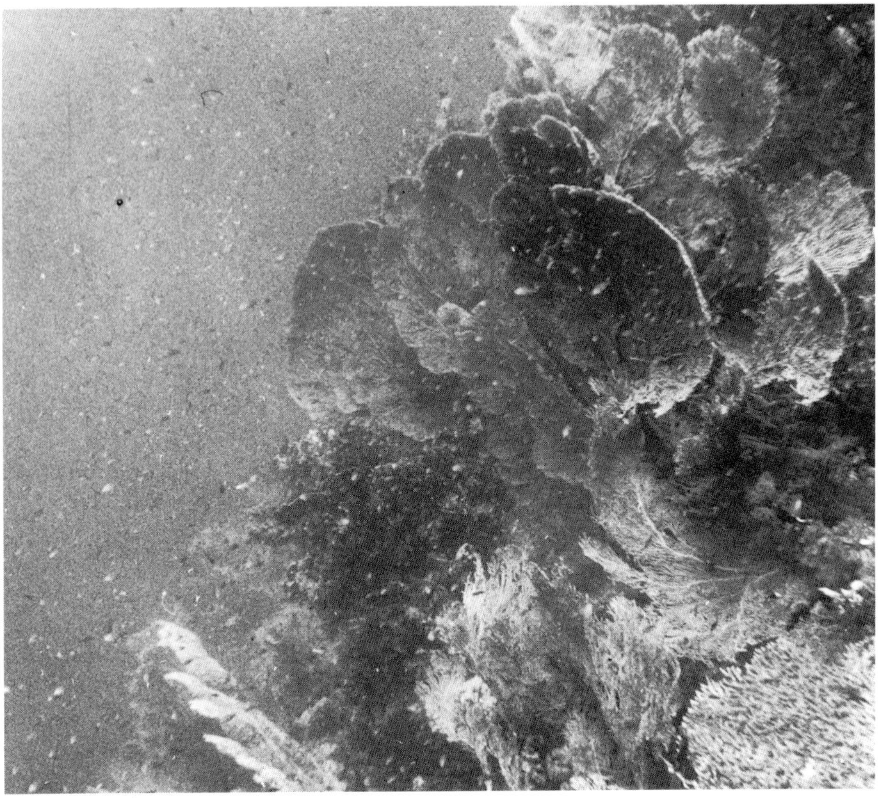

Fig. F.35. Belt of large gorgonians at 20 m depth on almost vertical wall. Single colonies about 2 m high. Ras Um Sid 7 km south of Marset el At

side of the Gulf around Tiran Island, where similar general oceanographic conditions prevail, i.e., two-sided exposure and higher fertility of the waters, resulting in higher plankton production.

In comparison to the reefs along the coast north of the Straits of Tiran, the Ras Muhammed reefs are exposed to much higher water energy. This is reflected by the appearance of distinct algal rims protecting the edge of the reef front from destruction by the waves as in larger barrier reefs in the open ocean. Where open lagoons develop behind reef crests, particularly on the eastern side of Ras Muhammed, the shallow basins are starved, with sediments trapped by sebkhas developing shorewards. In these shallow lagoons large meadows of *Cymodocea* develop, reflecting the turbulent water conditions generated by wave and tidal currents. As in Ras Burka, *Calcarina calcar* appears under these conditions.

The temporary violence of water movement is documented by the size of macroids reaching 10 cm in diameter (discovered by H. Fricke in the submarine GEO) in more than 100 m depth.

The comparatively rich plankton production is reflected by the development of a continuous belt of huge gorgonians (Fig. F.35) in about 20 m depth along the

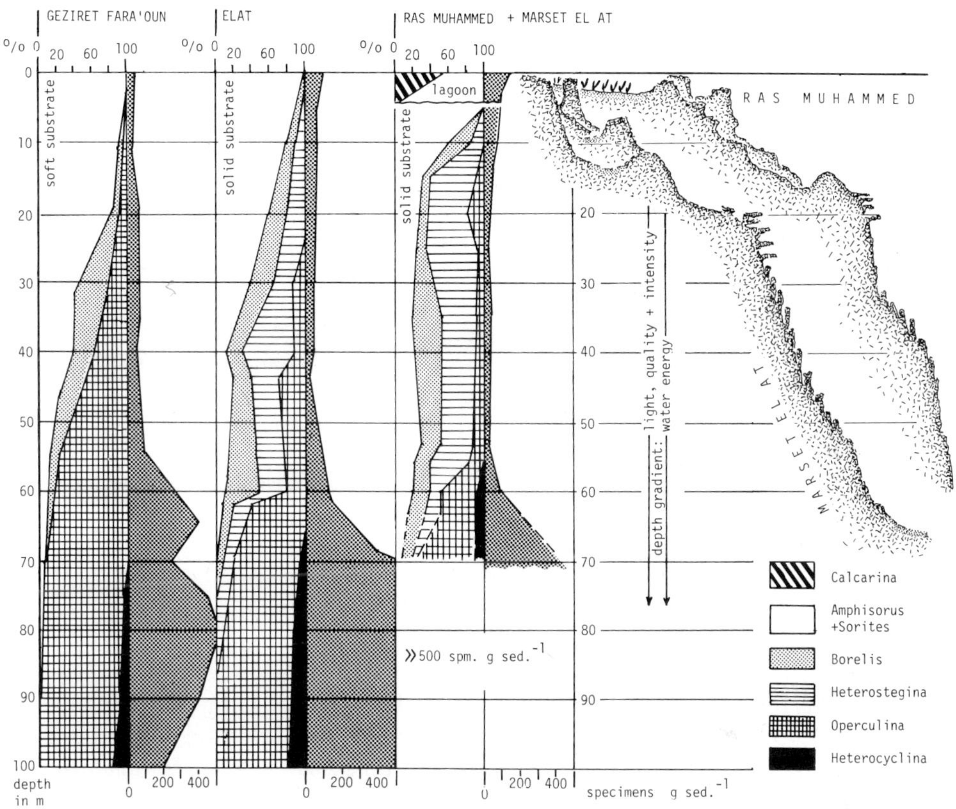

Fig. F.36. Depth distribution of larger Foraminifera at tip of the Sinai peninsula (Ras Muhammed) on almost vertical walls as compared to the distribution on less steep soft and hard substrates in the shelter of innermost Gulf of Aqaba

vertical walls. These suspension feeders reflect stratification of the plankton and indicate by the orientation of their fans the main current direction. On vertical walls exposed to the southeast, the fans are oriented horizontally, using either inconspicuous upwelling currents or, more probably, the alternating up and down movement of the water column generated by the waves. Similar dispositions to 40 m depth were observed in the Mediterranean on vertical walls of Île Maire, near Marseille. North of the Straits of Tiran, especially in the Dahab area, this particular reef zone is restricted to the most exposed reef spurs, but positioned in aproximately the same depth of 20 m, where there is also a characteristic break in the inclination of the slope. The gorgonians are oriented vertically, perpendicular to the main longshore current. They are associated to *Acropora* table colonies oriented horizontally. The filter-feeding mechanisms of these two groups of anthozoans must, therefore, be differentiated in a way which is not yet understood.

The foraminiferal associations as far as known today do not reflect this particular zone linked to plankton layering which indicates to some extent the independence of benthic Foraminifera from food provided by the planktic realm. One must, however, take into account that the environment described houses a community having matured over a long time, excluding most opportunistic species. This might be another reason why potential filter feeders, such as *Miniacina* and other arborescent Foraminifera, seem to be relegated into the background of cryptic life. On the other hand, the sediment does reflect this particular zone by its high contents of gorgonian spicules.

At greater depths, the photos taken by the GEO submersible indicate a development of bryozoan bioherms similar to those observed in the northern part of the Gulf, buth with shapes of superposed projecting roofs when attached to vertical substrates.

In spite of the peculiar situation of the benthos living on vertical substrates, the distribution of the larger Foraminifera is not changed and reflects the depth zonations on hard substrate as usual (Fig. F.36).

References: Cohen (1975), Fricke (1982 b), Gavish et al. (1978), Gvirtzman and Buchbinder (1978), Gvirtzman et al. (1977), Hottinger (1977 a, 1980, 1983 a), Hottinger and Leutenegger (1980).

F.2.8 Grafton Passage, Tiran

The reefs bordering the Grafton passage in the Straits of Tiran were explored in order to check the impact of tides and currents on the distribution of larger Foraminifera. Diving operations were supported by the research vessel Arnona, with J. Erez in charge, for several days in early June 1973. The sketch map locating the samples taken in this area (Fig. F.37) is based on Hall's (1975) new bathymetric chart of Tiran modified according to the detailed nautical and topographical maps available and to personal observations.

As the reefs are located on the saddle forming the straits, their basement is less steep than that of the Ras Muhammed fringing reefs. The Tiran reefs are therefore much broader, and develop into a barrier reef by transformation of the reef platform into shallow lagoons behind a typical barrier reef crest. The topographic disposition of barriers and lagoons is determined largely by the prevailing northerly winds and currents, accumulating moderate amounts of biogenic sediments in the open bay south of Johnson Point on Tiran Island. In this bay coral growth is restricted to 1–3-m-high patches covered mainly by soft corals and milleporids. The latter are particularly abundant along the exposed side of the barrier reef crest. The barrier reef slopes show the usual accumulation of *Acropora* tables and gorgonians between 15 and 25 m depth. The orientation of the gorgonians indicates the existence of currents parallel to the reef barrier. They produce spicules accumulating in the sediment. Deeper parts of the slope consist of bare coral rock, sparsely overgrown by scattered incrusting corals. Sandy bottoms appear deeper than 50 m. No macroids have been observed. There is neither a topographic break in the slope representing a possible equivalent of the 40-m-deep terrace observed farther north, nor does the bottom of the Grafton passage between 60 and 70 m in depth correspond to the 80-m terrace.

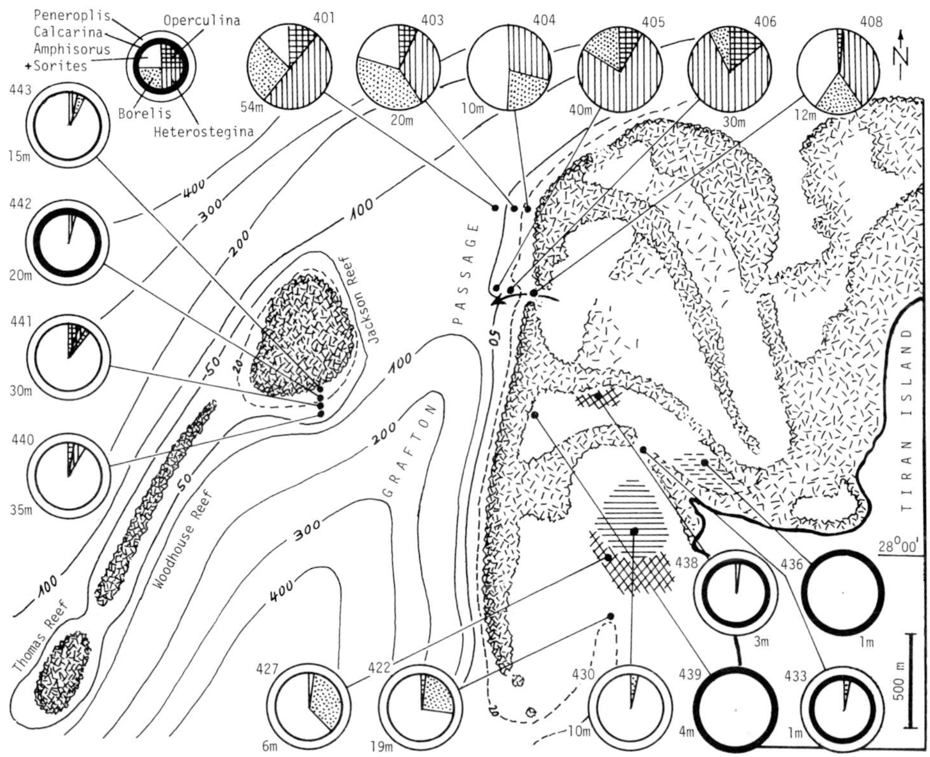

Fig. F.37. Frequency and distribution of larger Foraminifera, *Peneroplis* and *Calcarina* in the Grafton Passage, Straits of Tiran. High frequency of *Peneroplis* and/or *Calcarina* indicated by *white* and/or *black rings* when more abundant than all larger Foraminifera together. Substrate see legend Fig. F.15c. *Black arrow*: outflow of heavy hypersaline water during low tide

Jackson reef, blocking the straits, is a 70-m-high biogenic construction of cylindrical shape. Its upper part is penetrated by innumerable deep cavities where water circulates with high velocity. The walls of these cavities are usually smooth. Overgrowth is restricted to very small or smoothly incrusting organisms. It is, however, suspected that the cryptic fauna is rather productive in the interior of the reef complex, adding considerable quantities of food for the strikingly dense population of coral fishes by organisms swept out of the cavity system. This is supported by the unusually high percentage of *Miniacina* and *Homotrema* shell debris in the coral sand accumulating at the foot of the reef, below 30 m. These arborescent Foraminifera were not observed on the open slopes and surfaces of the reef and must consequently represent a part of the cryptic fauna in the cavity system.

Soft bottoms in the lagoons and in the protected bay are diversified in respect to soft bottoms farther north, but the lowest water energy levels are reflected as usual by *Halophila stipulacea* meadows. In highly oxygenated sand, moved by tidal currents, *Halophila ovalis* occurs in patches along the inner margin of the reef barrier. On open stretches of sand, large zones are covered by disseminated *Hali-*

meda plants deeply rooted in the mobile substrate. This facies strikingly corresponds to the *Halimeda-Penicillus* communities of South Florida, characterized by calcareous algae with a short life cycle on soft substrate. Where the sedimentation rate is lower and the water energy higher, *Cymodocea* occurs in rather localized patches, while on hard beach rock brown algae form fairly dense vegetation covers.

Evaporation in the shallow lagoons inserted into the reef platform is high enough to significantly alter the **salinity** in spite of the water circulation between the lagoons through numerous passages by tidal currents, driven by the northerly winds into the open bay (Fig. F.37). The importance of the circulation is reflected by the development of small secondary fringing reefs along the shores of Tiran Island and by the blossoming life on coral patches and bars in the lagoons. The water masses of higher salinity are heavy enough to flow out through the passages in the reef barrier in front of Jackson reef, falling over the reef slope toward greater depths and producing a submarine waterfall. One diving team was caught by surprise in such a falling current and had to crawl back upslope over the bottom, against the current reaching a velocity of 0.5–1 ms^{-1}. These currents must be a fairly constant phenomenon, being reflected by a linear distribution of the fire coral *Millepora* to a depth of at least 55 m in the ravine channeling the current.

In contrast to smaller forms, the **associations** of the **larger Foraminifera** (Fig. F.37) do not react to the differentiation of the vegetation cover in the lagoons and in the protected bay. An exception is a shift toward *Sorites* in the *Sorites*/*Amphisorus* ratio of the shallowest assemblage, reflecting the high water energy and/or perhaps the late season in which the samples were collected. The large number of *Peneroplis* present in many of the shallowest samples may be an indication of the high salinities in the lagoons. *Cymodocea*- and *Halimeda*-covered bottoms support large numbers of *Calcarina calcar*, most of them living at this period of the year, while *Challengerella*, *Ammonia*-related rotalids and *Cymbaloporetta* sp. are abundant in the sediments at a lower energy level with or without permanent vegetation cover. In some samples in the lagoon and on the upper reef slope, *Poroeponides* appears in high numbers, but these occurrences cannot yet be correlated with a particular environmental factor.

On the barrier reef slope the foraminiferal associations are dominated by *Heterostegina depressa*, while the number of *Borelis schlumbergeri* remains constantly very low, showing however, a small increase in its preferred depth, i.e., between 20 and 30 m, where patches of coral sand accumulate. The abundance of *Planorbulinella* in the sediments on the upper reef slope is striking. Most of their shells are deformed, indicating a quasi-permanent sessile life on uneven hard substrates. All samples carry many debris of the arborescent foraminifer *Homotrema*, living at the base of coral heads.

The associations of larger Foraminifera in the sediment collected on the southward exposed slope of Jackson reef are composed almost exclusively of *Sorites* shells, showing by their deformation their attached way of life on hard substrates in agitated water, while the number of *Heterostegina* and *Borelis* shells is unusually low. The soritid shells are associated with high numbers of *Planorbulinella* and, strikingly enough, with very abundant *Peneroplis pertusus*. The latter might have lived in ephemeral algal covers during early spring and disappeared completely in the period of our visit in June. Their abundance in shallow environments on the exposed Jackson reef can obviously not be linked to high salinities and must therefore have some nutritional reason.

References: Gvirtzman et al. (1977), Hall (1975).

F.3 Significance of the Depth Gradient

The survey of the "shelf" in front of the MBL, Elat demonstrates the **succession of communities** replacing each other with the **depth gradient.** In the absence of a biologically significant thermocline or halocline, two major factors must be responsible for selecting the dominant organisms in different depths, viz. **"light"** and **"water energy."** Their roughly parallel gradients are not uniform and may independently change their relation to absolute depth according to local or general oceanographic circumstances generated by regional or local exposure, regional or local sediment input, regional or local water energy, fertility of the water mass, etc.

The light gradient has an immediate and specific impact on the physiology and the energy budget of the **benthic species depending on symbiosis** and/or on **benthic plants as food organisms**. On solid substrate, below the euphotic (s. str.) zone, at 130 m depth and 0.5% surface light intensity, the benthic organisms depending on light nearly disappear. They are replaced by organisms often living a cryptic life in reef cavities with strong water circulation, where they constitute populations of comparatively low density kept under control by organisms with a long life span and covering the bottom in dense formations. At least 50% of the bryozoan species and all three species of arborescent Foraminifera are common to shallow and deep environments. Such species, particularly passive filter feeders, are relieved below the euphotic zone from competition pressure and lose their cryptic way of life by covering all available hard substrate exposed to currents.

With the change of the faunal composition below the euphotic zone and the disappearance of the most extreme K-strategists (stony corals and higher plants forming permanent bottom covers), the maturation process of the benthic hardbottom communities is drastically shortened. Consequently, below the euphotic zone the diversity of the hardbottom community drops significantly.

The infauna burrowing in mobile sediments is insufficiently known to even estimate their change with depth. The only argument in situ observations can provide is the relative density of larger **burrowing organisms** producing mounds and funnels of the *Callianassa* type. These are densest in *Halophila*-covered substrates between 20 and 35 m depth, where the benthic primary production is supposed to be highest, as estimated from photosynthetic maximum rates of foraminiferal and coral symbionts and from the density of the *Halophila* cover.

The **diel cycle of the plankton rise and fall** in the water column can be observed on the bottom by the reaction of the stony corals extending their tentacles during the day in depths below 40 m, while in the fringing reefs the corals are open and the echinoderm filter feeders appear in large numbers on exposed spots during the night. As the pellets produced in the plankton layer and the plankton organisms themselves are a source of nutrition for many benthic organisms, the 40-m depth level of the **upper plankton layer** during the day (and the **plankton layering** in general) may represent a **supplementary depth-dependent factor** for the distribution of benthic species. It is also reflected by the sudden increase of pelagic organic remains in the sediments.

Water energy, generally decreasing with depth, acts in different ways on benthic communities. While **tidal effects** are mainly restricted to larger lagoons in the southern part of the Gulf, the shallowest subtidal zone is exposed to **wave action** all along the Sinai coast down to 5–15 m according to the exposure of the shore and to the seaway of the waves.

The fringing reefs resisting to wave action provide shelter to larger organisms but medium to fine sediment fractions are carried away by the turbulent water so that appropriate substrate for many free, smaller organisms is lacking. The different **communities of stony corals** as reflected in Loya's or Mergner's reef zonations seem to have an impact on foraminiferal communities only in so far as they influence the **regime of local water and sediment movement** by their type of predominant colony **shape** and **reef morphology**. In particular, where closed reef fronts restrict turbulence and circulation on the reef platform, the washed-out sediment may accumulate in shallow lagoons and channels along the shore where sea grass covers can develop. These harbor the shallowest epiphytic and bottom-dwelling community adapted to extreme changes of temperature, light intensity, and salinity during diel and seasonal cycles.

In front of the fringing reef coarse sand is kept in perpetual movement by **wave turbulence** inhibiting the growth of vegetation covers. These stretches of bare sand are a lethal environment for all benthic organisms, with the exception of very few, specifically adapted groups, such as, e.g., the burrowing echinoid *Lovenia*. These sands have yielded up to now only dead and transported foraminiferal assemblages, providing therefore no support for Liebau's theory (1978) on heavy ornamentation in rotalid Foraminifera as a protection against abrasion.

Below wave base, **long-shore currents** determine the water energy regime. Current velocities change with weather and season, but diminish generally in intensity with depth. At 40 m depth they are still able to lift larger coral debris; down to approximately 100 m depth, the sediments are winnowed, keeping sedimentation rates low. The energy maxima, periodically reaching particular depths during short periods of time, determine the depth distribution of particular substrate types, as demonstrated by the mechanisms of macroid formation, while constance of direction and minimal current velocity govern the nutritional regime at depth. It is not known if the constant pressure kept up by the slow, deeper currents on the interstitial water in the relief-forming bryozoan build-ups is responsible for the processes of their cementation.

F.4 Significance of Substrates

The sediments in the Gulf of Aqaba are produced mainly by biological **fixation of carbonate in skeletons and shells** and by **their destruction by other organisms**. As the net production of sediments in the water is comparatively low and the input of terrigenous material is very localized, the distribution of the mobile sediments depends primarily on the **topography** (inclination and exposure) of the sea bottom, on local net production rates relative to **growth rates of frame-building organisms**, and on the **rates of sediment fixation by early diagenesis**. The result is a narrowly meshed pattern of surfaces providing soft or solid substrates. The latter support respectively different benthic organisms in each depth zone.

Where light is available, vegetation covers develop on mobile, and coral covers on solid substrate. **Solid substrate** has a mechanical role supporting sessile organisms in high water energy (as in the fringing reefs) or on unproductive bottoms (as in coral carpets). **Soft substrates** support vegetation covers in areas of low water energy where the bottom is productive but where sedimentation rates hamper coral growth. Both kinds of covers are **permanent during the seasonal cycle** and **determine the composition of the benthic communities**. As the water body in the Gulf is characterized by very low nutrient contents, both ecosystems are designed to **recycle the nutrients**. On hard substrate, the corals are using the **symbionts** as recycling devices. On mobile substrate, the **burrowing infauna remineralizes** the organic matter produced by the plants. Both cover-forming groups of organisms, corals and vascular plants, are long-lived, using **K-strategies** to survive in an environment starved of nutrients. Whereas the two parallel nutrient cycles are almost totally closed, the energy flow of the system fuelled by the sun has its sink in the fixation of carbonate deposited on the sea bottom as mobile sediment or in the bioconstructions.

Most **larger** and doubtless many **smaller Foraminifera** adopt a **strategy of nutrient recycling by symbiosis** and live either on hard substrate (*Heterostegina*) or as epiphytes in the soft-substrate ecosystem (*Sorites-Amphisorus*). The populations of many smaller Foraminifera producing restricted amounts of biomass live on or in the soft bottons, participating in the nutritional cycle of the soft bottom ecosystem. The alveolinids and the particularly successful, omnipresent and abundant amphisteginids seem to profit from both strategies possessing symbionts and **feeding actively** on the benthic primary producers. This corresponds to their distribution on both kinds of substrates.

Below the euphotic zone, the ecosystems of hard and soft bottom seem to depend more closely to the sparse plankton production. The mechanical role of the hard substrate is particularly important, supporting all the passive filter feeders, while on soft substrate primary consumption seems to be reduced to scavenging.

F.5 Seasonality

Seasonal changes in a reef community were studied by Mergner and collaborators in Winter 1975–76 at the backside of the fringing reef platform near the Marine Science Station of Aqaba, in about 1 m depth on a 5 × 5-m-wide test surface of coral rock, with sandy patches and scattered coral heads. On the hard substrate, not occupied by living corals, the algal vegetation covering 25% of the total surface undergoes seasonal changes. Starting in December, each successive month provides highest densitites of *Padina pavonia, Cystoseira myrica* and *Sargassum dentifolium* respectively, while *Stypopodium zonale* and *Turbinaria elatensis* together reach their maximum coverage ratio in June (Fig. F.38). Soft corals, namely *Xenia* species and the stony coral *Stylophora pistillata* are submitted to similar changes within the same period.

In situ observations along the Sinai coast, where ephemeral algal covers are rare, provide additional qualitative data. *Padina pavonia* blooms were observed, for instance, in October on 20-m-deep terraces on the outer slope of the Dahab fringing reefs. The most conspicuous seasonal change, however, is the development of a community of *filamentous*, mostly *green algae*, covering loosely the permanent *Halophila* vegetation, as well as bare sand stretches around coral heads down to at least 40 m in depth during March. These clouds of algae house many **juvenile Foraminifera**, in particular amphisteginids, soritids and textulariids, **transported with the algal clouds** by accelerated currents during stormy days, parallel to the coast. This is probably the most important mechanism of dispersal for larger Foraminifera, as well as a disturbing factor for the distinction of hard and soft substrate associations in the same depth zone. As this mechanism works, however, for some weeks only during the year, its impact apparently does not obscure the lateral differences in foraminiferal associations.

No studies are available on changes in **seasonal production** by the permanent vegetation cover in the Gulf of Aqaba, but in temperate climates *Zostera marina* was observed to change its biomass ratio between rhizomes and leaves from 1:2 in winter to 1:1 in summer. The biomass of the rhizomes was doubled from March to September while that of leaves and flowering turion increased about fivefold during the warmer season. This would suggest not only seasonal increase in production rates but also the formation of reserves in the rhizomes during the productive season, to be consumed during the unproductive one. Similar mechanisms might play some role in the Gulf of Aqaba to equilibrate the changing availability of nutrients during the yearly cycle in the permanent vegetation.

The microflora, particularly **cyanobacteria** and their complex associations, constitute a permanent vegetation cover, where the **salinity** rises to values excluding all potential grazers. These systems were studied exclusively in the permanently hypersaline Solar Lake where seasonal changes linked to changing hydrographic conditions do occur. In the medieval harbor basin of Geziret Fara'oûn, closed by an artificial dam, salinity rises in the calm summer season to such an extent that cyanobacterial mats cover the bottom, while the rest of the benthic fauna is killed. The latter consists of a variety of gastropods and some byssate bivalves. The foraminiferal fauna is almost exclusively restricted to *Peneroplis*

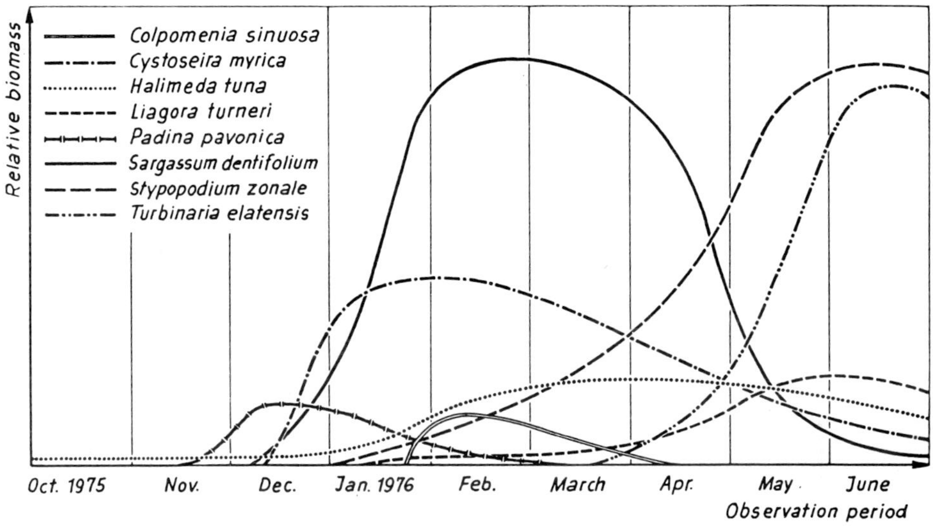

Fig. F.38. Seasonal distribution of benthic algae on solid substrate near Aqaba (Mergner and Svoboda 1977)

pertusus found in cores taken in the center of the harbor pool, in direct contact with the algal mats (Fig. F.39). This is in good agreement with the exclusive mass occurrences of *Peneroplis pertusus* in hypersaline lagoons of the Persian Gulf and with their abundance in the lagoons of Tiran with high salinities. Their occurrence on Jackson reef, however, indicates not a direct relation by exclusive physiological salinity tolerance, but perhaps a nutritional relationship. Thus, benthic primary production seems to be submitted to seasonal change in a complex way, whereas the nutrient concentration in the open water is only one of many factors regulating the process. The nutrient concentrations are likely to produce their most direct impact on the clouds of filamentous green algae and on parts of the benthic microflora. When and wherever intensive algal growth is observed by bright green and green-brownish colors in the underwater landscape, in small isolated patches or over larger areas, larger Foraminifera produce dense living populations covering the substrate to a large extent. The geometric and timely correspondence between **surfaces of active algal production** and the **density of living populations of Foraminifera** is so striking that there must be some close biological relationship. The latter is supposed to be a **nutritional one** at least for those benthic Foraminifera known to have diatom frustules in their protoplasm or in their feeding cyst.

Some benthic Foraminifera represent thus **first consumers** of the benthic primary production. Nothing is known about secondary consumers feeding on these Foraminifera. Although their shells have been found in the intestine of many different larger organisms, such as sea urchins or holothurians, it is not known whether these shells were taken up when alive or after death, nor if Foraminifera actually contribute a substantial part of the food used by these organisms. Nowhere did we ever observe under water a systematic grazing strategy on living fo-

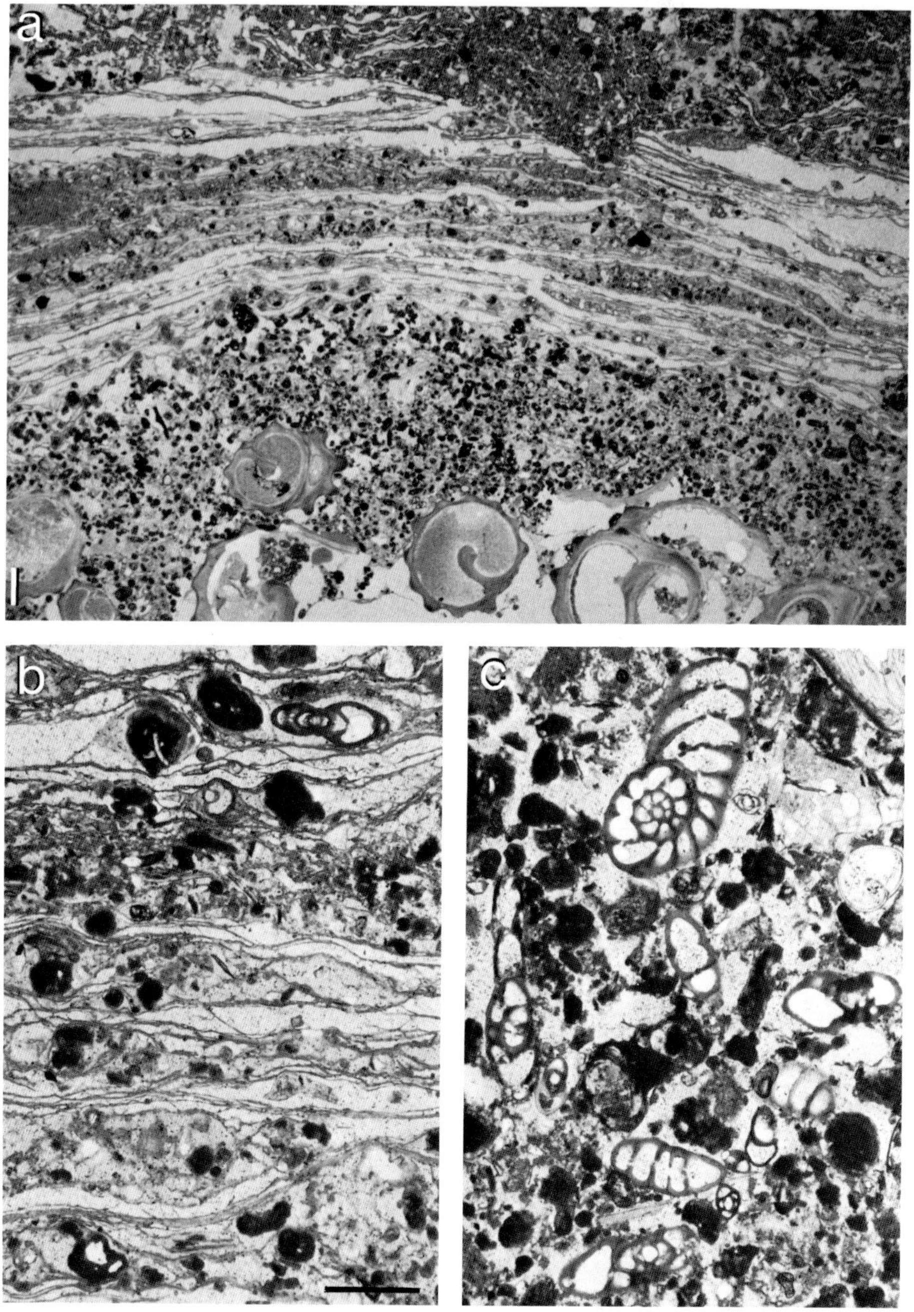

Fig. F.39 a–c. Seasonal deposits in Geziret Fara'oûn island's ancient harbour (site 029, Fig. F.15 c). Vertical thin section of artifically hardened sediment. Layers of cyanobacteria alternating with *Cerithium* shell deposits (**a**). **b** Detail of summer layer with single *Peneroplis pertusus* inbetween Cyanobacteria layers. **c** Detail of winter layer with numerous *P. pertusus* as almost exclusive representatives of foraminiferal assemblage. Scale bars 1 mm raminiferal populations, covering always in undisturbed, uniformly dense formations large patches on the bottom, in spite of their potential as a food source, because of the rich contents of storage products in their protoplasm. The striking **patchiness** observed in so many different foraminiferal populations seems thus to be independent of **predation pressure** and is likely to reflect an **uneven distribution of their food source.**

In contrast to planktic communities, most larger benthic organisms are thus less dependent on physicochemical factors in the water column and more closely related to mechanisms where abiotic and biotic processes interact on neighboring small surfaces in different ways. The benthic communities are thus subjected to an environment consisting of a mosaic of substrates evolving in the polarized field of the depth gradient and exposed to the rhythms of daily, yearly and long cycles of environmental change. An insight into the processes generating the different substrates and into the complex significance of the depth gradient is insufficient to understand the distribution pattern of a particular benthic population. **Each species or group of taxa reacts in a specific way to its specific environment**. These specific reactions and adaptations can be understood only when the relevant basic features of the **life history** of the respective group of organisms are known. In the Foraminifera, some relevant features of their life history are **nutrition, symbiosis, periodicity and mode of reproduction, shell construction** and **shell functions for motility, nutrition, protection and reproduction**.

In order to understand the specific composition of particular benthic communities, an additional category of problems must be considered. These are summarized by the notion of **provincialism** which cannot be investigated in the Gulf of Aqaba alone. The large-scale, regional distribution patterns of benthic species depend on **historical processes**, e.g., **long-term environmental change, regional maturation processes** and **diversification** in the communities and **barriers** for the propagation of species, which superimpose their effect on the relationships between populations and environment as observed in the Gulf of Aqaba.

References: Benayahu and Loya (1977), Fricke (1982a), Fricke and Hottinger (1983), Fricke and Schuhmacher (1984), Friedman (1968), Gabrie (1982), Hottinger (1980, 1983a, b, 1984a), Krumbein (1977), Krumbein and Cohen (1974), Krumbein et al. (1977), Loya and Slobodkin (1971), Mergner (1971, 1979, 1980), Mergner and Schuhmacher (1974), Mergner and Svoboda (1977), Muscatine and Porter (1977), Scheer and Pillai (1983).

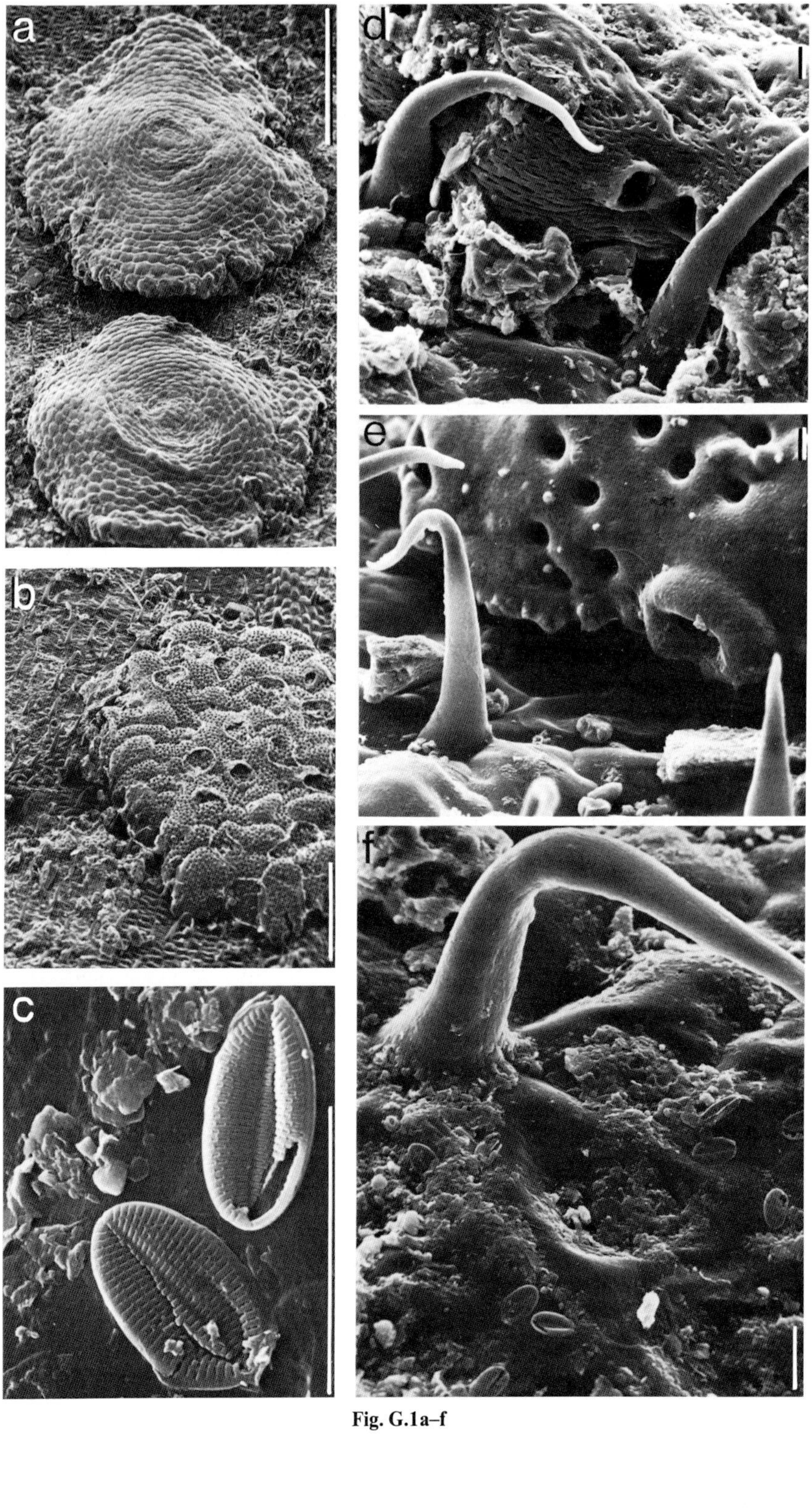

Fig. G.1a–f

G. Benthic Foraminifera: Response to Environment

As pointed out in the previous chapter, the response of each benthic species to the abiotic and biotic conditions in its environment is a specific one, reflected by its particular life history. An attempt is made in the following to summarize and integrate in situ observations, experimental evidence and assumptions based on circumstantial evidence separately for each species or for (informal) groups of species of benthic Foraminifera living in the Gulf of Aqaba.

Knowledge of each species or group of species varies widely from superficial shell morphology alone to more or less complete parts of their life history. Energy budgets, metabolic and storage rates, specific food or nutrient requirements, waste disposal and genetics are utterly unknown. In all groups, the morphology of dead shells is the best known item on which identification must be based. But even identification – of smaller benthic species in particular – raises problems which are not yet solved. In the following paragraphs, each group of species is treated according to the following sequence of items (if anything is known about them):

1. general characteristics and significance of shell morphology, symbiosis and soft parts common to all species in the group;
2. extant Red Sea species, their historical origin and biogeography as far as relevant to the understanding of their distribution;
3. distribution, biometry, seasonality, growth, calcification and reproduction as far as known in each particular species. Finally, some conclusions are drawn with regard to the significance of shell shape, symbiosis and stable isotope compositions of the shell minerals to the benthic Foraminifera in general and the significance of the latter's distribution in the Gulf of Aqaba as an actualistic model.

◄

Fig. G.1 a–f. Some epiphytic Foraminifera. **a, b** Neighbouring specimens of *Sorites variabilis* and a young specimen of *Acervulina inhaerens* on *Halophila* leaf. Scale bar 0.5 mm. **c** Diatoms (*Cocconeis* sp. see Lee et al. 1980 b) on leaf representing most important part of benthic primary production and foraminiferal food. **d** Detail of **a**, sector of apertural face in *S. variabilis* in between trichomes (leaf hairs) demonstrating their mechanical role in fixing the Foraminifera on the leaf. **e** Detail of **b**, *Acervulina* marginal apertures directed towards leaf surface. **f** Leaf surface with trichome supporting diatoms and coccoliths. Scale bars of details (**c–f**) 0.01 mm. SEM, washed leaf dried in air, Ras Burka, site 334, 20 m

G.1 Larger Foraminiferans

G.1.1 Soritines

Discoidal representatives of the subfamily Soritinae Ehrenberg are characterized by their obliquely overcrossing stolon system. They are the most common porcelaneous larger Foraminifera living in shallow water as epiphytes on *Halophila* or green algae (Fig. G.1). Their living body consists of protoplasm with a lacunar structure. In body parts corresponding to passages between chamber cavities (intercameral foramina and annular passages) of outer chamber rings, cell organelles of ectoplasmic character, coated vesicles and microtubular complexes are particularly frequent. Food particles, many of them empty diatomacean frustules of different types, point to the kind of diet taken up through the foramina. Feeding experiments with *Amphisorus* and *Amphistegina* confirm the uptake of algae as food, and selectivity in different species, but are inconclusive as to the effect of conditions of starvation in light or in dark before feeding.

Lateral parts of the chamber cavity delimited by the subepidermal partitions of the shell are filled with denser protoplasm containing many symbionts per chamber compartment. In the Red Sea species, the ultrastructure of these symbionts is similar to that of the dinophyte *Symbiodinium* (= *Zooxanthella*) *microadriaticum*, with the exception of a lower number of chromosomes in the nucleus and the presence of two small, free flagellae (Fig. G.2). The latter are capable of moving the symbiont through the lacunar system of the host from one side of the shell to the other. The location of the symbionts in the shell depends on the light intensity in their environment. At high light intensities, the symbionts move to the shaded, in low intensities to the illuminated side of the shell. A change of the symbiont's position, easily initiated by turning over the shell on its substrate, takes place in a few minutes. In addition, *Amphisorus hemprichii* shows negative phototaxis at high light intensities. Respirometry revealed its photocompensation level to be fixed around 10 klx.

There might be some relation between photoinhibition and the epiphytic way of life in soritines. Large *Amphisorus* specimens swept from their plant substrate to the sand by a heavy storm were observed to crawl back to their plant support within a few days. In tank experiments, the same phenomenon occurs. As the light reflection of the coral sand is very important, a shell lying flat on the bottom is illuminated strongly from both sides and has its light-regulation system blocked. Crawling back to the dark plant support restores the appropriate light conditions. The only exception to this rule was observed in the boulder field, between 6 and 20 m depth, in front of the MBL of Elat, where very large numbers of soritines live on and in between loose pebbles of dark crystalline rock. They are not found, however, on dark brown algae such as, e.g., *Padina*. The crawling back to plants and selecting the substrate implies the existence of some kind of orientation mechanism. Since differences in illumination and reflection in the tanks and jars were not taken into account, various experiments on crawling behavior in the laboratory did not furnish conclusive results as to the significance of dispersion of overcrowded populations on horizontal or vertical substrates, nor as to the kind of

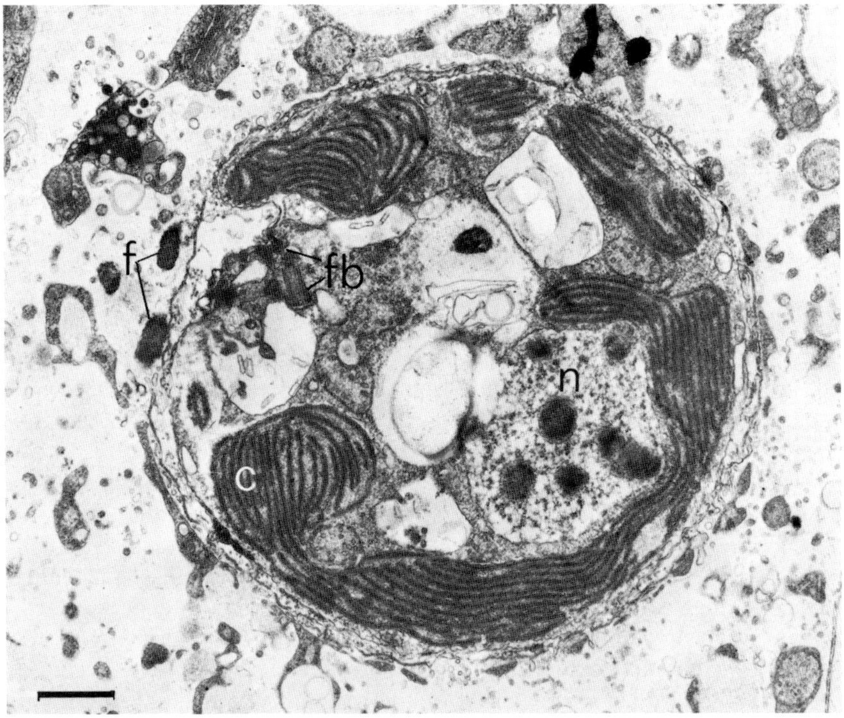

Fig. G.2. Motile dinophycean symbiont in *Amphisorus hemprichii*. *c* Chromatophores, *f* free flagellae, *fb* pair of flagellar bases, *n* nucleus. TEM, scale bar 1 μm (Leutenegger 1977a)

surface texture (smooth or frosted glass, etc.) preferred as a permanent substrate. In horizontal Petri dishes with uniform illumination and uniform distribution of diatoms used as food on the bottom, specimens of *Amphisorus* tend to keep a distance from each other corresponding to the radius of their extended pseudopods. This implies a device for recognition of the neighboring specimen. No studies are available on the interaction of specimens from different species.

Red Sea Species and Their Historical Background. Three soritine species are present in the Gulf of Aqaba: *Sorites orbiculus* Ehrenberg (1839), *Sorites "orbitolitoides"* (Hofker 1930) and *Amphisorus hemprichii* Ehrenberg (1839). A topotype specimen of true *S. orbitolitoides* provided by J. Hofker, Sr., from the Caribbean indicates that Lehmann (1961) has defined this species too broadly. The true *S. orbitolitoides* belongs to the *Puteolina-"Broeckina"* (Lévy 1977) group and has green, probably chlorophycean symbionts, whereas the Red Sea species has brown, dinoflagellate symbionts. Lacroix's name *S. variabilis* erected in 1941 on material from Aqaba is available for this latter species.

Other extant and widespread soritine species are absent from the Gulf of Aqaba and probably from the entire Red Sea–East African faunal province, in particular the Indopacific *Sorites marginalis* and *Marginopora vertebralis*. Citations of these two species in the Gulf of Aqaba are due to erroneous identifications. In the Caribbean, soritine species have also been misidentified. In particular, *Sorites orbiculus* was cited under the name *S. marginalis*. The latter has green symbionts, whereas the cosmopolitan *S. orbiculus* has brown ones wherever it has been found. Some *S. orbiculus* were mistaken for *A. hemprichii*. The absence of soritine species with green symbionts from the Red Sea–East African province may be caused by the absence from the bottom floras of the chlorophycean species necessary for recruitment after sexual reproduction of the host. The absence of *Marginopora vertebralis* with dinophycean symbionts may, however, be due to other factors which also prevented the largest and most complex representatives of other groups, such as *Alveolinella* or *Nummulites*, from penetrating into the faunistic border province since Miocene times.

Caribbean porcelaneous shallow-water faunas today are dominated by pillared shells with the radial stolon system of the *Archaias* and *Cyclorbiculina* groups, while Indopacific shallow faunas are characterized by soritine shells with a crosswise-oblique stolon system and by the absence of archaiasinids. Although a separation of Indopacific from Caribbean foraminiferal faunas has been evident since Late Cretaceous times, Miocene forms of archaiasinid structure are widespread all over the Tethys from Israel to Indonesia and occur also in central America. Either the larger porcelaneous Foraminifera developed independently *Archaias*-like new structures in both major provinces, such as, e.g., in early Tertiary conical groups, or the Caribbean pillared forms are the surviving representatives of a worldwide, Early Miocene population of archaiasinids. This question must be taken up by a careful revision of all fossil material available for comparison. However, the large Indopacific soritines have apparently no roots reaching beyond Miocene times and might be derived from simple and small *Sorites* species like those present in Early Miocene "Asmari" limestones in the Near East. The history of the Paleogene *Orbitolites* shows the same evolutionary trends.

Definition and delimitation of extant Indopacific species is difficult as we do not yet know how to interpret striking morphological differences observed in *Marginopora* and *Amphisorus* from isolated Island groups such as the Seychelles, the Maldives or New Caledonia. Do we find in the shallowest faunas true endemisms, similar to the unquestionably endemic animals and plants on land, or does a shift in depth-dependent environmental factors change the species associations of which the deeper-water ones have not been adequately sampled in all Island groups? There is, however, a shift of shell morphology in soritine populations from thick to thin discs with progressive depth of the habitat. This has been observed in *Marginopora* in New Caledonia, in *Amphisorus* on the Maldives and it is also evident in *Sorites* from the Gulf of Aqaba. The latter are interpreted as two species relaying each other with considerable overlap toward depth: *S. orbiculus* produces thick discs in shallow water, *S. variabilis* thin shells in the shade of plants or boulders and in deeper water. The distinction of the two species is difficult without laborious equatorial sectioning of their shell. For this reason a quantitative study of their respective distribution was not undertaken, but the intuitive interpretation adopted here is the same as the one for the two *Planorbulinella* species present in the Gulf and based on ample biometric data. *Planorbulinella* shows the same depth-dependent trend, viz. a much smaller megalosphere followed by narrower chamber rings in the deeper species. *Amphisorus hemprichii* shows no

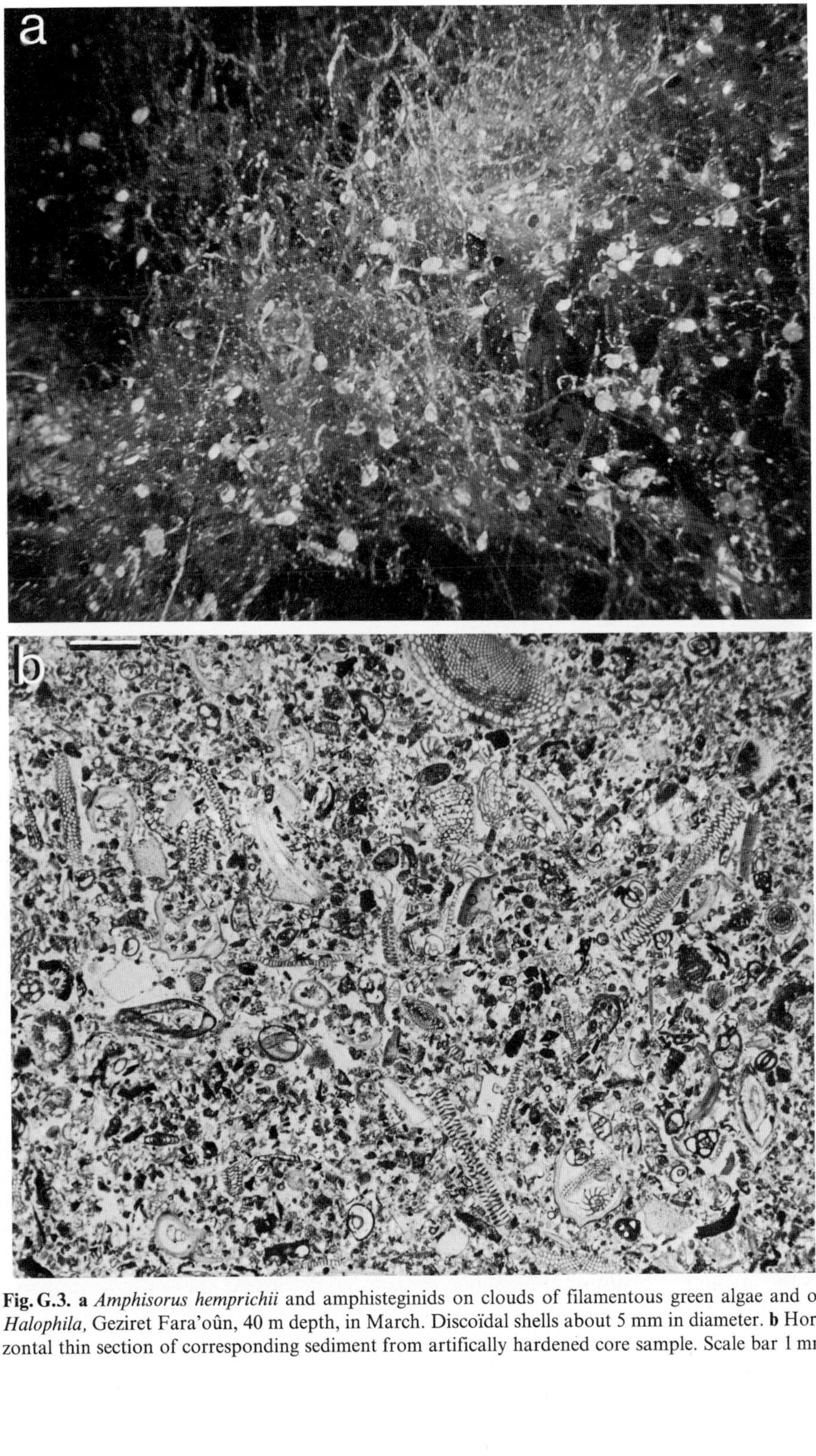

Fig. G.3. **a** *Amphisorus hemprichii* and amphisteginids on clouds of filamentous green algae and on *Halophila,* Geziret Fara'oûn, 40 m depth, in March. Discoïdal shells about 5 mm in diameter. **b** Horizontal thin section of corresponding sediment from artifically hardened core sample. Scale bar 1 mm

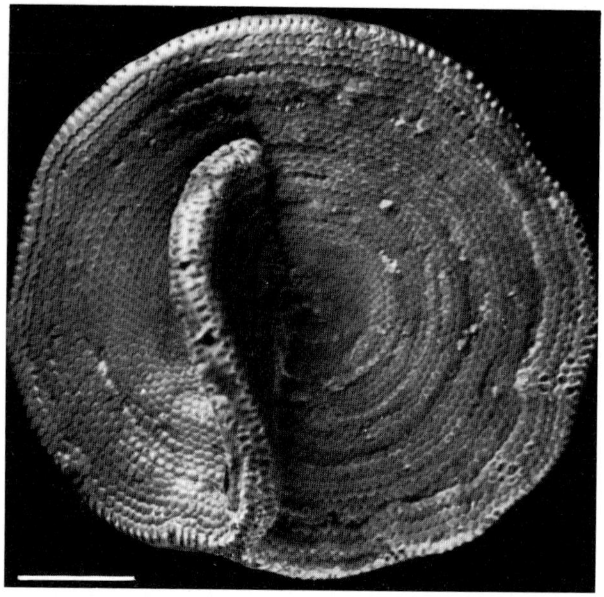

Fig. G.4. Irregular growth in *Amphisorus hemprichii.* Light microscope, scale bar 1 mm

obvious flattening trend with depth in the Gulf of Aqaba, whereas on the Maldives a large, very flat, new species seems to relay *A. hemprichii* in about 40 m depth. This interpretation is supported by shaking experiments showing no effect on growth, or on shell shape of *A. hemprichii* from Elat, while the growth of lamellar-perforate *Amphistegina lobifera* was enhanced and the shells grew thicker in artificial and natural high-water energy envirnonments.

Habitat and Distribution. Mapping soritines under water (see Fig. F.26, p. 175), mainly the large epiphytic *Amphisorus hemprichii* and microspheric *Sorites orbiculus* were registered (Fig. G.3). Their distribution of course follows the vegetation cover, but is limited by the 35-m isobath, clearly indicating the light requirements of these two species. *S. orbiculus* is more frequent in the very shallow *Diplanthera* meadows, while *A. hemprichii* occurs mainly on *Halophila* leaves or occasionally on accumulations of loose boulders.

In tidal pools of the reef platform, where temperature and salinity may temporarily rise to very high values, aberrant shells occur. Their deformation consists either of a development of a second shell plane, perpendicular to the original one (Fig. G.4), of twinned shells where adult growth rings enclose two embryos, or of a shift in growth direction deforming the discoid shell into a cylinder. Whereas similar features of irregular growth in megalospheric embryos produced in culture are lethal, shell growth continues in tidal pools, indicating the short duration of the adverse conditions. Fossil occurrences of deformed shells are quite common and have misled some authors into introducing new names for such deformations. In other cases, high frequencies of twinning or other irregularities are characteristic of particular species living in extremely shallow water and thus being submitted more often to short-term adverse conditions than species with a deeper habitat. These irregularities are not to be confused with mechanical damage repaired at all depths by regeneration of the shell part containing a vegetative, interphase nucleus.

Sorites variabilis occurs either as free or attached. Free specimens generate regular discoidal shells, attached specimens follow the surface relief of their substrate during growth. Some occur in the shade at the foot of *Halophila* plants, others below coral heads, boulders or coarse coated grains where they take part as a minor component in the construction of *Acervulina* macroids. They occur thus on hard or soft substrate in moderate numbers producing the soritine "tails" in the percentage distribution graphs of larger Foraminifera at a depth of between 35 and 70 m, particularly on hard, light-reflecting bottom.

Reproduction. The last chamber rings of microspheric soritines are thin-walled, slightly inflated and often irregularly subdivided. Such final growth stages in fossil material have often been interpreted as "senile" ontogenesis. They are in fact brood chambers. After division (probably during the night) of the mother cell, the young resorb the organic cell envelope (organic lining) and the mineralized shell in the equatorial zone of the parental disc, fitting snugly into shell compartments widened by resorption (Fig. G.5).

A similar sequence of events inferred from shell geometry was observed by Eggink and Baumfalk (1983) in *Orbitoides gensacicus* (Maestrichtian), where the young used the original chamberlets of the mother shell for subdividing their giant embryo by enveloping a bunch of mother chamberlets with their own, thickened embryo wall in a narrow space cut out by resorption.

Both megalo- and microspheric generations of adult *Amphisorus hemprichii* and *Sorites orbiculus* have numerous interphase nuclei characterized by their large size, by their lobate contour in section, their nucleolar bodies in peripheric position and their lack of a perinuclear reticulum. In addition to these vegetative nuclei destined to disintegrate during gametogenesis, two megalospheric specimens of *S. orbiculus* collected in September in Crete (Mediterranean) had reproductive micronuclei of three different types, supposed to represent successive stages of gamete differentiation. The latter are grouped in an inner ring around the embryonic apparatus of the mother shell and are small enough to escape, even as zygotes, through the apertures into the ambient environment without destroying the walls of the mother shell. We do not know if the copulation of the gametes takes place autogamously inside of the mother shell or by free, heterogamous gametes in the ambient environment.

The **population dynamics** of *Amphisorus hemprichii* were studied in the Gulf of Aqaba during one seasonal cycle by measuring the growth of living specimens collected from the plants and the bottom on uniform, neighboring test areas of the same, 4-m-deep *Halophila* meadow south of Elat. The accretion rates of the shell diameter closely reflect the growth rates measured as per weight increase (Fig. G.6). During autumn and winter of the first year, the population consisted of 94% megalospheric and 6% microspheric individuals. Until the reproduction period, the mean standing crop (150 specimens per 225 cm^2 leaf surface) and the

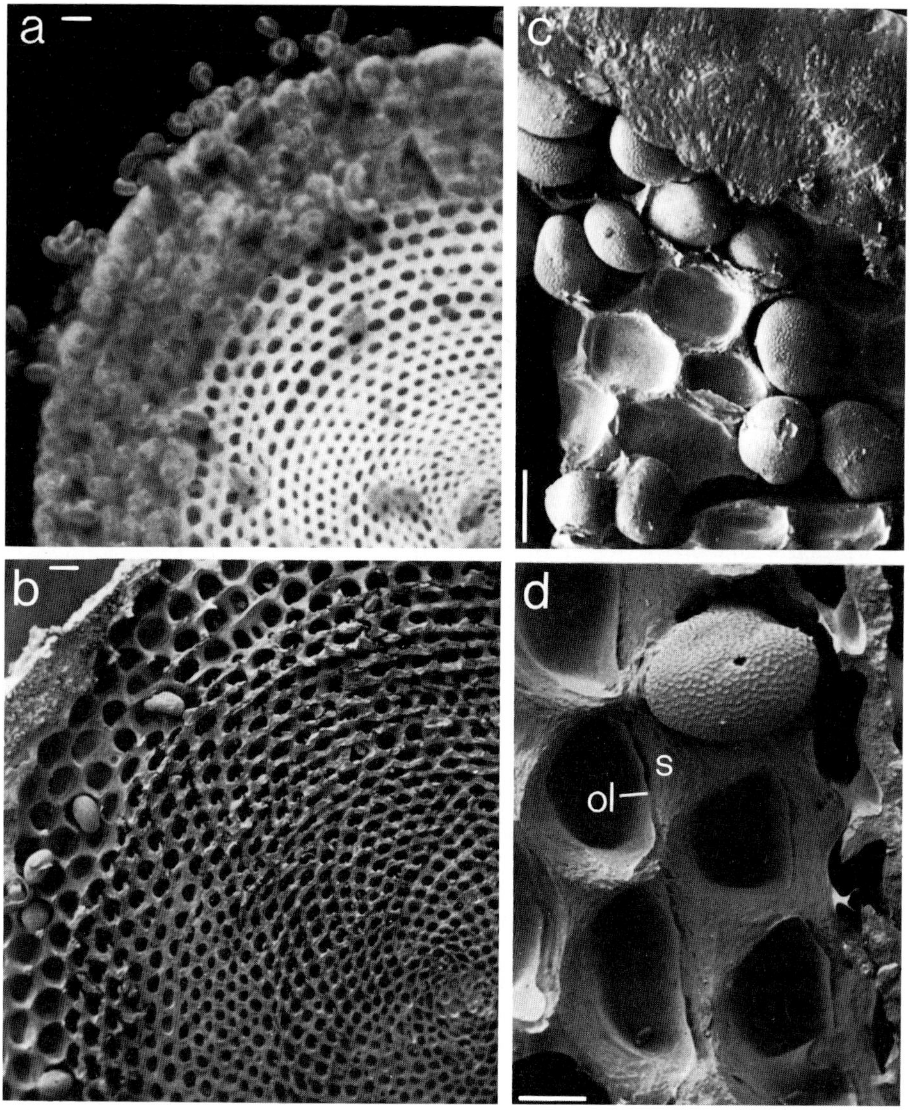

Fig. G.5 a–d. Asexual reproduction and brood chamber in microspheric *Sorites orbiculus*. **a** Megalospheric embryos on mother shell, light microscope, living specimens under water. Scale bar 0.1 mm. **b** Dry microspheric specimen broken in median plane of shell with some embryos remaining in the brood chamber. SEM, scale bar 0.1 mm. **c** Brood chamber with embryos. SEM, scale bar 0.1 mm. **d** Detail of brood chamber. Note resorption of equatorial part of mineralized shell *(s)* and organic cell envelope ("organic lining", *ol*). SEM, scale bar 0.05 mm

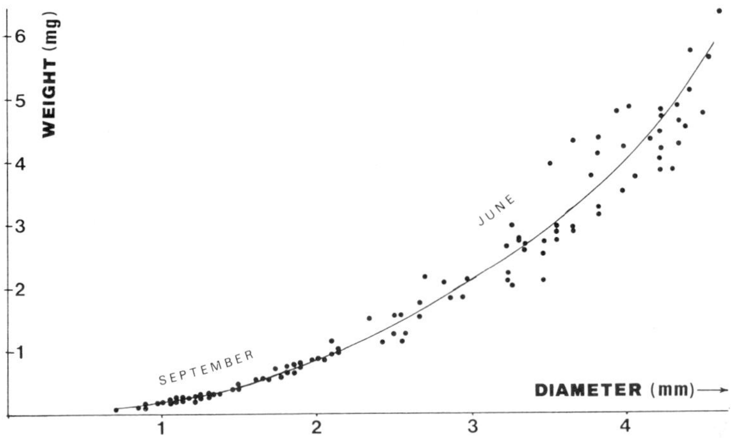

Fig. G.6. Growth in *Amphisorus hemprichii* expressed as relation between size and weight. *Line drawn* represents third-order regression fit to data. (Adapted from ter Kuile and Erez 1984)

growth rate of the shell (100 μm growth of equatorial shell diameter per week) equal to about 14% weight increase, were stable (Fig. G.6). Before reproduction, growth is declining. Asexual reproduction takes place during April and May, apparently triggered by the rise of the temperatures in surface waters. After reproduction, the standing crop rises by one to two orders of magnitude, but an important mortality of the juveniles reduces the density of the population back to about normal again within 2 months. However, the second generation observed in the test meadow was 100% megalospheric, its standing crop being stabilized at 350 specimens per 225 cm^2 and having significantly larger megalospheres (Fig. G.7). While the higher values for standing crop are to be related to a simultaneous, sharp decline of the population density of *Sorites orbiculus* in competition for food and/or space on their common substrate, the larger size of the megalospheres and the absence of microspheric individuals in *A. hemprichii* might be explained by a paratrimorphic life cycle (as postulated for other larger Foraminifera), but extending over 2 years (Fig. G.8). In the first year, megalospheric A_1 and (minoritarian) microspheric B specimens would produce for the following year a uniformly megalospheric population consisting of A_1 schizonts and A_2 gamonts. In order to reduce the sexual offspring to a 6% minority of the adult population, predation on the gamont's products must be very heavy. As no very young microspheric shells are found in the sediment (in contrast to dead megalospheric juveniles), their loss must concern mainly gametes and/or zygotes in the first stages of gamogony, probably prior to the construction of the microsphere.

Postulating a 1 yr-cycle of paratrimorphic nature for soritines would imply the expatriation of all sexual products out of the population area and a mechanism to reimport the young microspheric specimens at irregular yearly intervals. Mechanisms of pseudoplanktonic export of sexual products are known so far only in the benthic group of *Rosalina–Tretomphalus–Cymbaloporetta* with small, thin-walled shells where a final floating chamber is formed by the mother specimen prior to reproduction (see Rückert-Hilbig 1983). Dense populations ("blooms")

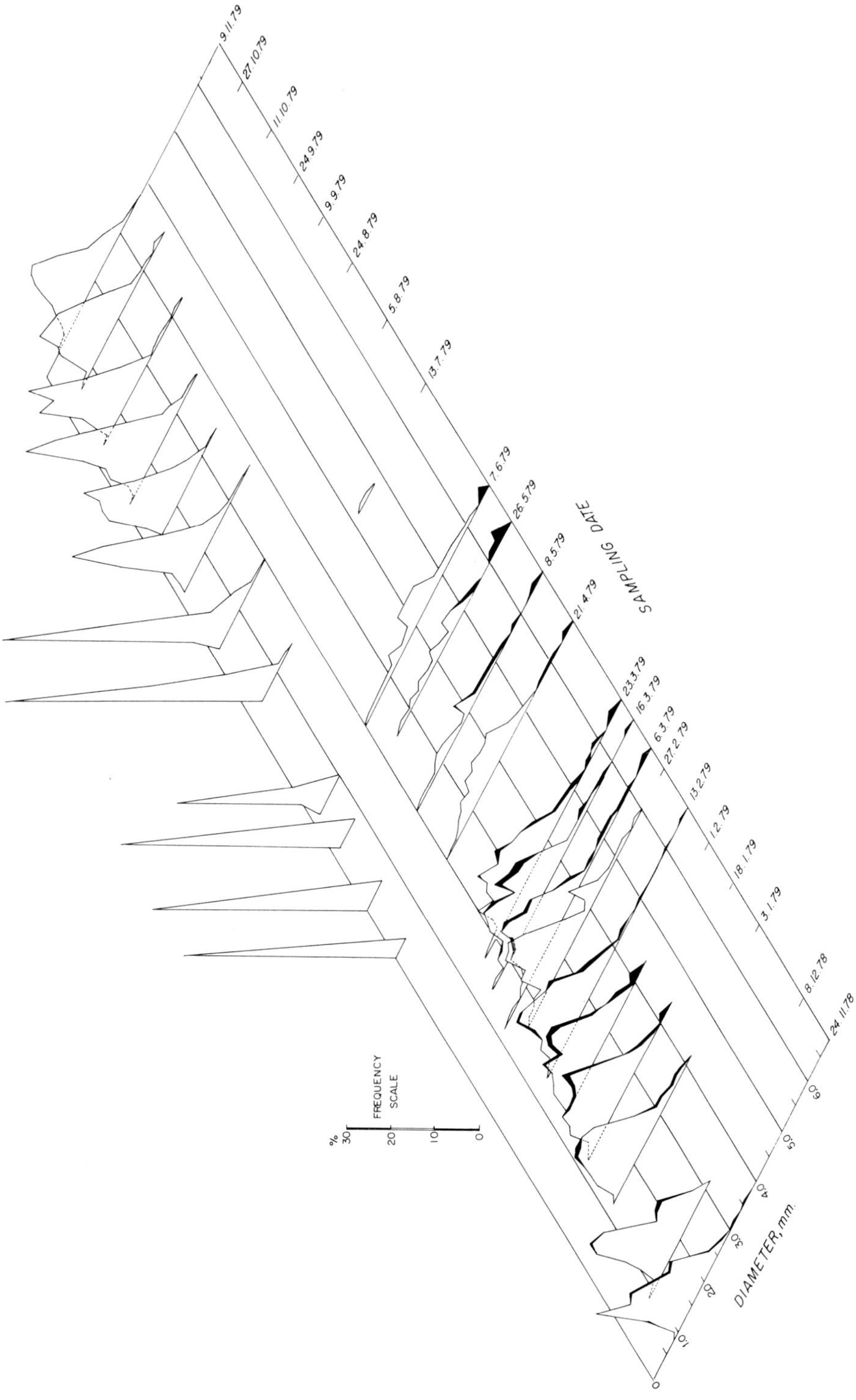

Fig. G.7. Size frequenzy distribution in population of *Amphisorus hemprichii* sampled over a 1-yr period. *Black areas*: microspheric specimens. Note reproduction period starting in April (Zohary et al. 1980)

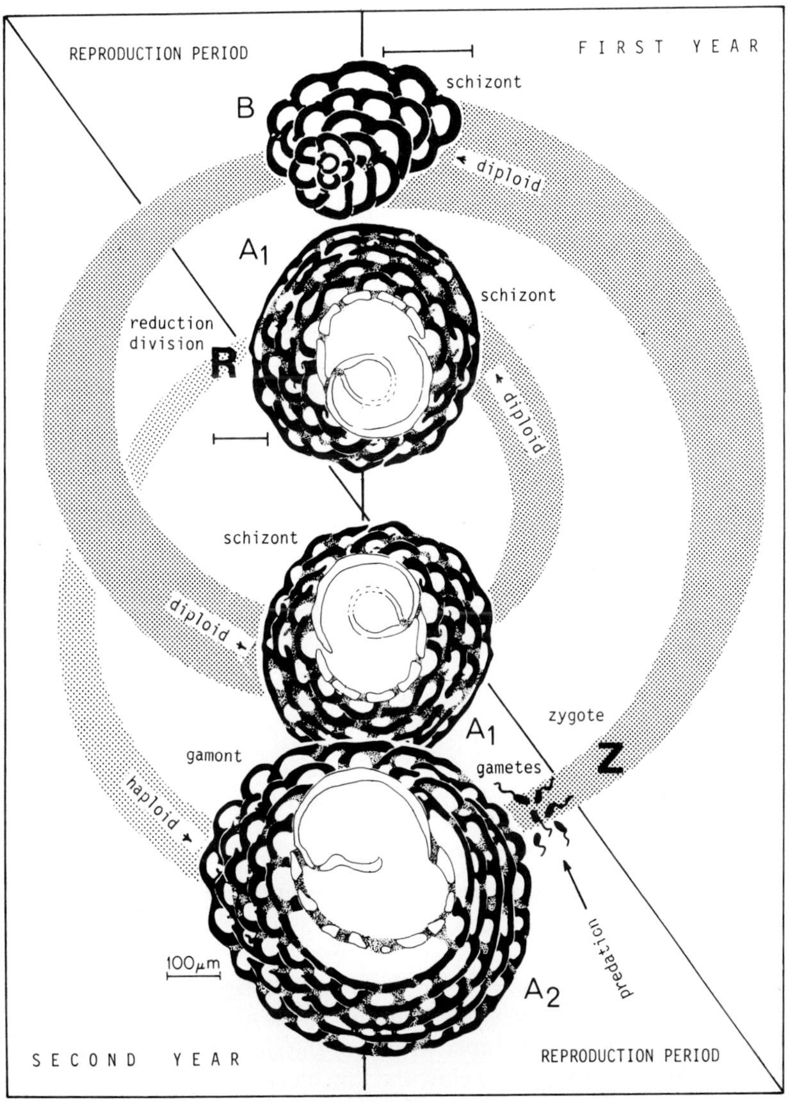

Fig. G.8. Reproduction cycle of *Amphisorus hemprichii* reinterpreted as paratrimorphic and coordinated with the seasonal cycles of two successive years (Hottinger 1983 b)

of large, adult *Amphisorus hemprichii*, easily detectable by the diver, were observed locally also in autumn. At this time of the year, the "blooms" are restricted to comparably small and isolated patches where short-lived, filamentous green algae are growing locally as they do everywhere in spring. From this observation it is presumed that the synchroneity of the life cycle in soritines is not absolute, and that growth is correlated to some extent with productivity of food, particularly of diatoms, reacting quickly to the amount of nutrients in the water. Nutrients are known to be recycled below the bottom–water interface and to be used up by the vegetation cover. Thus, autumn blooms of actively feeding benthic Foraminifera, such as soritines and amphisteginids, might depend on local processes producing nutrients in the subsurface of the sea bottom, whereas the overall production of shallow-water larger Foraminifera would be governed by the availability of food, depending on the concentration of nutrients in the water body.

Calcium carbonate production from this latter cycle was estimated 160 g, Ca $CO_3 m^{-2} yr^{-1}$ for *Amphisorus hemprichii* alone. Attempts to collect kilogram-sized samples of larger Foraminifera in spring from boulder-covered surfaces at a depth between 5 and 7 m by means of an underwater "vacuumcleaner" device yielded an estimated 80:20 wet weight ratio of *Amphisorus* to *Sorites* and a 60:40 ratio of soritines to amphisteginids and heterosteginids. Bias resulting from the harvesting method by suction devices are at least 10% in favor of the soritines. Therefore, the total production of foraminiferal calcium carbonate may reach 400 g m^{-2} yr^{-1} as the most conservative estimate in shallow waters of the northern part of the Gulf of Aqaba. This estimate matches those on foraminiferal carbonate production in Hawaii, in spite of the low primary production in the Gulf, and emphasizes the importance of nutrient recycling in systems where light energy is available to symbionts in far from limiting quantities.

References: Eggink and Baumfalk (1983), Frenkel (1974), Halicz and Reiss (1981), Hansen and Dalberg (1979), Hottinger (1977a, 1978a, 1980, 1981a, 1982a, 1983b), ter Kuile and Erez (1983), Lacroix (1941), Lee (1980), Lee et al. (1980a,c), Lehmann (1961), Leutenegger (1977a,d, 1984), Lévy (1977), Noy (1982), Reiss et al. (1961), Rückert-Hilbig (1983), Said (1949), Zohary et al. (1980).

G.1.2 Alveolinids

The living body of the alveolinids consists of protoplasm with a lacunar system similar to that in other porcelaneous Foraminifera. The ultrastructural elements characterizing the ectoplasm are concentrated in the preseptal passages and around the intercameral foramina of the ultimate and penultimate shell whorl. The protoplasm in the chamberlets of the inner whorls is rather vacuolated and houses symbionts of diatomacean structure. Under high irradiation by light the symbionts are retracted to the inner whorls of the shell. Under conditions of dim light, the symbionts are aligned below the thin chamberlet roof of the last whorl. The shell is planispiral and involute. Its elongated-involute, subglobular to fusiform shape represents a device to shorten communication distances from the innermost shell whorls to the ambient environment through the preseptal passages and the polar intercameral foramina. The latter are multiplied in large, elongate forms and oriented parallel to the axis of coiling by a spiral twist of the apertural face at the poles (Fig. G.9). Such elongate forms were observed to bury themselves in coral sand, dragging the shell one pole first in the shade of one to several layers of sand grains by extending the pseudopods from one of the polar apertural faces. On hard bottom or plants, the shells may be fixed to the substrate by pseudopodial protoplasm extruding through both polar apertural faces. The apertures in the equatorial zone of the shell are often closed by plugs made of organic material.

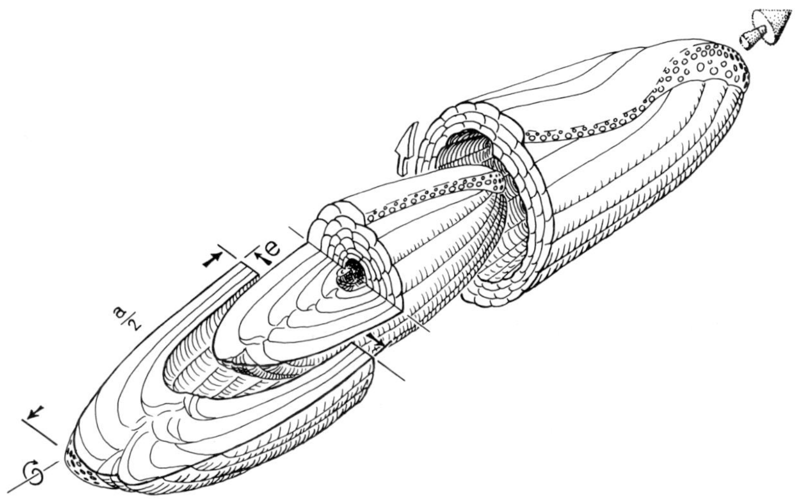

Fig. G.9. Direction of growth *(white arrow)* and of movement *(stippled arrow)* in a planispiral-fusiform, alveolinid shell. *a* Axial diameter of shell; *e* equatorial diameter (Hottinger 1981c)

In the Red Sea and along the East African coast to Mauritius Island a single species, *Borelis schlumbergeri* (Reichel 1937) seems to represent the alveolinid organisational type of porcelaneous Foraminifera. On the Maldive Islands in the open Indian Ocean, three species occur together: *Borelis schlumbergeri* Reichel, *B.* cf. *pulchra* (d'Orb.) and *Alveolinella quoii* (d'Orb). In New Caledonia, *A. quoii* only was found, whereas in the Carbibbean, *B. pulchra* is the only alveolinid representative. Extant alveolinids have thus a clear provincial distribution going back to Miocene times when *B. schlumbergeri* occurred in Tiran, while *Flosculinella* (as a very close relative and predecessor of *Alveolinella*) lived in Sri Lanka (Ceylon), Indonesia and in Eastern Africa.

Where *Borelis* is associated with *Alveolinella*, its comparatively wider, mainly shallower distribution matches the ecological part played by *Glomalveolina* or *Alveolina boscii* compared to *Alveolina* s.str. in Late Paleocene to Eocene times. Whereas the Eocene species have identical structural patterns in the simplified, small-sized and the complex, large-sized organizational types and go back to a common phylogenetic root, i.e., *A.(Glomalveolina) primaeva* Reichel, the extant representatives *Borelis* and *Alveolinella* have different structural patterns and separate phyletic origins: *Alveolinella quoii* derives from small, subglobular *Flosculinella reicheli* Mohler (Early Miocene), whereas the earliest, small and globular *Borelis vonderschmitti* Schweighauser has a late Middle Eocene (Biarritzian) age.

Curiously enough, all extant *Borelis* ssp. have bifid teeth blocking part of the foramina and becoming resorbed when a new chamber is formed (Fig. G.10). Such teeth have never been observed in other alveolinids. Their presence in *Borelis* supports the interpretation of this genus' independent origin from some undifferentiated miliolid, in spite of its basically similar, alveoliniform arrangement and subdivision of the chambers and of its similar symbionts. In the Gulf of Aqaba, where *B. schlumbergeri* alone represents the alveolinids, the intraspecific

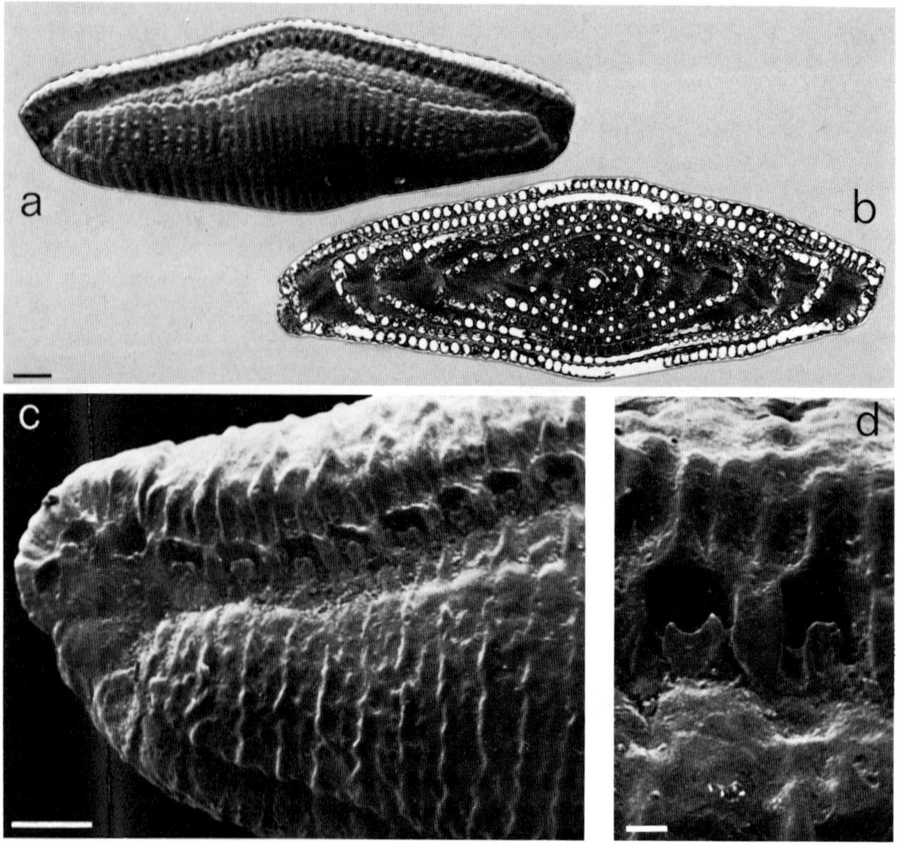

Fig. G.10 a–d. *Borelis schlumbergeri.* **a** Apertural view, SEM. **b** Axial thin section, light microscope. Scale bars 0.1 mm. **c** Polar area of shell. Note axial direction of enlarged polar aperture. **d** Detail showing bifid tooth in aperture. SEM, scale bars 0.01 mm. (After Hottinger 1977a)

variation in axial elongation is much wider, ranging from oval to fusiform shell shape (index of elongation 2.2–5, mean values around 3), than on the Maldive Islands where three alveolinid species are associated. The two *Borelis* species in particular, a subspherical to ovoid and a fusiform-elongate one, are clearly distinguished by their respective shell shape. Although their areas of distribution overlap almost completely, there are some samples where only one of the species is present.

The *Borelis* spp. shells being too small to be accurately registered by the diver under water, their distribution has to be reconstructed by counting the shells in the sediment samples. Their highest ratio relative to soritines and nummulitids reaches, irrespective of hard or soft substrate, about 50% in depths between 25 and 35 m. Above 10 and below 60 m depth *B. schlumbergeri* represents less than 10% of the shells of larger Foraminifera. On the Maldives, the common depth distribution of the three alveolinid species is approximately the same.

References: Hottinger (1977a, 1978b, 1980), Leutenegger (1984), Reiss and Gvirtzman (1966), Said (1949).

G.1.3 Amphisteginids

In contrast to imperforate, nonlamellar Foraminifera, the living body of amphisteginids is characterized by vacuolated protoplasm. It contains numerous food particles. Some of them can be identified as empty diatom frustules. Symbionts are lodged in deep internal pore depressions ("pore pits", a structure similar to egg holders) below the pore chimney used for gas exchange as demonstrated by radiotracers. The particularly thick organic cell envelope lining the chamber walls, the peculiar pore plug below the sieve plate at the bottom of the pore and the organic fill of the pore space, concentrating in electron-dense layers at each level of lamellar construction of the shell, apparently are no barrier for small molecules exchanged through the pores.

In the Gulf of Aqaba, four species representing the genus *Amphistegina* occur, viz. *A. lessonii* d'Orbigny (1826), *A. papillosa* Said (1949), *A. lobifera* Larsen (1976) and *A. bicirculata* Larsen (1976). None of them is endemic to the Red Sea. The widespread, Indopacific species *A. radiata* (Fichtel and Moll) is absent.

The phylogenetic background of extant *Amphistegina* species may reach back to Eocene times, but is insufficiently known to explain the genesis of present-day faunistic provinces. With the exception of *A. lessonii* and "*A. quoii*" (= *A. radiata*), no detailed structural analysis is available to compare species groups and to assess the postulated two major lineages ("phyletic groups I and II" of Larsen), characterized respectively by high or low numbers of chambers per whorl in their earliest representatives.

All extant amphisteginid species live in symbiosis with diatoms (Table G.1). The symbiont's structures identified in situ are uniform B 2a-types in *A. lobifera, A. lessonii* and *A. bicirculata*, slightly distinct B 2b-types in *A. papillosa*, but fundamentally different B 4-types in *A. radiata*. This distribution of the symbionts does not reflect the major morphological subdivision of extant amphisteginid species, but might explain the absence of *A. radiata* from the Gulf of Aqaba by a possible lack of B 4-type diatoms in the bottom flora which may provide symbionts to *A. radiata* after sexual reproduction.

Distribution. Two thick-walled, subglobular species, issued from phyletic group II, i.e., *A. lessonii* and *A. lobifera*, occur in shallow waters. They are gradually replaced by comparatively thin-walled, lenticular *A. papillosa* and *A. bicirculata* at depths between 40 and 80 m (Fig. G.11). The shallow-water species *A. lobifera* occurs in environments of higher energy, below fringing reef fronts for instance, whereas *A. lessonii* dominates the faunas of quieter lagoons or channels, as observed in particular at Ras Abu Galum. The shallow-water species are living as epiphytes on *Halophila* or other, sometimes emphemeral plant covers, but they share also the habitat of *Heterostegina depressa* on hard bottom.

On boulder-covered and current-exposed slopes, as present for instance in front of the MBL, Elat at a depth between 10 and 20 m, large numbers of *A. lobifera* and *A. lessonii* live on the lower, shaded surface of the boulders. Their withdrawal into the largely open pore space of the coarse sediment is interpreted here as a negative-phototropic reaction linked to the photoinhibition of their symbionts in shallow depths, as determined experimentally by measuring the rates of

Table G.1. Amphisteginids, ultrastructure of symbionts in situ compared to ultrastructure of isolates from some host species. Secondary isolates interpreted as contamination of cultures (Leutenegger 1983)

MORPHOLOGICAL TYPE: Pyrenoid: □ simple, ▨ compound; frustule: ○ absent, **O** present; size: x 500; ? unknown

FORAMINIFER	LOCALITY MONTH and YEAR of COLLECTION and (→) date of later processing of specimens kept in raw culture	m DEPTH at collecting site	ALGAE: IN HOST	ALGAE: ISOLATED dominant	ALGAE: ISOLATED rare	TAXONOMIC IDENTITY
Amphistegina lobifera	Crete 9→10.74	38	○			Diatoms
	Rhodos 10.73	10	(?)			Diatoms
	Red Sea 9.76	10	(?)			Diatoms
	5.73	1-10	○			Diatoms
	3→ 4.74	10	○			Diatoms
	3.78	3	○			Diatoms
	3.78	10-40		**O**		Fragilaria shiloi
					+ ▨	Nitzschia frustulum
					+ C	Chlorophyceans
					+ ○	N.panduriformis
Amphistegina lessonii	New Caledonia 3.81	20	○			Diatoms
	3.81	55	○			Diatoms
	Hawaii ? →	10	○			Diatoms
	? →	2-6		(?)		Nitzschia laevis
	Red Sea 6.71	10	○			Diatoms
	9.76	10	○?			Diatoms
	3→ 9.78	20	○			Diatoms
	3.78	3	○			Diatoms
	3.78	45	○			Diatoms
	3.78	10-40		**O**		Fragilaria shiloi
					+ C	Chlorophyceans
					+ ▨	Nitzschia frustulum
					+ ○	N.panduriformis
Amphistegina bicirculata	Maldives 3.77	?	○			Diatoms
	3.80	50	○			Diatoms
	Red Sea 9.76	60	○?			Diatoms
	3.78	38	○			Diatoms
	3.78	68	○			Diatoms
	3→ 9.78	20	○			Diatoms
Amphistegina papillosa	Red Sea 10.72/9.76	60	(?)			Diatoms
	3.78	68	○			Diatoms
	3.78	35		**O**		Fragilaria shiloi
					+ (?)	Navicula reissii
					+ ▨	Nitzschia frustulum
					+ C	Chlorophyceans

photosynthesis in relation to different light intensities. Under conditions of low light intensities, however, amphisteginids show positive photo- and thermotaxis.

The deep-water species *A. bicirculata* and *A. papillosa*, issued from phyletic group I, live both on soft and hard substrates of the open shelf, between 80 and 130 m depth at least.

Growth Rates and Shell Thickness. The trend from thick to thin shells with increasing depth within the photic zone is not particular to amphisteginids, nor even to lamellar Foraminifera. Related species with successively flatter shells substitute for each other with progressive depths in overlapping bathymetric zones, while intraspecific variability of adult shell shape shows similar trends. Substitution of species and intraspecific flattening trends occur in foraminiferal groups with or without symbionts at all depths of the photic zone. In amphisteginids (Fig. G.12), the shell thickness measured as ratio of the largest equatorial diameter to the axial diameter reflects (1) the spiral characteristics defined by the increase of radial chamber height during ontogeny; (2) the number of chambers (=growth steps) per whorl determining the number of outer lamellae enveloping the shell at each volution; and (3) the average thickness of the lamellae in the last volution at the periphery and in all volutions at the shell poles. While spiral characteristics depend especially upon the diameter of the proloculus and the growth rate of the soft body in all following growth stages, the thickness of individual lamellae may depend on calcification rates after each growth step. Umbilical lamellae in amphisteginids from Elat are 8–12 μm thick at 10 m depth, 2 μm thick at 90 m depth. The thickness of dorsal lamellae ranges from 4–8 μm in 10 m depth to 1 μm in 90 m. These measures constitute a kind of light-meter for paleoenvironments. It should be kept in mind, however, that the extension and thickness of the lamellae are predetermined to a large extent by the disposition of organic matter involved in the process of biomineralization.

The relation between the largest equatorial diameter and the dry weight increase during growth in *Amphistegina lobifera* follows mathematical laws (Fig. G.13) permitting the use of the shell diameter as a meaningful measure of growth in terms of $CaCO_3$ weight added per unit of time. The weight of dry organic matter is within the range of 2–3% of the total shell weight. In situ and laboratory experiments show the growth rates to depend on light, food and water energy at comparable temperature intervals. Maximum growth rates are observed in 20–30 m depth with a 6% weight increase per day, while in fed specimens kept in the dark the growth rates are 30% of the optimum values. Without food the dark controls stop growing. Growth rates of animals kept in cages near the bottom are about 1.5 times higher than those in cages suspended in the open water at comparable depths. This indicates the importance of benthic food and feeding rates for foraminiferal growth rates. Suspending cages in shallow depths, where wave movements rock the containers, or keeping the animals in a shaker-incubator enhanced the thickening of the shell with respect to controls kept unter quite identical conditions of feeding and illumination.

Irradiation intensity and/or quality of light affect the growth rates of amphisteginids by the effects of light on the photosynthetic process, determining metabolic rates of the symbionts by the effects of free diatoms and/or other algae in

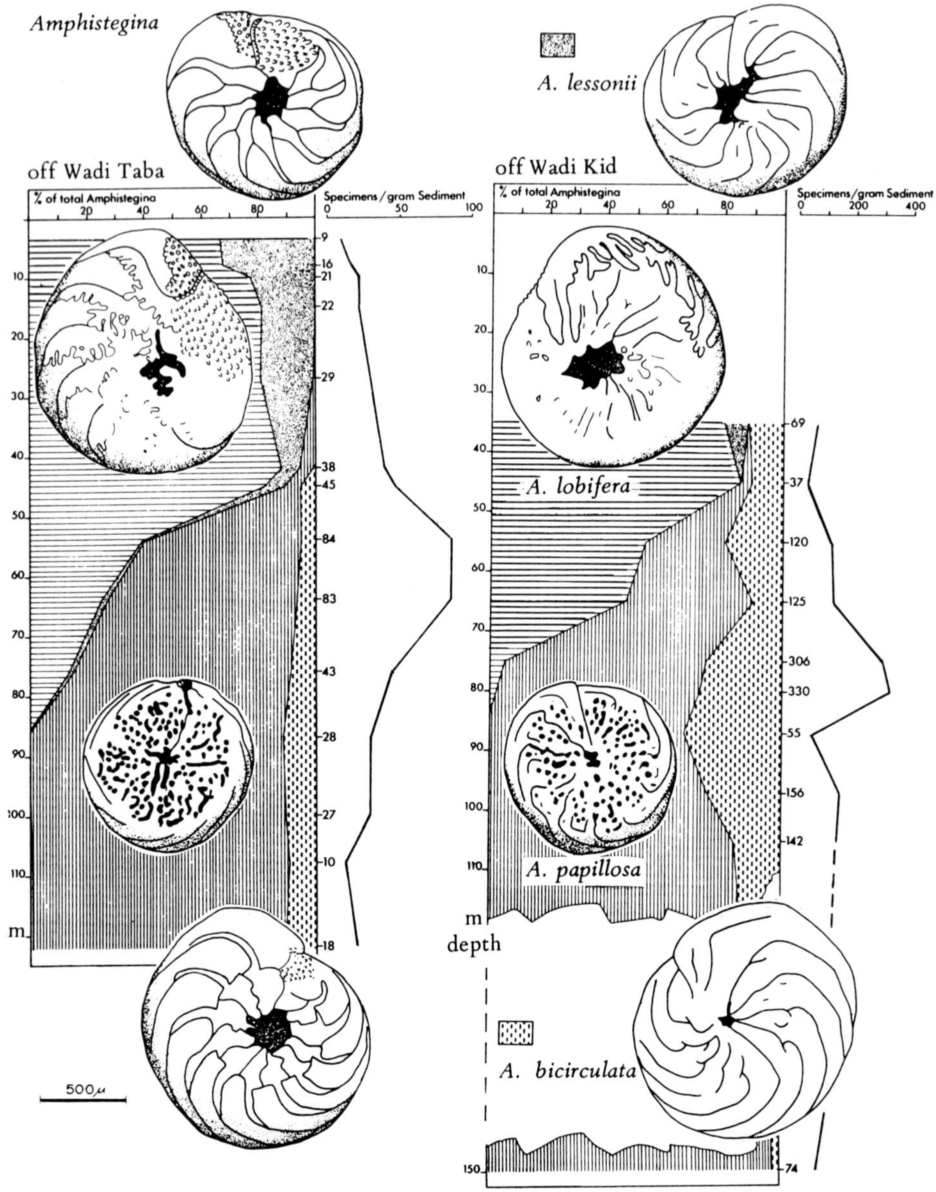

Fig. G.11. Depth distribution of *Amphistegina* species in the Gulf of Aqaba and external aspect of shell, ventral view *(left)*, dorsal view *(right)*. (Combined from Hansen and Buchardt 1977, Larsen and Drooger 1977)

Fig. G.12 a–i. *Amphistegina* species from the Gulf of Aqaba, horizontal and vertical sections, camera lucida drawings. **a–c** *A. lobifera,* 20 m; **d, e** *A. lessonii,* 28 m; **f, g** *A. papillosa,* 70 m; **h, i** *A. bicirculata,* 135 m. All specimens megalospheric except **b**. Note flattening of vertical sections in deep species in spite of similar number of septa per shell whorl (= number of outer lamellas)

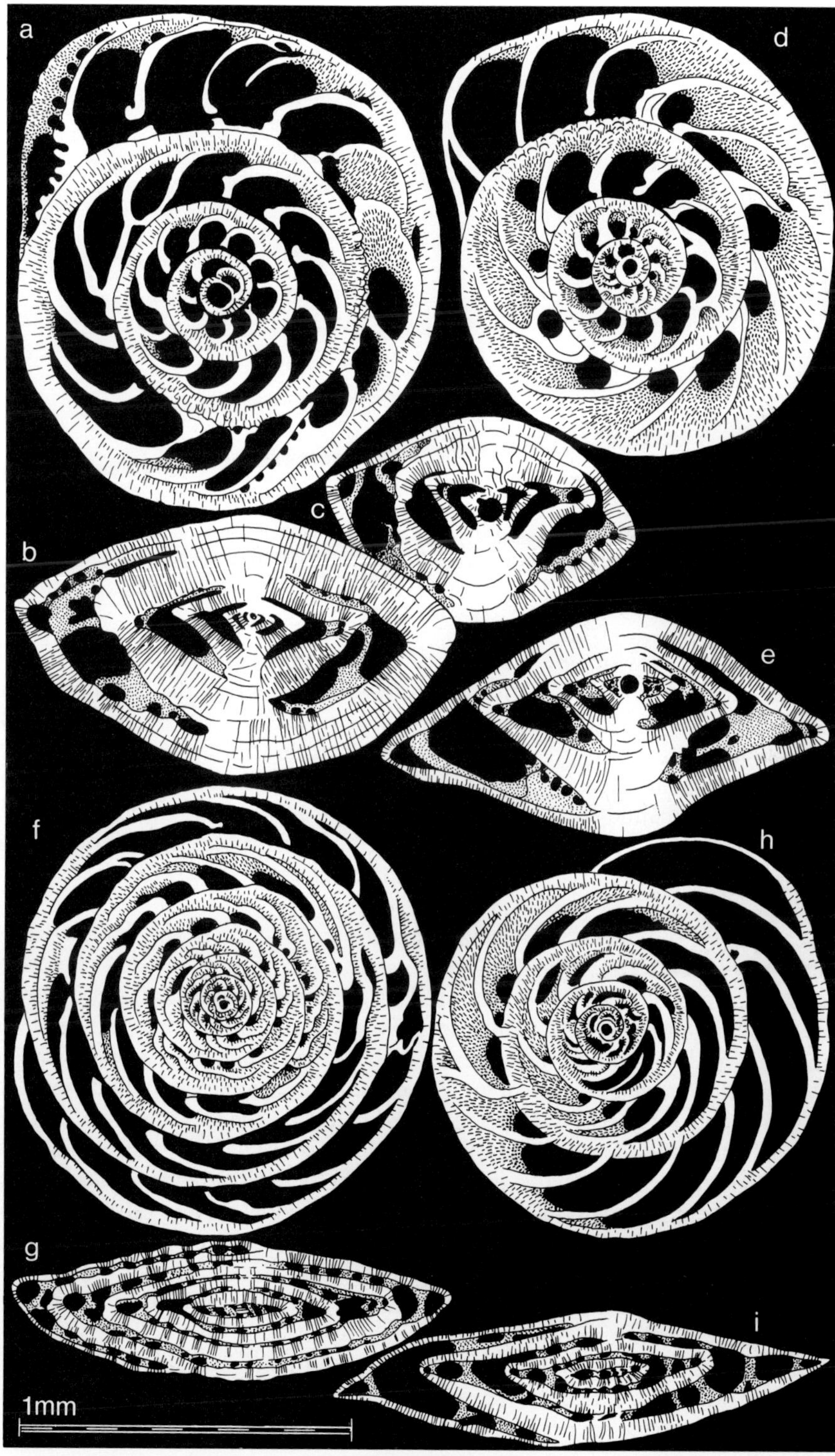

Fig. G.12 a–i

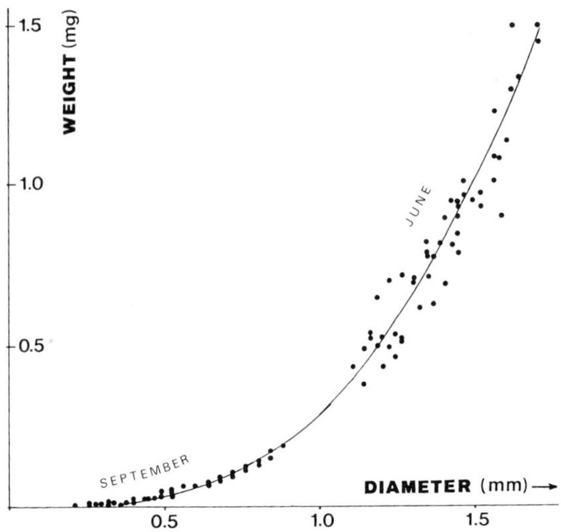

Fig. G.13. Growth in *Amphistegina lobifera* expressed as size-weight relationship. *Line drawn* represents third-order regression fit to data. (Adapted from ter Kuile and Erez 1984)

the environment providing the food or by a direct photobiological effect on biomineralization processes. Experiments designed to separate the effects of light and food provide contradicting results which may be explained by synergistic effects of light and food on growth rates. In order to test this hypothesis, a model curve derived from enzyme kinetics was successfully fitted to the experimental data on amphisteginid growth considering light as the growth-limiting "substrate" (Fig. G.14). This indicates that one common process might control the three factors involved in the relationship between light and growth.

Seasonality. In the Gulf of Aqaba, asexual reproduction of the shallow-water species *Amphistegina lessonii* and *A. lobifera* are controlled by the seasonal cycle. The growth rates in *A. lobifera* measured on specimens from Elat and observations under water during the sampling activities demonstrate that maturity size is reached within 6 months. This corresponds to the results of clone culturing of Hawaiian *A. lobifera* and *A. lessonii.* The biannual reproductive periods seem to be January and July for *A. lobifera* in Elat. In Hawaii, the schizogony of *A. lobifera* lags behind the reproduction of *A. lessonii.* It is noteworthy that in cultures, *A. lessonii* reproduced asexually at all available light levels, whereas *A. lobifera* reproduced only at the highest light levels. Sexual reproduction and ratios of microspheric to megalospheric specimens in populations are presently unknown.

Coiling Direction. Amphisteginids are the only trochospiral larger Foraminifera present in the Gulf of Aqaba available for analysis of their coiling direction. *A. lessonii, A. lobifera* and *A. bicirculata* are predominantly sinistral while *A. papillosa* is dextral. There are, however, specific changes in proportions of coiling directions within particular populations.

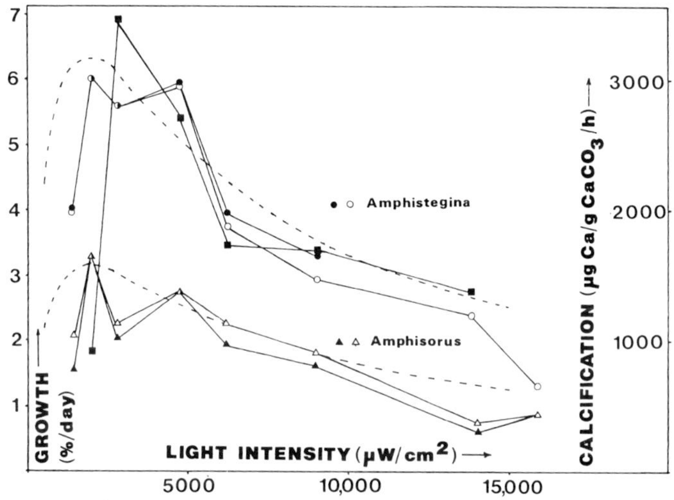

Fig. G.14. Growth rates of *Amphistegina lobifera* and *Amphisorus hemprichii* plotted as function of light intensity at incubation depth. *Closed symbols* represent growth rate values based on real final weight, *open symbols* represent values of final weight calculated from final size. *Closed squares* represent growth rates in *Amphistegina lobifera* based on calcification rates measured by J. Erez (1978) using ^{45}Ca labels. *Dotted lines* based on equation describing growth on substrate inhibiting growth at high concentrations as derived from enzyme kinetics. (After ter Kuile and Erez 1984)

Minority (dextral) coiling direction increases in the larger size classes of *A. lessonii* and *A. lobifera* reaching almost 50% of the largest (microspheric?) individuals. In cultures kept at 24°–26 °C, the normal field temperature of surface water, schizogony produced sinistral *A. lessonii* and dextral *A. lobifera*. The proportion in coiling direction in the clones produced in culture corresponds to the one in the field populations from Hawaii from where the parents were selected. The minorities in Hawaiian populations are smallest in juveniles from field samples, taken shortly after the reproduction period in March–April, for *A. lobifera* and January–February for *A. lessonii*. The proportion of minority *A. lobifera* rises significantly in a thermal effluent plume of a Hawaiian power plant producing a +4 °C warmer habitat. Effects of temperature on coiling directions are known in planktonic Foraminifera. In amphisteginids, some factor associated with reduced temperature may inhibit the development of the majority reproducing earlier in the year and having a higher mortality of the juveniles. Minority ratios in coiling direction of amphisteginids are therefore to be considered as a useful tool to distinguish winter and summer schizonts.

Symbiosis. There is apparently contradicting evidence as to the specificity of the symbiont–host relationship, particularly in amphisteginids. In all *Amphistegina* species, the ultrastructure of symbionts observed in situ is identical in all symbionts occuring in shells of the same species from different depths, different seasons, and different provenances in the Gulf of Aqaba or in other seas (Table G.1). The ultrastructure of the symbionts in *A. lobifera, A. lessonii* and *A. bicir-*

culata is identical with that of the diatom *Fragilaria shiloi* Lee, extracted artificially from the hosts and kept in culture. These cultured isolates produce frustules permitting a taxonomic identification of the symbiont at species level.

The symbionts of *A. papillosa* in situ have the same ultrastructure as seen in the isolated *F. shiloi*, but their pyrenoid is significantly larger. Other, so-called "secondary" or "rare" isolates from amphisteginid species have distinct ultrastructures, particularly in their pyrenoid, which have never been found in situ. Therefore, these "secondary" isolates might represent contaminations by epizoan algae overlooked while cleaning the living amphisteginids prior to extraction of the symbionts, or by food organisms. Some of the latter can be recognized as empty diatomacean frustules in the last chambers of *Amphistegina* or in its food cyst.

Pigment extracts from *Amphistegina lobifera* and *Amphisorus hemprichii* showed similar absorption spectra where chlorophyll a is clearly dominant. Carotenoids are present in both species, while small amounts of chlorophyll c were found only in *A. lobifera*. Minor differences in the spectra are observed in the ranges below 400 nm. The spectra strongly resemble those from extracts of *Archaias angulatus*. The pigment extract spectra seem thus not to reflect the different ultrastructural types of the symbionts, representing three different classes of algae with different colors.

Chlorophyll a content in respect to shell weight was found constant from 5 to 15 m, but increasing with depth from 15 to 25 m. Röttger et al. (1980) report chlorophyll a contents per unit protein to be highest at light intensities corresponding to optimal growth rates. This might reflect the density of the symbionts in the host rather than an adaptation of the single cell to diminishing light intensity. As the symbiont cells were seen to divide in the host, their number per unit of the host's weight are a measure of the symbiont's metabolic activity depending on photosynthesis.

References: Erez (1978), Hallock (1979), Hallock and Hansen (1979), Hallock and Larsen (1979), Hansen and Buchardt (1977), ter Kuile and Erez (1983), Larsen (1976, 1978), Larsen and Drooger (1977), Lee (1980, 1983), Lee et al. (1979 a,b, 1980 b,c), Leutenegger (1977 a,b,c, 1983, 1984), O'Herne (1974), Reiss et al. (1961), Röttger et al. (1980), Said (1949), Zmiri et al. (1972, 1974).

G.1.4 Nummulitids

Common to all nummulitids is a complex system of marginal canals and marginal furrows ("marginal cord") representing a peculiar type of marginal pseudopodial flow (Fig. G.15). Even in cyclical forms, where the apertural face replaces the marginal chamber wall in forming the shell margin, the deep furrows of the margin are taken over by the apertural face, while lateral, intraseptal canals substitute marginal canals below the furrowed margin wall. The change from spiral to cyclical growth offers thus an opportunity to witness a substitution of structural elements in the shell, where an element responsible for radial distribution of protoplasm takes over functions of marginal distribution and shelter of the pseudopods (Fig. G.16). This is an indication that marginal cords are not only

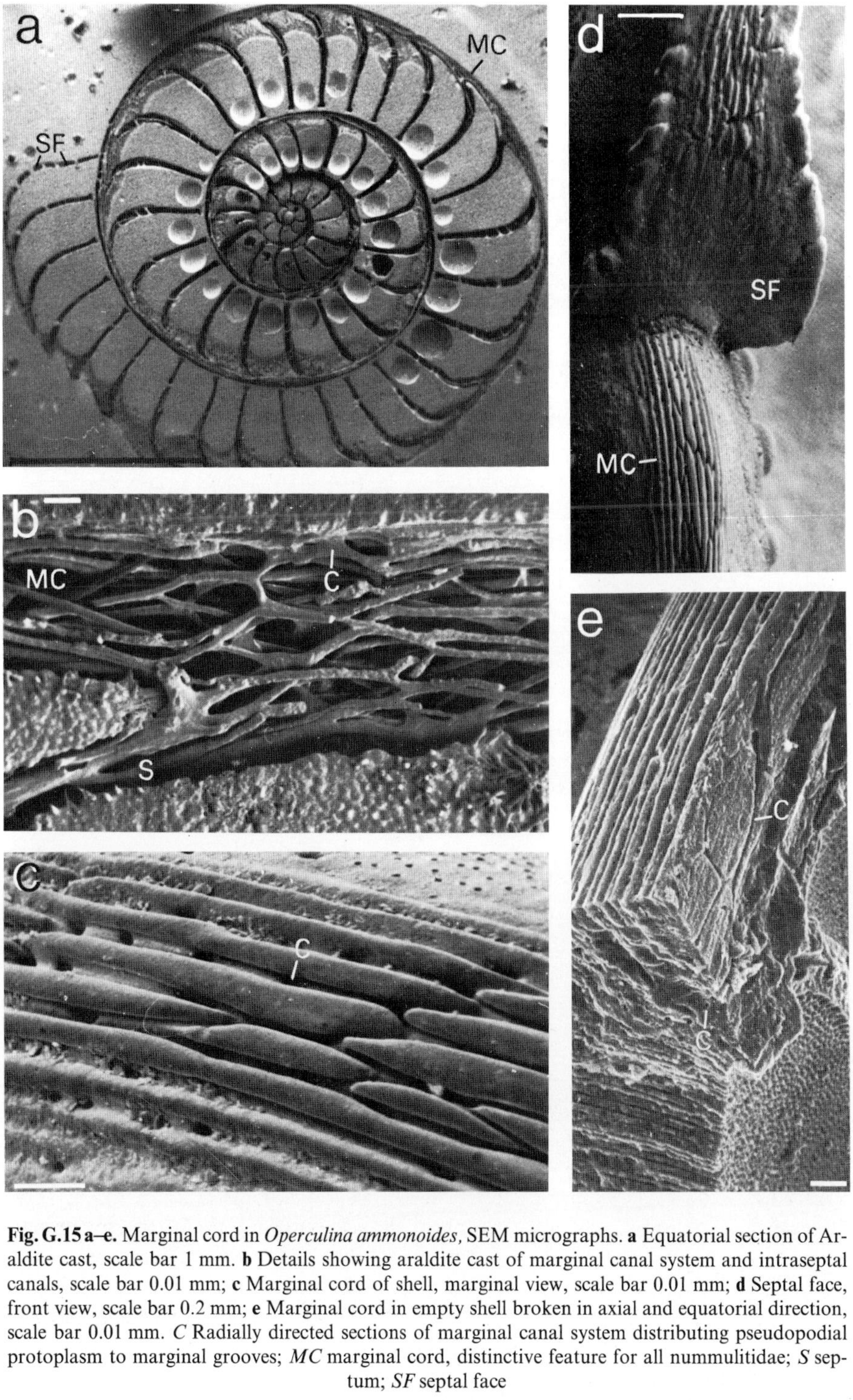

Fig. G.15 a–e. Marginal cord in *Operculina ammonoides*, SEM micrographs. **a** Equatorial section of Araldite cast, scale bar 1 mm. **b** Details showing araldite cast of marginal canal system and intraseptal canals, scale bar 0.01 mm; **c** Marginal cord of shell, marginal view, scale bar 0.01 mm; **d** Septal face, front view, scale bar 0.2 mm; **e** Marginal cord in empty shell broken in axial and equatorial direction, scale bar 0.01 mm. *C* Radially directed sections of marginal canal system distributing pseudopodial protoplasm to marginal grooves; *MC* marginal cord, distinctive feature for all nummulitidae; *S* septum; *SF* septal face

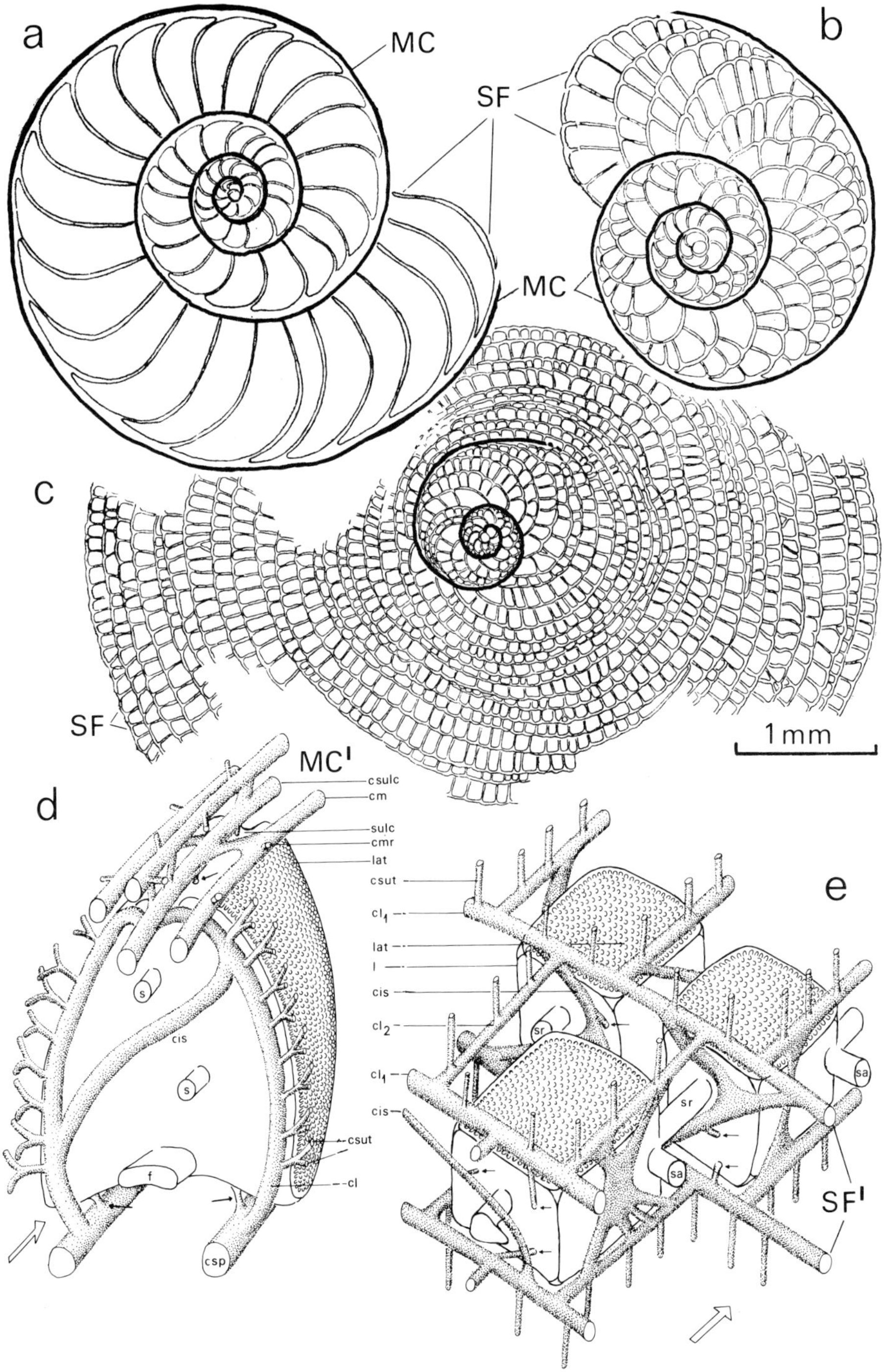
a
MC
b
SF
MC
c
SF
1mm
d
MC'
csulc
cm
sulc
cmr
lat
csut
cl1
lat
l
cis
cl2
cl1
cis
s
cis
s
f
csut
cl
csp
e
sr
sr
sa
sa
SF'

Fig. G.16 a–e

a taxonomic feature of family rank, but also a reflection of at least some common shell functions related to pseudopodial protoplasm responsible for motility and food uptake.

In Foraminifera with canal systems, generated by infolding of the outer shell surface, the chamber and the canal cavities house different kinds of protoplasm. In the chamber plasma protected by the organic cell envelope, electron-dense or -transparent, large vacuoles, symbionts and the large vegetative nucleus hint at the predominance of vegetative-physiological functions located in the chamber lumina. In the canals, the protoplasm is not protected by an organic envelope and presents, in retracted position an ultrastructure similar to freely expanded pseudopods, viz. numerous microtubuli, small mitochondria, small vacuoles and osmiophilic granules, indicating functions related to motility and transport. Apparently, the canaliculate Foraminifera keep their pseudopodial protoplasm permanently differentiated in their canal system, whereas noncanaliculate species transform the ultrastructure of their pseudopods when retracted into the last chambers of their shell.

The symbionts of the three nummulitid species present in the Gulf of Aqaba are diatoms of B3d-type (see Fig. G.41) identical in their ultrastructure analyzed in situ and identified as *Nitzschia panduriformis* in culture. In foraminiferal species from other families with or without canal system, diatom symbionts have different ultrastructures.

Red Sea Species and Their Historical Background. Three species of nummulitids are present in the Gulf of Aqaba: *Operculina ammonoides* (Gronovius 1781), *Heterostegina depressa* d'Orbigny (1826) and *Heterocyclina tuberculata* (Moebius 1880). Their different generic designation is based on differences in their shell structure and reflects their separate historical roots within the nummulitid family: *O. ammonoides* is a descendant from Indopacific Miocene *O. bikiniensis* (Cole) and related species; *H. depressa* doubtfully from late Eocene *H. aequatoria* Cole; whereas for *Heterocyclina tuberculata* no predecessor with cyclical chamber arrangement and comparable stolon system is known.

In the Red Sea, all other extant Indopacific nummulitid species, *Planoperculina heterosteginoides* (Hofker), *Heterostegina operculinoides* Hofker, *Cycloclypeus carpenteri* Brady and *Nummulites (=Operculinella) cummingii* Carpenter (=*venosa* Fichtel and Moll) are absent. Whereas *O. ammonoides* and *H. depressa*

◄

Fig. G.16 a–e. Functional replacement of marginal canals distributing pseudopodial protoplasm around shell margin by intraseptal canals in concentrically growing nummulitids. **a** *Operculina ammonoides;* **b** *Heterostegina depressa;* **c** *Heterocyclina tuberculata.* Equatorial sections, camera lucida drawings. **d** Canal system in *Operculina,* schematic, not to scale. **e** Canal system in adult *Heterocyclina,* schematic, not to scale. *MC* Marginal cord; *MC'* Marginal canals below shell margin. *SF* Septal face; *SF'* Intraseptal, mainly primary lateral canals below septal face taking over new function. *cis* Intraseptal canals in primary septum; *cl* lateral (subsutural) intraseptal canal in primary septum; *cl* lateral canal in secondary septum; *cm* marginal canal; *cmr* radial or oblique mouth of marginal canal system; *csp* spiral canal; *csulc* sulcus canal; *csut* sutural canal; *f* foramen; *l* chamber; *lat* lateral chamber surface bearing marks of perforation in lateral shell wall; *s* stolo; *sa* annular stolo; *sr* radial stolo; *sulc* sulcus; *white arrows* direction of shell growth; *small black arrows* communication between chamber cavity and canal system. Note equatorial plane vertical in **d** and horizontal in **e** (Hottinger 1977 b)

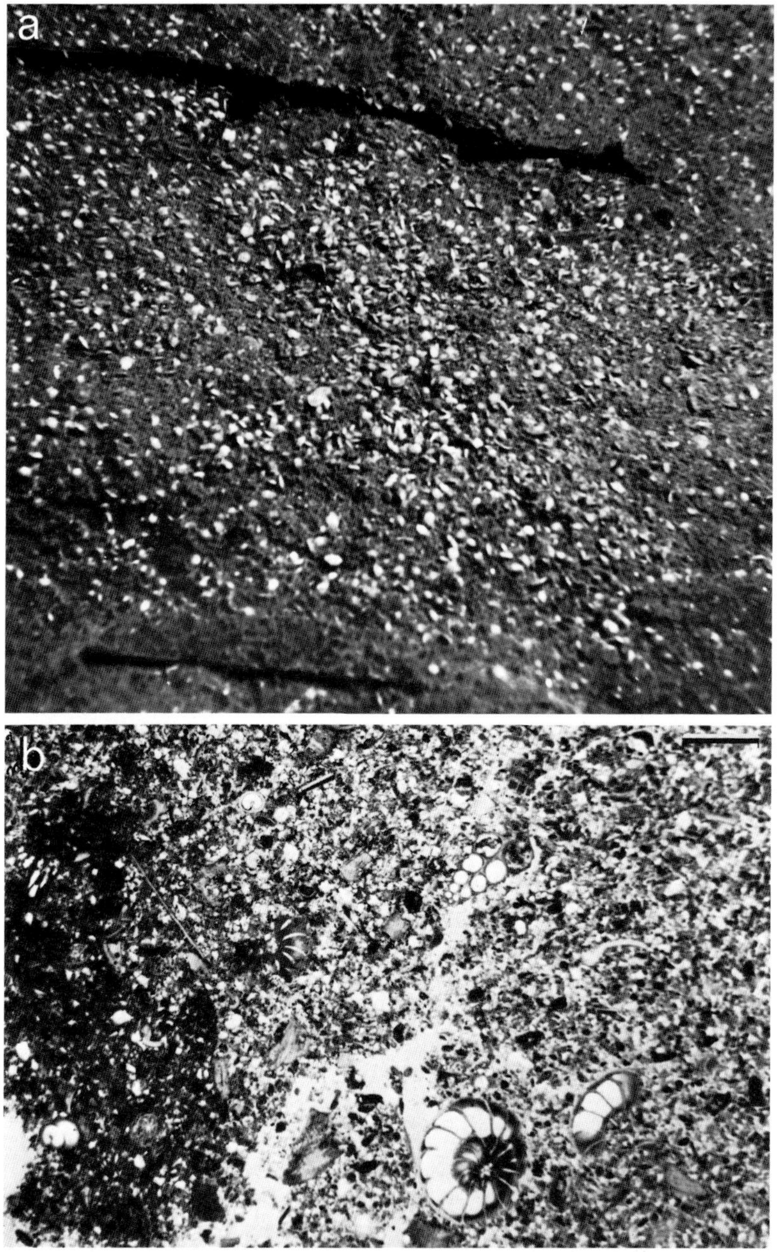

Fig. G.17 a, b. Living *Operculina ammonoides* on soft substrate **(a)** between dead *Halophila* leaves in 35 m depth, Geziret Fara'oûn area, site 172–173. Horizontal thinsection **(b)** of corresponding, fine-grained and bioturbated sediment with *Operculina* shells from artificially hardened core sample, scale bar 1 mm

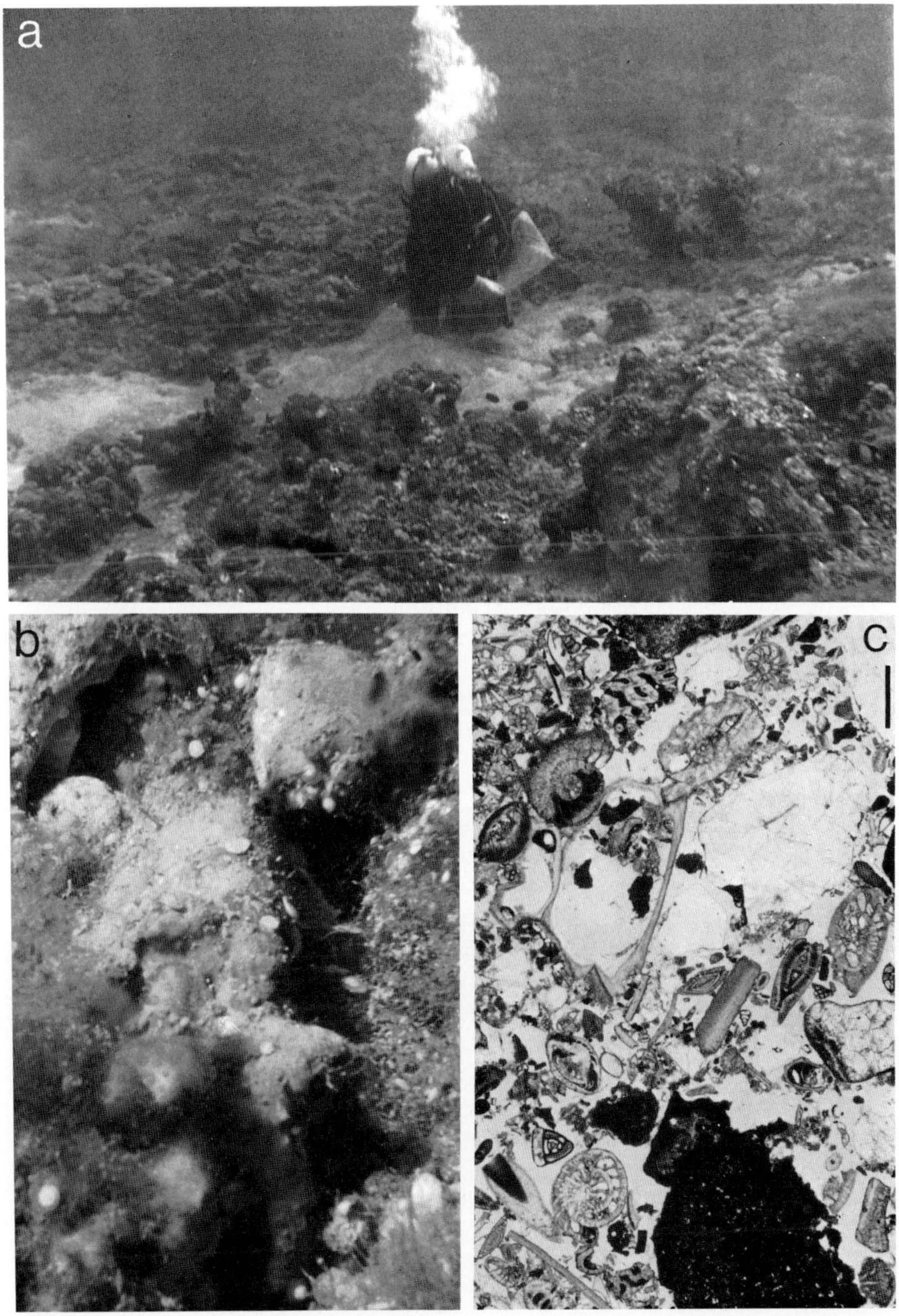

Fig. G.18 a–c. *Heterostegina depressa* on hard substrate. **a** Coral carpet off MBL, Elat, 40 m, with patches of bare sand where *Heterostegina* shells produced in the carpet accumulate. **b** Dead coral debris overgrown by small algal thalli and numerous living *Heterostegina*, about natural size. **c** Vertical thin section of corresponding sediment from artificially hardened core sample, off MBL, Elat, 52 m; scale bar 1 mm are cosmopolitan-tropical to subtropical species occurring in the Gulf of Aqaba as isolated representatives of their organizational type in their respective communities, the cyclical *H. tuberculata* replaces *Cycloclypeus carpenteri* in deeper environments and is considered here as a species truly endemic to the Red Sea–East African faunal province.

Distribution. The three nummulitid species are usually bottom dwellers throughout their life. Although specimens of *O. ammonoides* were found occasionally on *Halophila* or in clouds of filamentous algae, the species is not epiphytic. Often *O. ammonoides* shells are found attached to *Halophila* roots collected from a few centimeters depth below the sediment–water interface. It is not yet clear whether these specimens do live below the sediment surface or if the roots of the plants fix themselves on dead or dying shells, buried by the ever-present, intensive burrowing of the infauna.

In the Gulf of Aqaba, *O. ammonoides* and *H. depressa* share roughly the same depth zone, between 20 and 130 m, but are separated by the characteristics of their substrate: *O. ammonoides* lives on soft sediment in between *Halophila* plants (Fig. G.17), where water energy is low, while *H. depressa* crawls on hard bottom, at the base of corals, on boulders or coarse coral debris (Fig. G.18). Near boundaries of ecosystems corresponding to hard- or soft-bottom communities, the distribution of the two species overlaps, particularly in the deeper part of their depth range, where sedimentation rates are low. However, their separation is sufficiently characteristic to be easily mapped under water (Fig. F.26, p. 175). In the open Indopacific, where other nummulitid species are present in the same types of environments, a similar distribution is found, but *H. depressa* occurs often in shaded areas in shallower waters, at a few meters depth. Either *H. depressa* has been overlooked in shallow depths in the Gulf of Aqaba or the small fringing reefs do not provide shaded areas large enough to permit a significant development of this species, noted for its photoinhibition if exposed to high light intensities. In the Mahakam Delta (Kalimantan, Borneo) the distribution of *O. ammonoides* reaches much shallower waters than around Elat. This might be due to the high turbidity around deltas absorbing light in the water column. Therefore, *O. ammonoides* is also expected to react to high light intensities with photoinhibition, a conclusion supported by a reduced "vital effect" in $\delta^{18}O$ of shells grown in waters of less than 40 m depth in Elat.

Heterostegina depressa is the only nummulitid species known to produce a cyst of organic material enveloping the shell completely prior to building a new chamber. By rotating the shell in the cyst the space for a new chamber is generated. The ecological significance of this peculiar process is not known at present.

Below the growth of dense coral carpets on hard substrates or of permanent vegetation covers on soft substrate, limited to a depth of 70 m, *Heterocyclina tuberculata* is found regularly in densities which are 1–3 orders of magnitude lower than those of *O. ammonoides*. The microspheric generation, documented by a single specimen, represents 1% of the sectioned specimens of which all but one were megalospheric. *H. tuberculata* is found not only on soft substrates with low sedimentation rates, but also on hard bottoms down to the limit of the euphotic

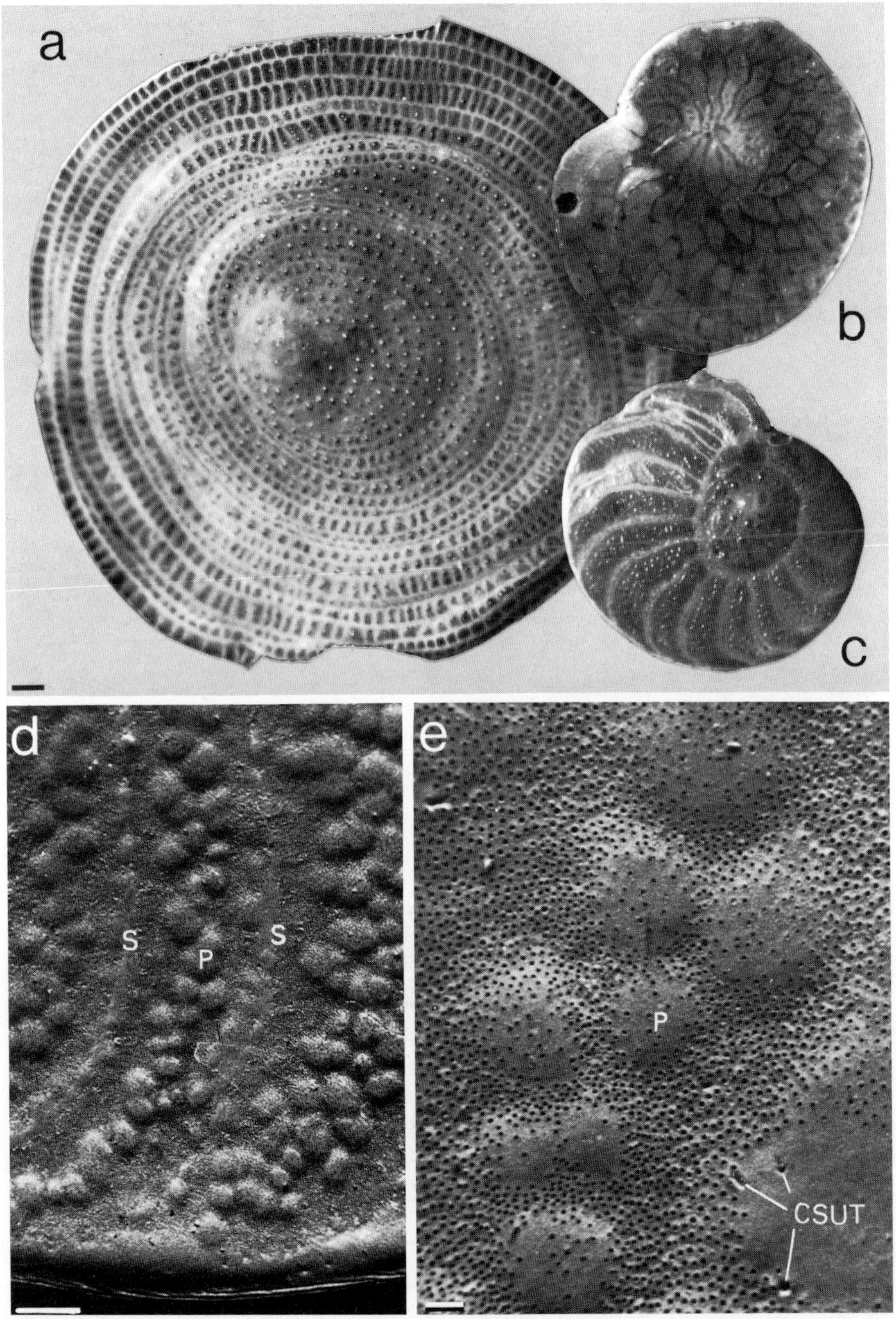

Fig. G.19 a–e. Lateral views of nummulitid shells. **a** *Heterocyclina tuberculata;* **b** *Heterostegina depressa;* **c** *Operculina ammonoides.* Megalospheric specimens, scale bar 0,1 mm. Note light reflections on intraseptal pillars suspected to function as light lenses. **d** Detail from **c** showing pillars *(P)* between septal sutures *(S)*. SEM, scale bar 0,1 mm. **e** Detail from **d** showing spacing of pores on interseptal pillars *(P)* and sutural canal mouths *(CSUT)*

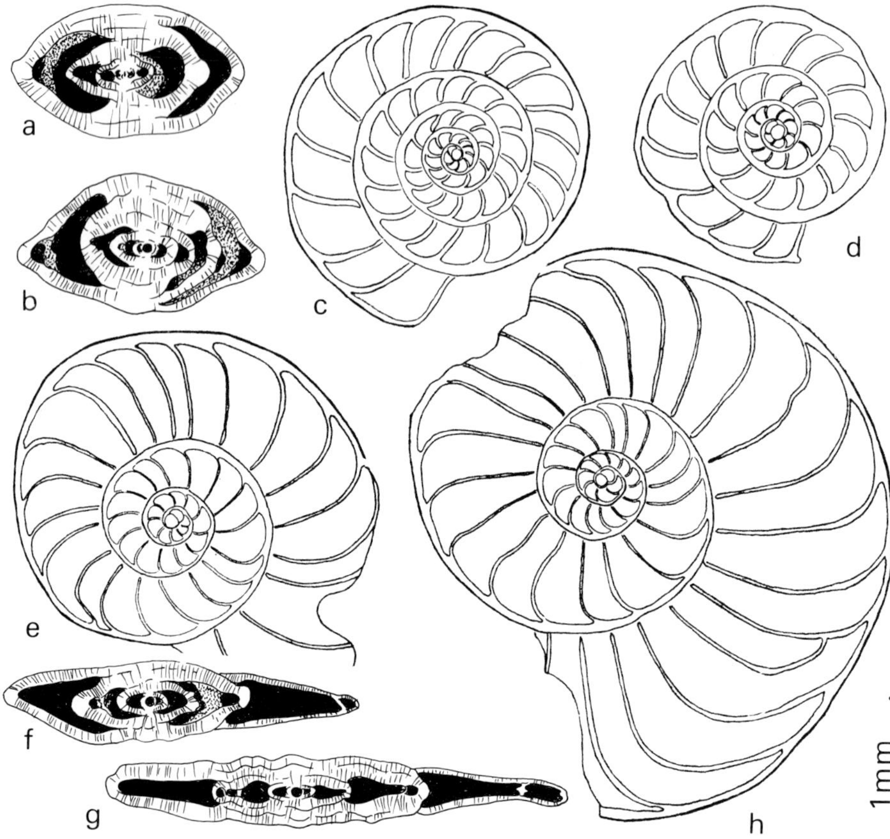

Fig. G.20 a–h. Shell flattening with depth in *Operculina ammonoides*. Camera lucida drawings of megalospheric specimens sectioned in axial and equatorial direction. **a–d** Involute specimens from 40 m depth; **e–f** semiinvolute specimens from 50 m; **g–h** evolute specimens from 90 m. Note variation of embryo and early whorls independant from adult shape. (Equatorial sections from Hottinger 1977 b)

zone. *Cycloclypeus carpenteri* has a similar depth distribution in the Indopacific Ocean. It was also found living on hard bottom on the frontal slopes of New Caledonia's barrier reef.

The specimens of the three nummulitid species living below 60 m depth are characterized by their single or multiple interseptal "pillars", forming half-globular humps on the lateral shell surface (Fig. G.19). These "ornaments" are interpreted here as lenses collecting the dwindling light to the profit of the symbionts, a hypothesis still to be tested by systematic investigations of the optical properties of the mineralized shell. The general trend of shell flattening with depth observed in the successive species and in the intraspecific variation of operculinids and heterosteginids seems to support a general hypothesis of shell-shape modification as an adaptation to light, as was postulated also for corals. The distribution of the symbionts below the upper, light-exposed **and** the lower, shaded-lateral shell surface, however, points to at least one supplementary factor playing some part in the system, i.e., gas exchange through the pores.

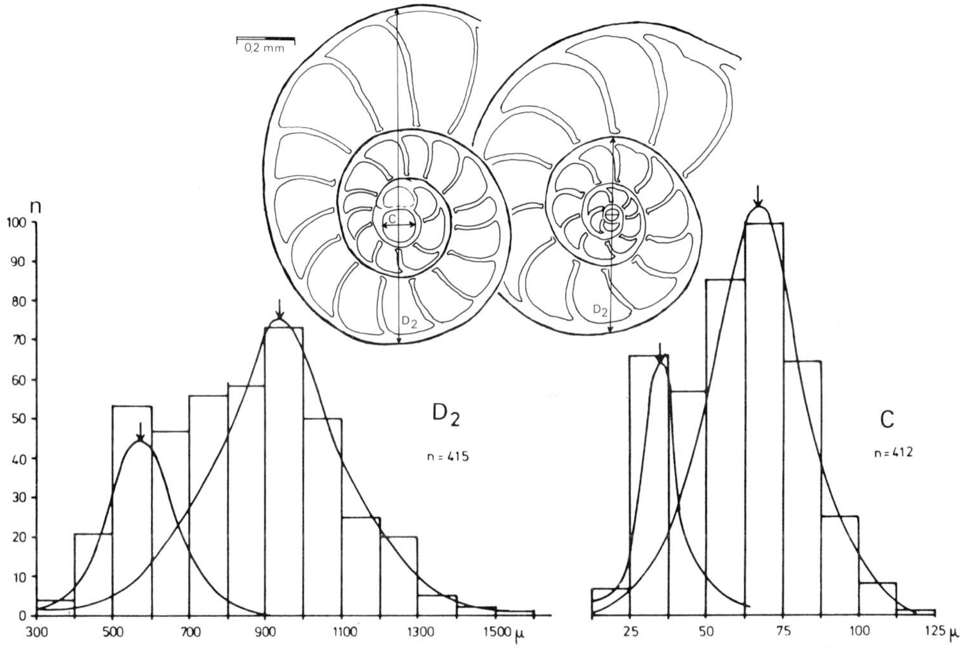

Fig. G.21. Intraspecific variation in early whorls of megalospheric *Operculina ammonoides* from all depths. Frequency distribution reflects heterogeneous population. (Histograms from Fermont 1977 a)

Biometry. Biometric studies on *Operculina ammonoides* based on comparatively simple linear measurements and chamber counts yielded chaotic results as long as involute and evolute specimens were not separated. In evolute, flat and involute, thick specimens, however, (Fig. G.20) the inner diameter of the megalosphere is positively correlated with the equatorial outer diameter of the first two whorls (Fig. G.21). This means that the larger the initial volume of protoplasm after asexual reproduction, the larger is the volume increase of synthesized new protoplasm in each growth step, within the ranges specific to generation and taxonomic unit.

The diameter of the first two whorls and the number of corresponding chambers are less well, but still clearly positively correlated in evolute forms, but apparently not correlated in involute ones. This is because the volume increase of the chamber with each growth step is better reflected by the equatorial diameter of the shell in evolute than in involute specimens, where the lateral parts of the chamber lumen house a substantial part of the total chamber protoplasm, which is not taken into account by measurements restricted to the equatorial plane.

Shell frequency distribution of involute *Operculina ammonoides* shows highest values at a depth between 50 and 80 m and of evolute types at a depth between 70 and 100 m, reflecting depth-dependent production rates in this species. The differences between frequency maxima and minima are of three orders of magnitude which cannot be explained by different dilution rates of the shells in the sediment. Microspheric specimens occur almost exclusively in the most productive zone. The frequency distribution of megalosphere size classes according to the depth gradient follow closely the curves of shell frequency per gram sediment and are thus correlated with productivity. Differences in megalosphere diameter reflect different initial volumes of protoplasm in the juveniles resulting from the ratio of the total protoplasmic volume in the parent shell to the number of offspring produced by asexual reproduction. The volume of the final protoplasmic mass in the parent generation may be related to rates of food uptake determined by the relation of food production to population density (availability of food) and by metabolic rates. The latter have been shown to be optimal in light conditions prevailing in the depth zones of highest propulation density in larger Foraminifera from shallower water, particularly in amphisteginids, as measured by highest vital effects in $\delta^{18}O$ values. In operculinids, highest vital effects coincide also with the highest shell frequencies per gram sediment. It may, therefore, be assumed that optimum light conditions for operculinid species prevail at depths between 50 and 100 m, where productivity is highest.

Frequency distribution of the megalosphere diameter from all depths is bimodal for evolute and unimodal for involute forms. The large overlap in involute forms (Fig. G.21) may result from bias due to uneven spacing of the samples in the depth gradient. Elimination of this bias would accentuate the heterogeneity of the population of evolute specimens. The number of living nummulitid specimens as defined by the greenish-brown color of their symbionts, conserved even in dried shells or registered underwater by the diver, is highest during the early spring months and drops strikingly during the late summer and autumn on both hard and soft substrates. The grab samples taken at depths over 60 m yielded living nummulitids throughout the year. Thus, the reproduction of nummulitids seems to be less dependent on the seasonal cycle in deeper than in shallow waters. The ratios of megalospheric to microspheric shells suggest that all nummulitids reproduce in paratrimorphic cycles. If successive megalospheric generations are different in their morphology, as suggested by differences measured in *Amphisorus hemprichii* and by negative correlations of megalospheric diameter and final shell size in *Operculina ammonoides,* a population consisting of a synchronous, uniform generation would have mean biometric values differing from those of a population of asynchronous and mixed generations. Other mixing factors such as burrowing intensity or biological factors such as, e.g., shell destruction rates by boring microorganisms, of preferential selection of juveniles during mass mortality periods after reproduction, as suggested by minority coiling direction ratios in *Amphistegina,* may further accentuate or obscure biometric heterogeneity of the samples.

Why consider involute and evolute *O. ammonoides* as one species? Biometric measurements of internal characters of specimens from all depths reveal no significant differences, particularly in the ratios megalosphere diameter to equatorial diameter of the second whorl. The number of chambers per whorl shows significant differences only in late adult growth stages (29 chambers) where the effect of equatorial radius increase on chamber volume is greatest. Moreover only one (evolute) type of microspheric form has been found. Thus, involute thick and evolute thin types of *O. ammonoides* are interpreted as mere ecological variants replacing each other at the 70 m isobath. This limit corresponds to the limit of sub-

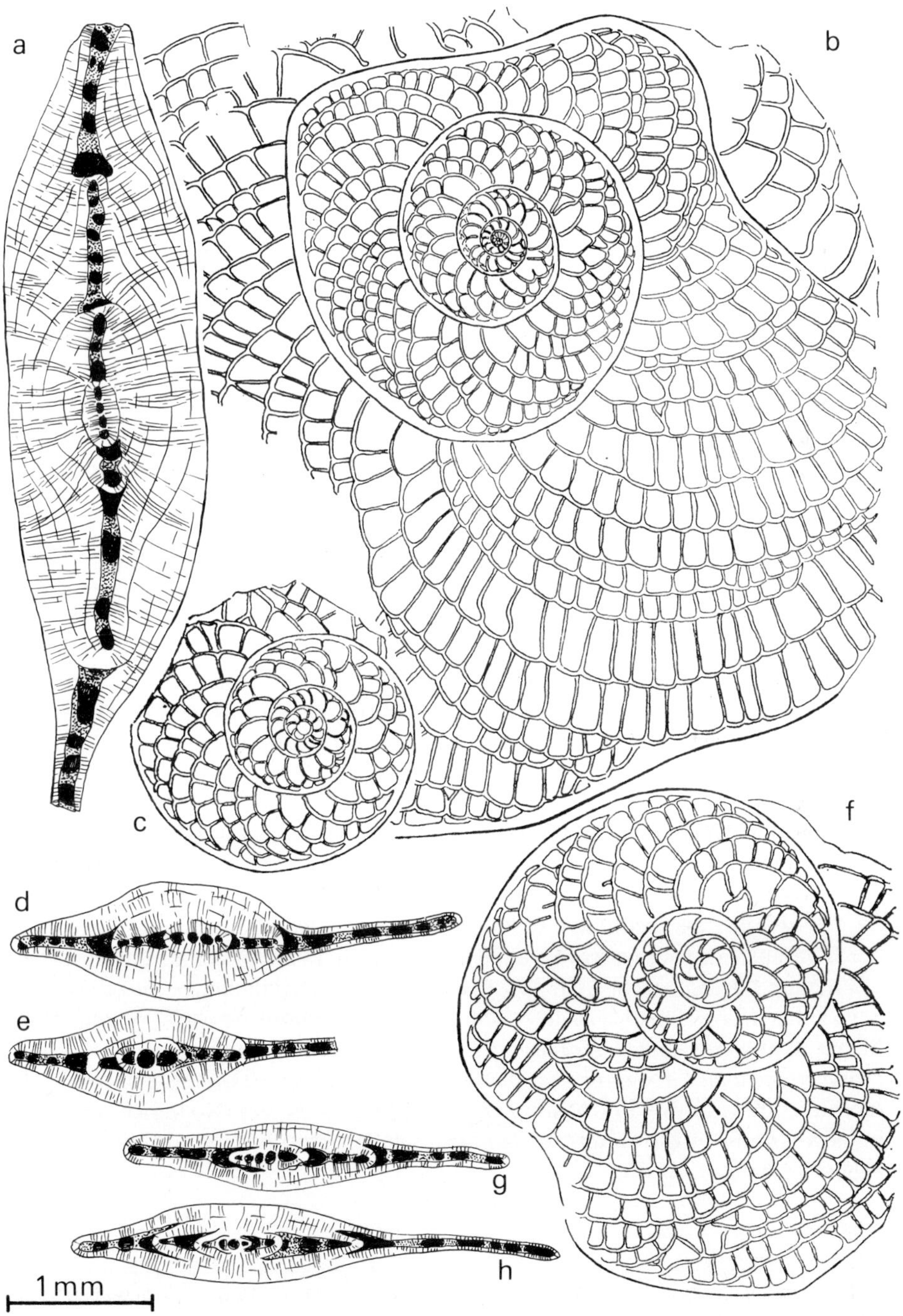

Fig. G.22 a–h. Shell flattening with depth in *Heterostegina depressa*. Camera lucida drawings of specimens sectioned in equatorial and axial direction. **a, b** Microspheric specimens, from 40 m depth; **c–e** megalospheric specimens, 40 m; **f–h** megalospheric specimens, 70 m. (Equatorial sections from Hottinger 1977b)

strate types as discussed in Chapter F. The response to the depth gradient in this case might be an indirect one as involute and evolute shells, independent of their depth provenance, are distinguished by their different $\delta^{13}C$ value reflecting their main carbon source (see below). The latter depends on feeding habits and food quality assumed to be different on different substrate types. It may be noted that the intraspecific variability in *O. ammonoides* is much larger in the Gulf of Aqaba than on the Maldives in the open Indian Ocean, where *Nummulites cummingii* and *Heterostegina operculinoides* coexist with *O. ammonoides*. The involute nummulite and the evolute heterosteginid might fill at least partially the nutritional niches left wide open for involute and evolute extreme variants of the only nummulitic occupant in the Red Sea.

Heterostegina depressa is uniformly involute although specimens from shallow waters are thicker than specimens from deep water (Fig. G.22). Average thickness of the lateral lamellae and growth spiral (measured as equatorial radius) contribute to the change of shell shape with depth. First whorl diameter and number of chamberlets in the long, curved chambers of late growth stages are poor characters to describe the flaring adult whorl of the shell reacting in particular to the depth gradient: the first whorl diameter represents too early ontogenetic stages and the length of the chambers adjusts to changing values of the growth spiral radius by different degrees of backward bending.

In the microspheric generation, the depth-dependent trends in adult whorls of the shell are most conspicuous. The frequency of this generation (relative to megalospheric generations) is about 5% at depth, much less in shallow waters, but still higher than in *O. ammonoides* where the frequency of B-forms is estimated to be less than 0.1%.

As in *Operculina ammonoides,* the variation of the diameter of the megalosphere is positively correlated with the one of the deuteroconch and of the equatorial diameter of the first whorl (Fig. G.23). The larger the protoplasmic volume in the first growth steps, the less operculine, unfolded septa are present.

The number of secondary septa per unit of apertural face length does not much diminish during ontogeny, a pattern common to all fossil and extant species of involute *Heterostegina* s.str. The diameter of the megalosphere and the correlated other characteristics of the shell change with depth between 30 and 70 m. Below 70 m, no systematic trend in the biometric data was observed. As in *Operculina ammonoides,* the nonlinear depth trend may reflect relative synchroneity in reproduction with or without supplementary mixing factors. No indications were found that any of the *Heterostegina* assemblages from Elat would be heterogeneous.

Fig. G.23 a–d. Intraspecific variation in *Heterostegina depressa*. **a** Mean values of morphological parameters in relation to depth. S_{20} number of secondary septa in the 20th chamber. S_{15} in the 15th chamber; X number of chambers without secondary septa; h diameter first whorl, $c_{1,2}$ diameter of embryo consisting of protoconch and deuteroconch; c_1 diameter of proloculus. Note close relation of variation curves for h, $c_{1,2}$ and c_1 and independence of S_{15} and S_{20} from the latter. X is inversely related to $S_{15,20}$. **b–c** Histograms for X and C, of specimens from all depths reflecting homogenous population. **d** Measurements in early whorls of *H. depressa*. (**a–c** after Fermont 1977 b)

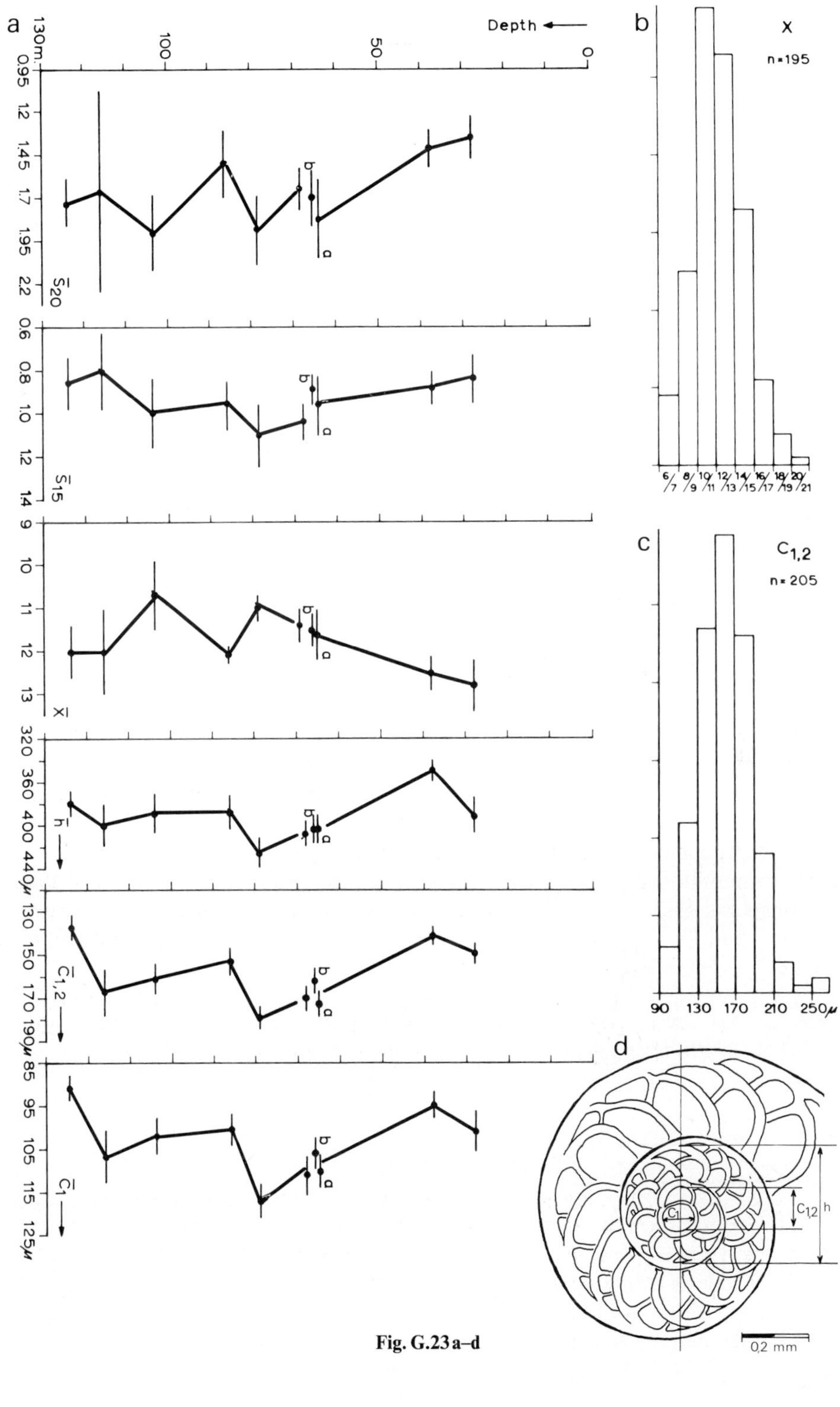

Fig. G.23 a–d

In *Heterocyclina tuberculata,* megalosphere and deuteroconch diameter, as well as number of spiral precyclic chambers were measured in 65 specimens from 86 to 116 m depth. The number of operculinid chambers is either 3 or 4, occasionally 5 including proto- and deuteroconch. The negative correlation between megalosphere or deuteroconch diameter and the number of spiral chambers indicates a weak, but clearly recognizable influence of the volumes in initial growth on the 26–38 following stages. This influence is kept up in *H. tuberculata* much longer than in *O. ammonoides* where the ecological response of adult growth stages overwhelms the influence of the initial volumes after about one and a half whorls, corresponding to less than 20 growth stages. This long-lasting influence of initial volumes during ontogeny in *H. tuberculata* can be measured exceptionally well in this species because its shell is extremely compressed, reflecting small differences in chamber volume in the pattern offered by equatorial sections. The weak and late ecological response might be due to the comparatively restricted depth range of the investigated samples.

Comparing the three extant nummulitid species in the Gulf of Aqaba to fossil nummulitid lineages by characters empirically used to describe "evolutionary advance" through time (megalosphere diameter, number of unfolded, operculinoid or precyclical chambers), all three species would have to be defined as "early" or "primitive", belonging to more or less initial stages of evolutionary lineages. *O. ammonoides* is comparable to the Middle Eocene *O. parva,* second or third evolutionary stage in *O. alpina* group, having a very similar habit in shell shape, chamber shape and spiral growth pattern. *Heterostegina depressa* is fairly similar to *H. helvetica,* the earliest known, Late Eocene representative of involute *Heterostegina* s.l. The number of precyclical chambers in *H. tuberculata* corresponds to the one of the earliest, Oligocene representatives of *Cycloclypeus*. This is taken as an indication that the phyletic history of the nummulitid species present in the Gulf of Aqaba might be short. The endemic character of *Heterocyclina tuberculata* supports this view. It may well be that the attempts to bridge the gaps in the fossil record around Mid-Miocene times in order to reconstruct the phylogeny of extant nummulitids are purely artificial and must be revised when more Late Miocene and Pliocene material from permanently tropical regions will be available for study.

References: Fermont (1977a, b), Fermont et al. (1983), Hansen and Reiss (1971), Hottinger (1977a, b, 1980, 1981b), Hottinger and Dreher (1974), Lee et al. (1979a, b, 1980a, b), Leutenegger (1977a, b, c, 1984), Meulenkamp (1977), Said (1949), Spindler and Röttger (1973).

G.2 "Smaller" Benthic Foraminifera

The delimitation of "larger" and "smaller" benthic Foraminifera is an arbitrary one. Here, large-sized and symbiont-bearing forms on which a large body of knowledge is available are included in a first chapter above, while forms of large sizes without symbionts are combined with small-sized benthics in the present chapter, mostly because the restricted knowledge of these groups does not

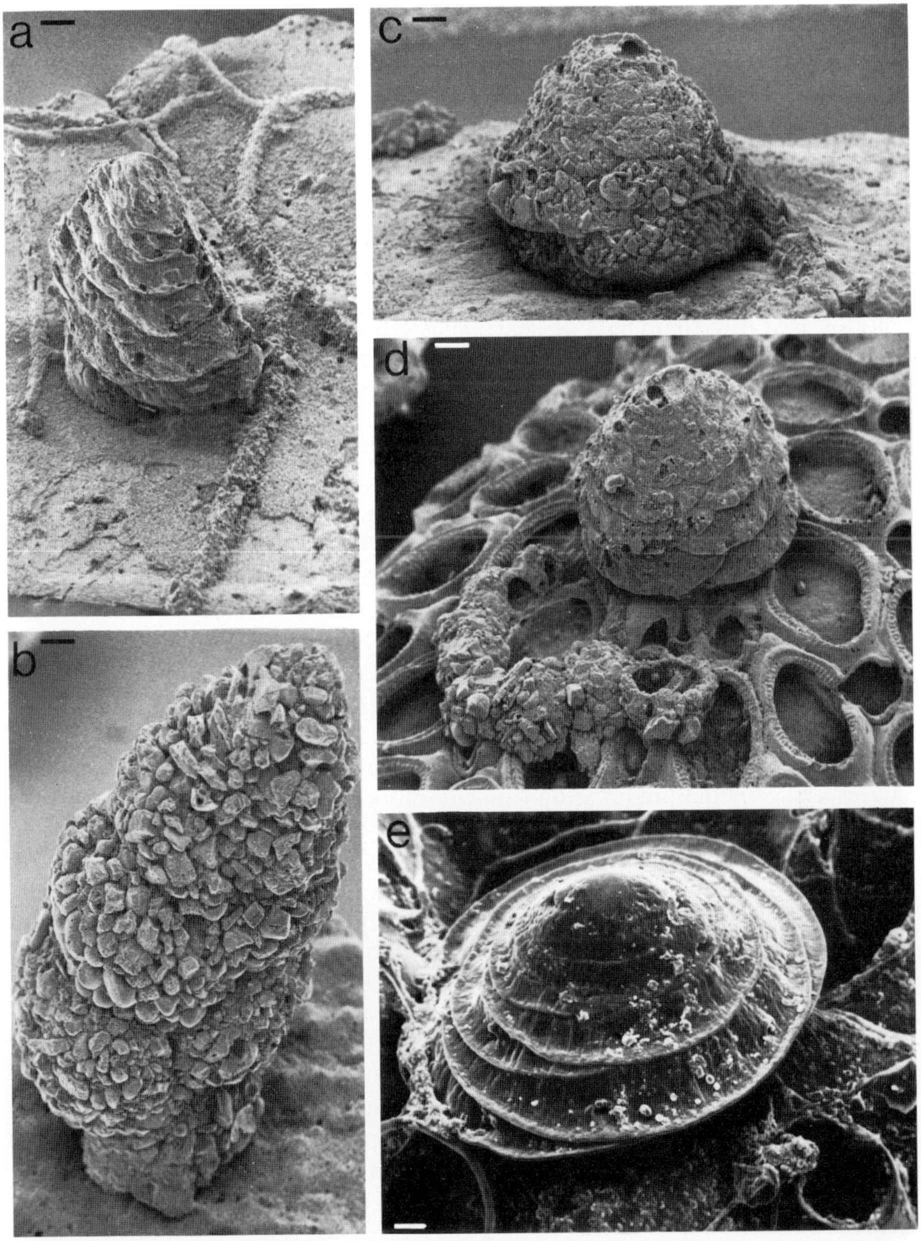

Fig. G.24 a–e. Benthic smaller foraminifera on solid substrate, off MBL, Elat, 150 m. SEM, scale bars **a–d** 0.1 mm, **e** 0.01 mm. **a** *Spiroplectammina kerimbaensis*, oblique apical view, with *Rhizammina* sp.; **b** *Textularia foliacea*, oblique lateral view; **c** *Textularia barkeri*, lateral view; **d** *T. barkeri*, oblique apical view (note broken, biserial embryo at apex) with attached species of *Rheophax* on bryozoan colony (*Conopeum* sp.); **e** *Patellina corrugata* on bryozoan colony. Note importance of apertural face for fixing shell on substrate merit single subchapters. Modern taxonomic revisions of the "smaller" benthic Foraminifera present in the Gulf of Aqaba are available only for agglutinated textulariids s.l. (Halicz and Reiss 1979), for perforate-lamellar Cibicididae (Perelis and Reiss 1975), for Bolivinitidae (Zweig-Strykowski and Reiss 1975) and for a number of selected species from various families. A complete inventory on the species level is still to be worked out in order to revise and complete Said's fine work from 1949.

Agglutinated Foraminifera occur in the Gulf of Aqaba at all depths and on all types of substrate. They represent up to 20% of the total number of benthic Foraminifera in the sediment samples. Incrusting saccaminids, particularly *Sagenina* sp. and tolypamminids are particularly frequent on larger shell debris in zones of low sedimentation rates and on hard substrate on the outer shelf, where they contribute substantially to the bioconstructions. They are often in competition of space with small bryozoan pioneers and seem to belong to the second-generation settlers following the arborescent perforate Foraminifera (see below). Some textulariids are also permanently fixed on shell debris (Fig. G.24). They are so tightly glued to their substrate that the shells do not fall off during the processing of the sediment samples. *Textularia barkeri* was found so often on dead bryozoan colonies that one may suspect this particular species to feed on these animals.

The lack of any transparency of their shell causes great difficulty in distinguishing the living from the dead agglutinated Foraminifera. However, populations of homogeneous shell size found in great density in the clouds of filamentous green algae in spring in depths between 20 and 50 m were certainly alive. Their great number is not reflected in the sediment samples collected in autumn. This suggests a restricted preservation capacity of at least some agglutinated species in the sediments.

The textulariids present in the Gulf of Aqaba show intraspecific variation of grain size and grain composition of the shell wall. Very small grains in between larger ones appear in the SEM to be of skeletal origin, but it is not yet proven whether the smallest mineralized elements of the shell are also detritus collected by the animal or if they were secreted or mineralized on the organic material present in between grains. All species classified here within the genera *Textularia* and *Spiroplectammina* possess blindly ending tubular structures coated with organic matter in their shell walls. These structures were called pseudopores by Reiss (1963). Their opening toward the chamber lumen is covered by the organic envelope of the protoplasmic body (inner lining) as in true pores of perforate-lamellar shells. The true pores have been demonstrated to serve gas exchange through the shell walls. Similar functions are ascribed here to the pseudopores in agglutinated walls.

Among the species present in the Gulf of Aqaba, *Textularia* cf. *aegyptiaea* Said (1949), *T.* cf. *agglutinans* d'Orb. (1839), *T. cushmani* Said (1949), and *T. barkeri* Hofker (1978) have tiny trochospiral early growth stages, while *T. foliacea* Heron-Allan and Earland (1915) *T. fungiformis* Fornasini (1887), and *T. conica* Mateu (1970) (*non* d'Orb.) are biserial throughout their ontogeny. The initial stages of *Spiroplectammina kerimbaensis* Said (1949) and of *S.* cf. *taiwanica* Chang (1956) are planispiral. *Neoseptigerina floridana* (Cushman 1922) possesses

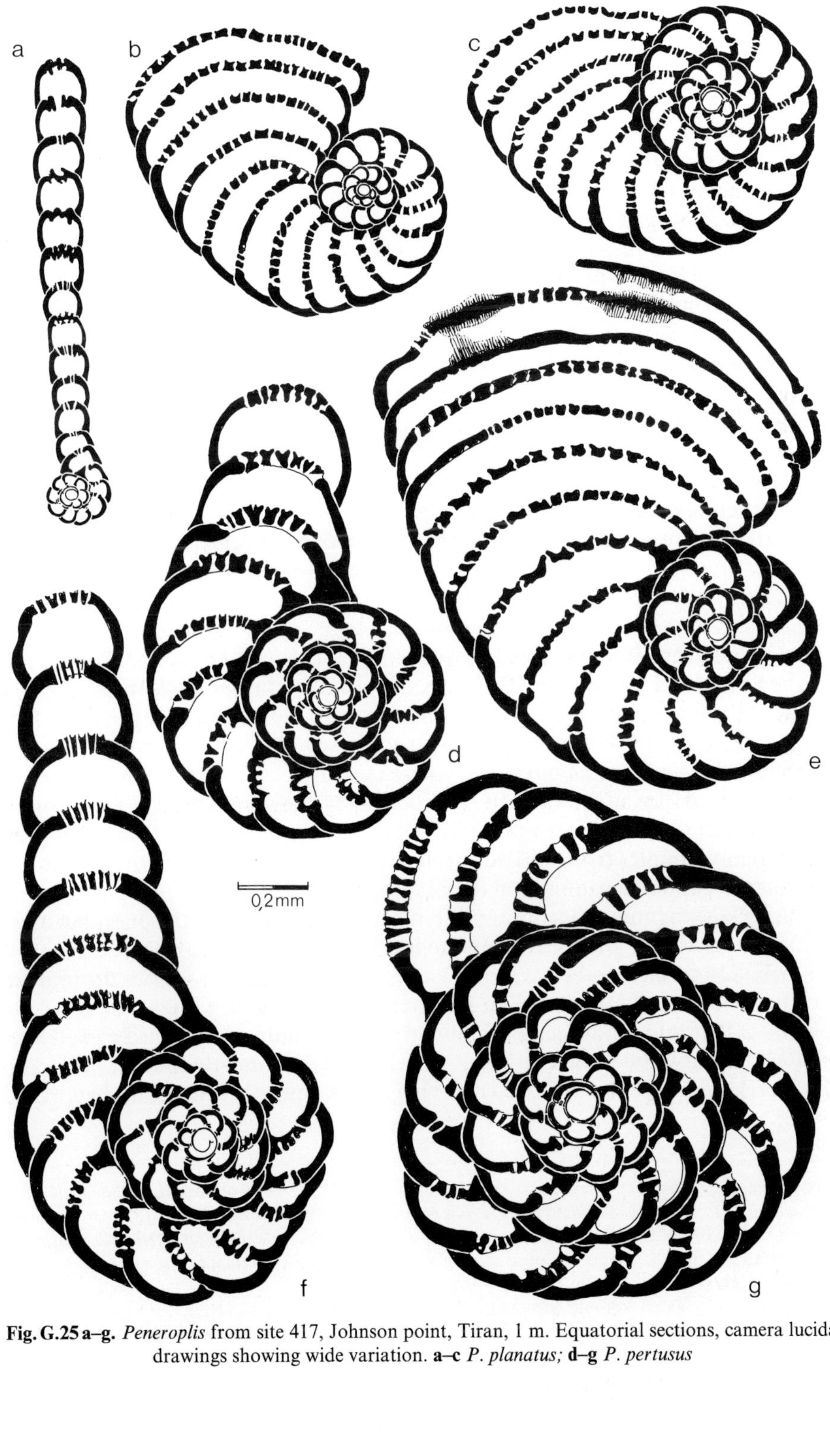

Fig. G.25 a–g. *Peneroplis* from site 417, Johnson point, Tiran, 1 m. Equatorial sections, camera lucida drawings showing wide variation. **a–c** *P. planatus;* **d–g** *P. pertusus*

internal chamber partitions, but lacks pseudopores. This seems to be the only agglutinated species systematically absent from hard substrates. The depth distribution of these species is rather uniform, most of them occurring regularly between 20 and 300 m. *T. fungiformis* and *T. barkeri* were not found in samples from depths greater than 70 m, while *S.* cf. *taiwanica* is present only in samples collected below 170 m.

The **porcelaneous** genus *Peneroplis* is represented in the Gulf of Aqaba by the species *P. pertusus* Forskal (1775) and *P. planatus* (Fichtel and Moll 1798). They are characterized respectively by coarse and fine longitudinal ribbing of the shell surface combined with different pitting patterns in the depressions in between ribs. As in miliolids, the pits are supposed to enhance the exchange of gases through the shell wall. Modes of growth in the early stages and details of the apertural morphology are also different in the two species (Fig. G.25). Both species have an intraspecific variability noted for its wide range from *Spirolina*- to *Dendritina*-like habit. The opening angle of the fan produced by the adult chambers varies widely. No depth relation was observed so far in intraspecific variability. Both species possess rhodophycean symbionts characterized in situ by R 1-type ultrastructures of their chromatophores.

P. pertusus is a shallow-water form restricted to the uppermost 40 m. In contrast to conditions in the Persian Gulf, where this species dominates frequently lagoonal foraminiferal associations, mass occurrences of *P. pertusus* are rare in the Gulf of Aqaba. Associations dominated by *P. pertusus* are, however, not restricted to hypersaline lagoons, but were observed also at the foot of Jackson reef in the Straits of Tiran (Chap. F). There, the intraspecific variation in the opening angle of the fan is much restricted in comparison to monospecific lagoonal associations. This may be due to effects similar to those noted above in *Borelis schlumbergeri* and *Operculina ammonoides,* where missing related concurrent species are "compensated" by larger variants.

P. planatus also presents a rather restricted intraspecific variability and is present in many samples from soft substrates to 70 m depth as a very minor constituent of the benthic foraminiferal fauna.

Porcelaneous milioline Foraminifera characterized by their particular, two-chamber-per-whorl growth rhythm are not restricted to shallow water, but their contribution to the number of specimens in the sediment associations drops drastically from up to 50% to less than 10% at the 50 m isobath. Kitazato (1981) demonstrates that different species of miliolids live as epiphytes, on hard or soft substrates or below the sediment–water interface in the interstitial spaces (see also Chap. H). Some species are known to have symbionts. Shell shape and apertural

Fig. G.26 a–o. Some porcelaneous smaller Foraminifera from the Gulf of Aqaba. **a** *Quinqueloculina pseudoreticulata;* **b** *Quinqueloculina parkeri;* **c, d** *Massilina corrugata* (sensu Collins 1958); **e** *Spiroloculina corrugata;* **f** *Spiroloculina communis attenuata;* **g, h** *Quinqueloculina oblonga;* **i** *Triloculina littoralis;* **j** *Articulina queenslandica,* earliest spiral growth stage missing; **k** *Vertebralina striata;* **l** *Edentostomina rupertiana,* apertural face; **m–o** *Articulina pacifica* (note change of sagittal plane of peristome after spiral stage). Elongated, smooth forms **(g, i)** living in the sediment, ornamented forms (all others) on plant substrate or on bottom. Significance of miliolid teeth and necked peristomes or teethless, large foramina are not yet understood

Fig. G.26 a–o

characteristics must respond to such different modes of life for simple mechanical reasons: burrowing forms in fine-grained sediments have flush, ovoid to elongate shells easy to slip through the sediment, while others found on hard bottoms have almost discoidal-flattened shells. Heavy ornamentation producing irregular outlines of the shell are found in species occurring in coarse-grained mobile sediment (Fig. G.26). Similar relations to moving on or in the sediment may explain the extreme modifications observed in the morphology of apertural faces, varying from elongated bottlenecks to large openings with flaring peristomes.

The **perforate-lamellar** Nodosariacea, in particular *Lenticulina* spp. and *Marginulina* spp. reveal a strikingly constant upper limit of distribution between 60–70 m. Below this limit they occur in small quantities at all depths.

The **Buliminacea** are represented by species of bolivinitids, buliminids and uvigerinids (Fig. G.27). An inventory of the bolivinitids in the Gulf of Aqaba reveals the presence of the following species providing up to about 20% of the total thanatocoenoses of benthic Foraminifera. Exclusively shallow-water species are heavily ornamented with peripheric spines [*Brizalina (Br.) simpsoni* (Herron-Allan and Earland 1915) and *Br. (Parabrizalina) convallaria* (Millett 1900)] or a lobate peripheric outline of the shell [*Br. (Pseudobrizalina) lobata* (Brady 1881)]. The spinose species *Br. simpsoni* is particularly frequent in coarse coral sands lacking a vegetation cover. The more lightly ornamented or smooth species *Bolivina hebes* (Macfadyen 1930), *Brizalina (Br.) italica* (Cushman 1936), *Br. (Parabrizalina) amygdalaeformis* (Brady 1881), *Br. (P.) africana* (Smitter 1955), and *Br. (P.)* sp. A7 occur in all depths, *Brizalina (Br.)* sp. B7 and *Brizalina (Parabrizalina)* cf. *subargentea* (Uchio 1960) mainly below the shelf edge.

The buliminids and the uvigerinids are increasingly abundant below the shelf edge reaching about 20% of the benthic assemblages at 500 m depth. In samples collected by submersible from 150 m depth, off the MBL Elat, *Reussella aequa* Cushman and McCulloch (1948), *Siphogenerina virgula* (Brady 1890), *Siphouvigerina ampullacea* (Brady 1884), and *S. porrecta* (Brady 1884), were identified (Fig. G.27). These species occur in association with *Brizalina (Parabrizalina) amygdalaeformis* (Brady) in fully oxygenated sediments.

The **rotaliids** are present as a very minor compound of the benthic foraminiferal assemblages, with the exception of *Challengerella bradyi* Billman, Hottinger, and Oesterle (1980), found in comparatively large numbers between 20–45 m depth living on *Halophila*-covered soft substrate (Fig. G.28). *Asterorotalia gaimardii* (d'Orb. 1906), and *Ammonia convexa* (Collins 1958) were occasionally encountered. Unidentified species of *Ammonia* occur frequently in the shallowest environments around mangroves. Species of *Pseudorotalia* (linked most probably to low salinities) have never been observed in the Red Sea.

Fig. G.27 a–p. Some buliminacean Foraminifera from the Gulf of Aqaba. **a, b** *Siphogenerina virgula*, apertural face and microspheric specimen; **c–e** *Bulimina marginata*, front view, lateral view and oblique front view; **f** *Reussella* cf. *spinulosa*; **g** *Reussella aequa*; **h** *Brizalina (Brizalina)* cf. *italica*; **i** *Brizalina (Parabrizalina) amygdalaeformis*; **j** *Brizalina (Parabrizalina)* cf. *africana*; **k, l** *Brizalina (Parabrizalina)* sp. A (Zweig-Strykowski 1973) with its apertural face; **m** *Brizalina (Brizalina) simpsoni*; **n, o** *Siphouvigerina porrecta*, apertural face and complete specimen; **p** *Siphouvigerina ampullacea*. SEM, scale bars on details showing apertural face 0.01 mm. *TP* tooth plate; *O* organic wall (temporarely?) closing part of aperture

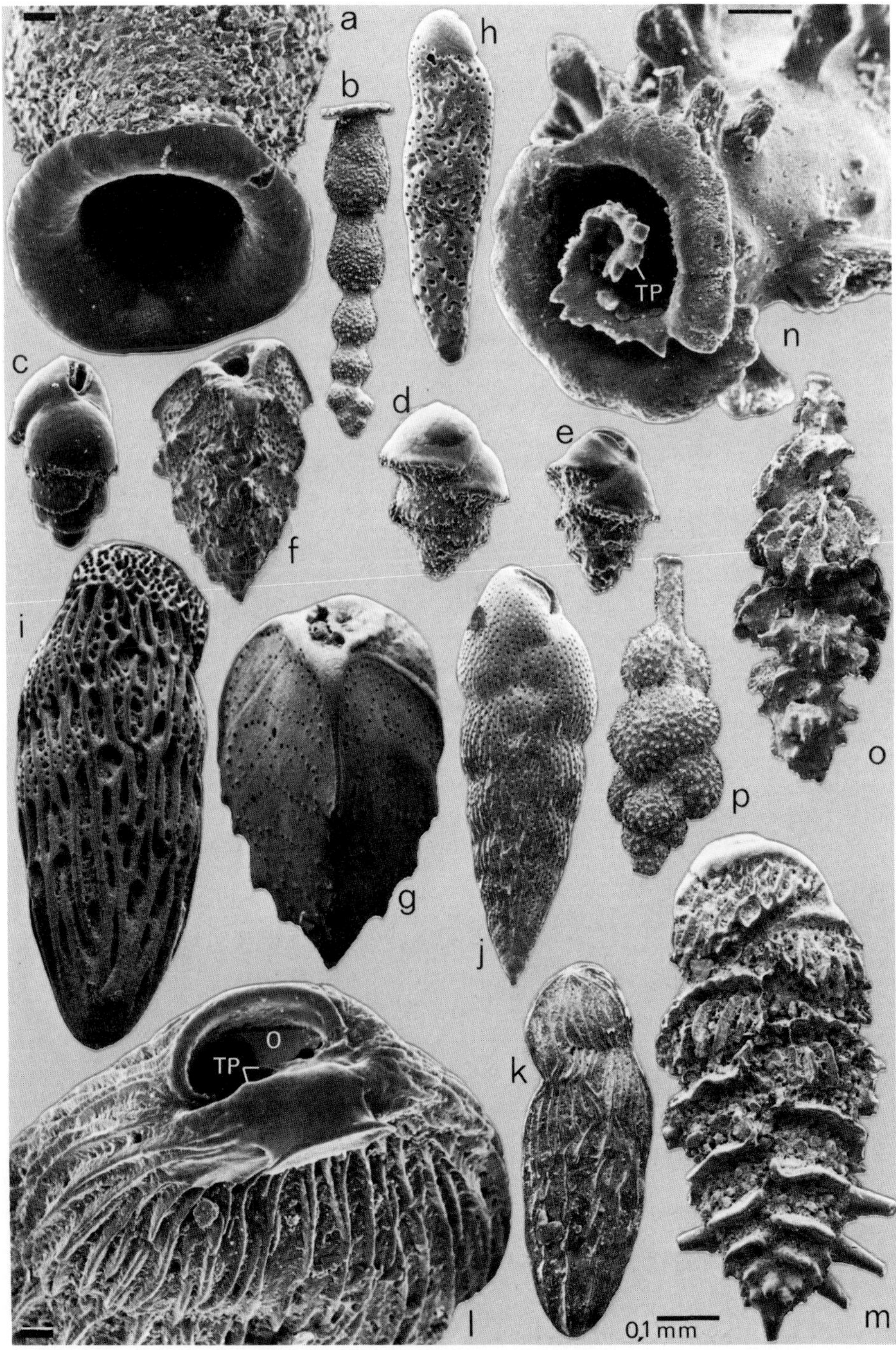

Fig. G.27 a–p

The **calcarinids** are represented in the Gulf of Aqaba exclusively by *Calcarina (=Pararotalia) calcar* d'Orb. (1839) in shallow environments of high water energy (Fig. G.28, while all larger calcarinids including *C. spengleri), Baculogypsina sphaerulata* and *Schlumbergerella floresiana* are absent from the "Red Sea marginal province."

Like other calcarinids, *C. calcar* is characterized by an extremely thick organic lining, forming massive pore plugs, and by a permanent differentiation of the protoplasm in the enveloping canal system. The single vegetative nucleus contains numerous spherical nucleoles. The symbiontic algae are subspherical diatoms similar to those of *Amphistegina lobifera* with two to three chloroplasts, but with a structure of the pyrenoid (B-1 type of Leutenegger) particular to symbionts in calcarinids.

"Eponidids," particularly of the *Poroeponides lateralis* (Terquem 1878) group, are a constant, but very minor component of the benthic assemblages in all depths. *Neoeponides schreibersii* (d'Orb. 1846) occurs very rarely in depths around 200 m.

The discorbid *Epistomaria punctata* Said 1949 (Fig. G.28 a) is regularly present in very low numbers in samples from *Halophila* covered substrate between 20 and 60 m depth.

The **cibicidids** as a group (Fig. G.29) constitute an important component of the benthic assemblages in all depths with a significant drop of their relative abundance below 200 m. This might be due to a reduction of the percentage of hard bottom surface below the shelf edge. A particular investigation of the ultrastructure of the shell and of their systematics revealed the presence of the following species which are temporarily or permanently fixed on solid substrate: *Cibicides lobatulus* (Walker and Jacob 1798), *C. mayori* (Cushman 1924), *C. pseudolobatulus* Perelis and Reiss (1975), *C. tenuimargo* (Brady 1884), and *Caribeanella elatensis* Perelis and Reiss (1975). *Cibicides* (?) *tabaensis* Perelis and Reiss (1975), *C. mabaheti* Said (1949), *C. sp. K4* Perelis and Reiss (1975), *C. aravaensis* Perelis and Reiss (1975) and *Paracibicides edomi* Perelis and Reiss (1975) have no obvious surfaces of attachment and are therefore interpreted as free-living. The genera *Paracibicides* and *Caribeanella* are defined by supplementary sutural apertures in different positions. All species have chamber walls of "indistinctly radial" nature with the exception of *Caribeanella elatensis* and *Cibicides* (?) *tabaensis,* which are distinctly radial. The many new species described from the Gulf of Aqaba are

Fig. G.28 a–m. Rotaliid Foraminifera. **a–d** *Challengerella bradyi,* megalospheric specimens; **a** dorsal view; **b, c** ventral view; **d** araldite cast, horizontal section, ventral view, showing umbilical canal system. **e–j** *Ammonia convexa,* megalospheric specimens; **e, f** umbilical view; **g, h** lateral view; **i, j** spiral view. **k, l** *Calcarina calcar,* megalospheric specimens, **k** spiral view, **l** umbilical view. Note open interlocular space between ultimate and penultimate chambers transformed to canal system mouths in between earlier chambers. Scale bars 0.1 mm. **m** Lamellation in interlocular space of *Calcarina,* specimen filled with epoxy resin, sectioned in horizontal direction, polished and etched according to methods of Hansen and Reiss (1971). Scale bar 0.01 mm. *b* passage chamber lumen-canal system; *ch* chamber; *f* foramen; *il* inner lining (of primary wall); *isp* interlocular space transformed to enveloping canal system in *Calcarina; ol* outer lamella (of primary wall); *p* pores; *sol* subsequent outer lamellae covering interlocular space; *sp* interlocular space in spiral suture covered by subsequent shell whorl and transformed to spiral canal; *u* umbilical interlocular space transformed to umbilical canal (cf. Figs. G.38 and G.40)

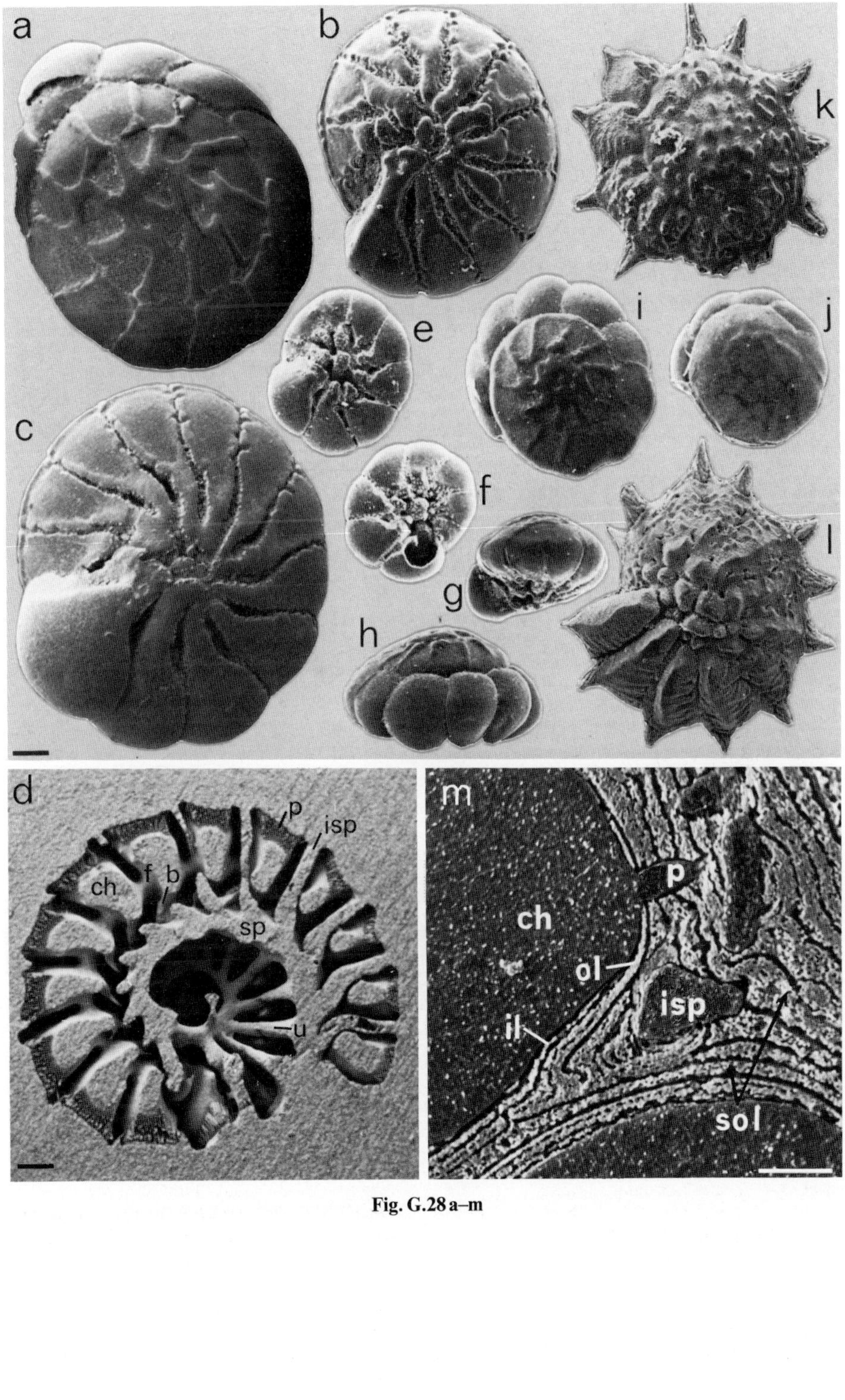

Fig. G.28 a–m

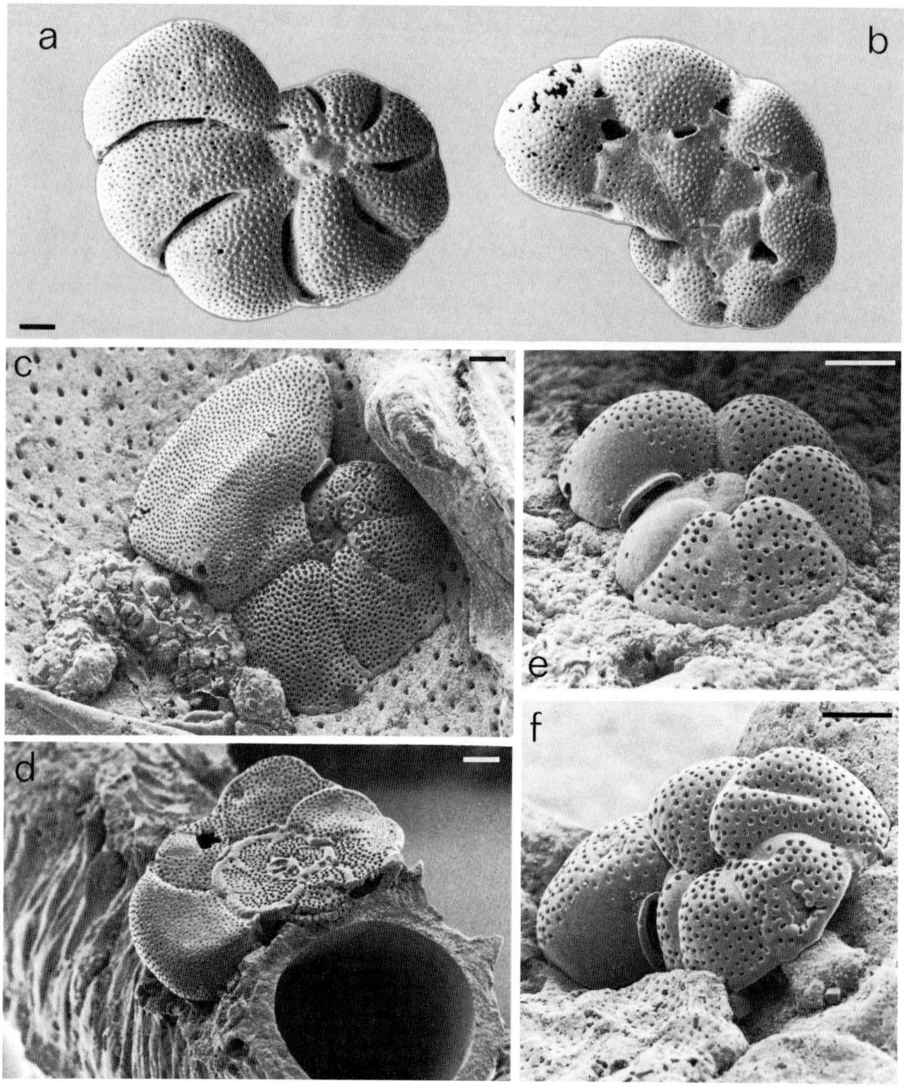

Fig. G.29 a–f. Some free-living or attached smaller perforated Foraminifera. **a, b** *Epistomaroides punctata* free on plant substrate, from Wadi Taba, 45 m. **c** *Caribeanella elatensis* with attached species of *Rheophax* on brachiopod shell. Note sutural apertures near periphery. **d** *Cibicides lobatulus* exceptionally attached with umbilical side to substrate. **e, f** *Paracibicides edomica* attached to comparably smooth and rough substrate with spiral side. **c–f** From samples off MBL, Elat, 150 m. SEM, scale bars 0.1 mm

probably not endemic, but a consequence of a higher level of morphological analysis helped by SEM methods in this particular group of benthic foraminiferans.

Planorbulinella is currently regarded as the closest extant relative of the extinct **orbitoids.** The shell is constructed by a single layer of concentric chamber rings communicating with a single layer of cross-wise oblique stolons. Perforation is particularly coarse and the heavy ornamentation is clearly differentiated in a coarser (ventral) and finer (dorsal) coat of pustules or pillars covering the lateral surfaces of the disc and suggesting a trochospiral origin of the group. *Planorbulinella* species have no symbionts. Like *Sorites variabilis,* they live as free specimens on hard substrates or on coarse grains on soft substrates. A large number of specimens are permanently fixed on the substrate adjusting their shell to irregular surfaces by distorting the disc. Frequently, *Planorbulinella* shells are trapped by encrusting Foraminifera in acervulinid macroids or by bryozoans coating hard surfaces on the deeper part of the shelf.

In free specimens, parts of single chamber rings are frequently disposed at low angles in respect to the equatorial plane of the shell. In equatorial sections of the shell, such oblique chamberlets seem to alter the regular structure of the stolo system, faking a relapse into the spiral arrangement of the first few nepionic chambers. Biometric measurements including the "relapse" chambers in the count of spiral chambers (Y and R in Fig. G.30) thus measure the frequency of disc-deformation in early post-nepionic growth stages together with nepionic acceleration.

Biometric studies were carried out in order to check eventual depth dependence of internal parameters in megalospheric early growth stages used to define time-dependent (evolutionary) changes in Miocene Mediterranean planorbulinellids. In the Gulf of Aqaba, two distinct, fairly homogeneous groups of assemblages were found in shallow and deep water respectively. Assemblages from intermediate depths between 80 and 125 m are interpreted as a mixture of shallow and deep forms. The two groups are distinguished mainly by the diameter of the megalospheric nepionic chambers and by the number of nepionic spiral chambers (Fig. G.30). The volume of adult chambers at comparable growth stages is also clearly different in megalospheric and microspheric forms. The two groups are therefore interpreted as separate species: *P. larvata* (Parker and Jones 1865) between 0 and 125 m depth with larger, voluminous shell cavities, and *P. elatensis* (Thomas 1977) between 90 and 240 m with smaller, less voluminous cavities. From their external aspect alone, the two species are difficult to distinguish. The intraspecific variability shows a weak trend of embryo volume increase with depth in both species. The proportion of the microspheric specimens in the samples varies from 0 to 30% and seems to be unrelated to depth.

In contrast to evolutionary trends observed in most larger orbitoid Foraminifera, the larger megalospheres in *P. larvata* are followed by a higher number (usually two) of spiral chambers than the smaller ones (with usually one) in *P. elatensis*. Thus, the two species present in the Gulf of Aqaba are not closely related to each other. Extrapolating the evolutionary trend in Miocene planorbulinellids, *P. elatensis* represents the extant end member of the *P. rokae-P. caneae* lineage, while *P. larvata* may belong to a different, Indopacific group (Fig. G.31).

Incrusting perforate Foraminifera form flattened discs programmed for rapid radial growth by orbitoidal or similar arrangement of chamber rings in the competition for space on hard substrate. Their soft parts lack symbionts and are protected by a particularly thick organic lining, revealing after dissolution of the shell in acid the communication pattern between their shell cavities. The shell has, in contrast to *Planorbulinella* spp., a clearly differentiated attachment surface similar to those of cibicidids. In cross-section the chamberlets of the attached layer

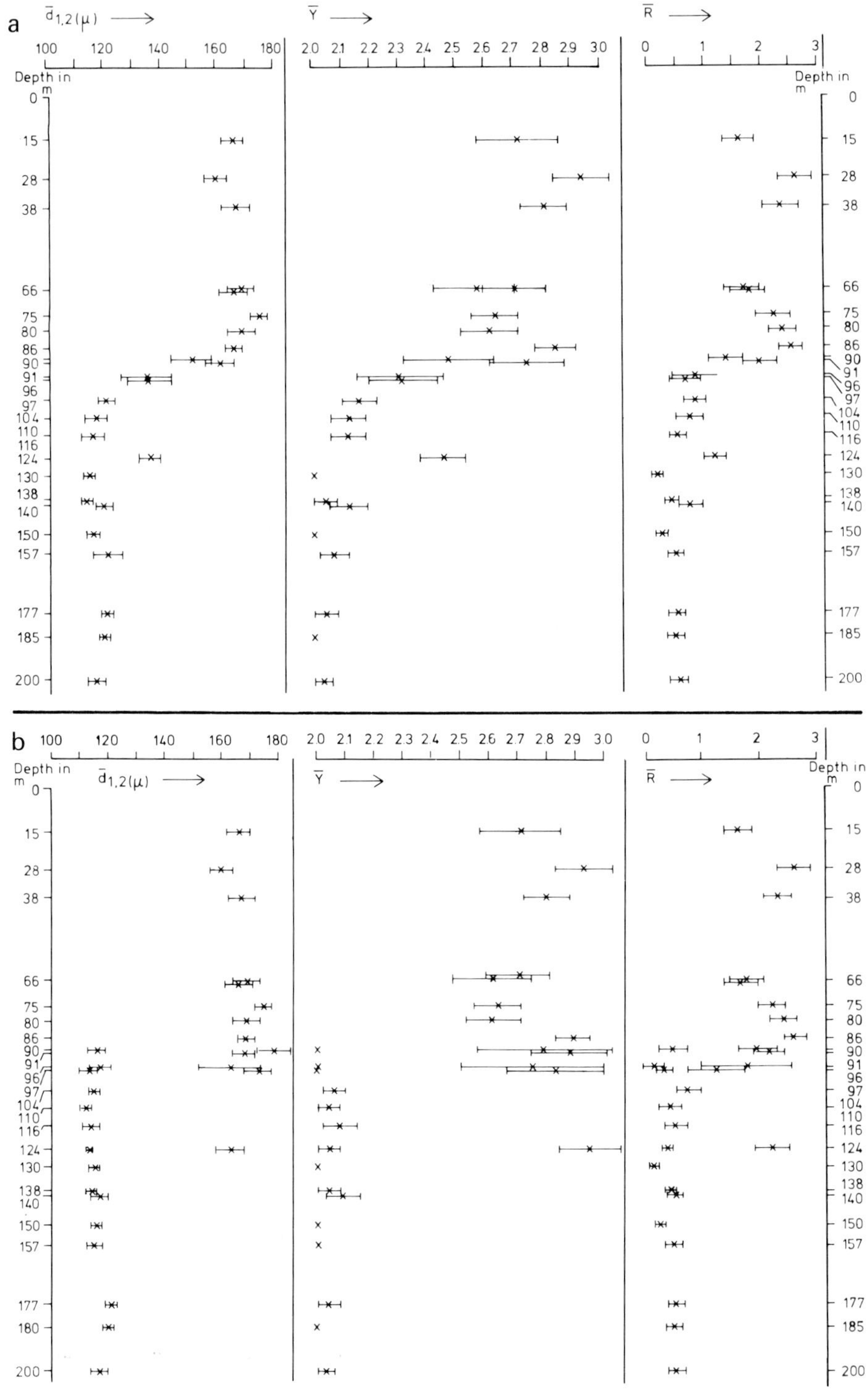
a
$\bar{d}_{1,2}(\mu)$
100 120 140 160 180
Depth in m
$\bar{Y}$
2.0 2.1 2.2 2.3 2.4 2.5 2.6 2.7 2.8 2.9 3.0
$\bar{R}$
0 1 2 3
Depth in m
0 15 28 38 66 75 80 86 90 91 96 97 104 110 116 124 130 138 140 150 157 177 185 200
b
100 120 140 160 180
Depth in m
$\bar{d}_{1,2}(\mu)$
2.0 2.1 2.2 2.3 2.4 2.5 2.6 2.7 2.8 2.9 3.0
$\bar{Y}$
0 1 2 3
$\bar{R}$
Depth in m
0 15 28 38 66 75 80 86 90 91 96 97 104 110 116 124 130 138 140 150 157 177 180 200
0 15 28 38 66 75 80 86 90 91 96 97 104 110 116 124 130 138 140 150 157 177 185 200

Fig. G.30 a, b

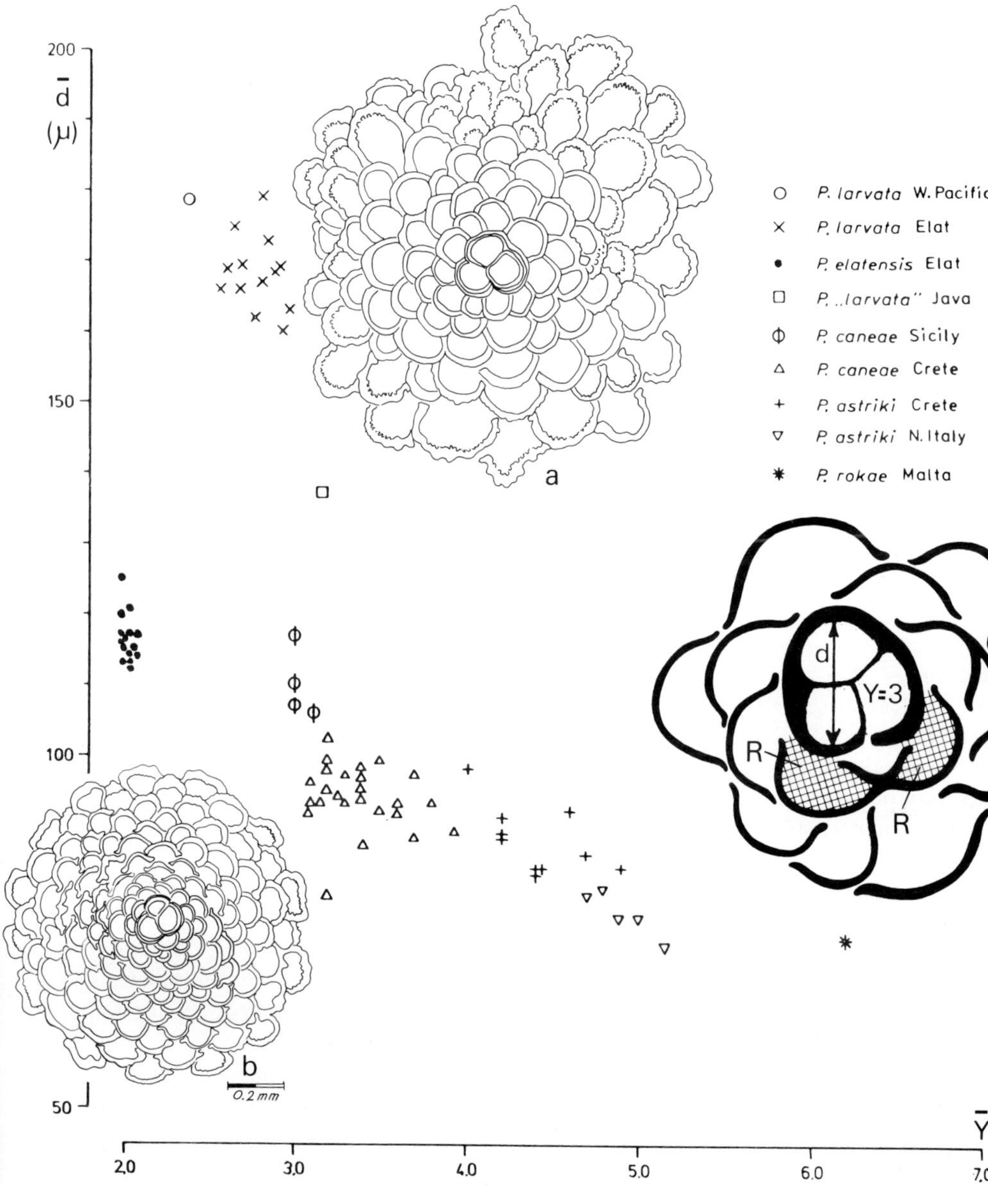

Fig. G.31 a, b. Recent and fossil *Planorbulinella*, d/Y scatter diagram (after Thomas 1977) and camera lucida drawings of equatorial sections in shells of *P. larvata*, 40 m **(a)** and *P. elatensis*, 138 m **(b)** from Gulf of Aqaba. Note on diagram *P. elatensis* in continuation of evolutionary trend in fossil species while *P. larvata* stays apart

Fig. G.30 a, b. *Planorbulinella* in the Gulf of Aqaba. Depth distribution of mean values for $d_{1,2}$ $(=d)$, *Y* and *R* (see Fig. G.31). **a** Values for all specimens; **b** values for specimens separated according to size of adult chambers into two distinct species with overlapping depth ranges, *P. larvata* and *P. elatensis*. Note perfect separation in the latter case for *d* and *Y* values, no separation for *R* values reflecting shell distortion. (After Thomas 1977) are evolute below the attached shell face and often slightly keeled at their margins, while they are involute and without keels at the free surface of the shell. In analogy to cibicidids, the attached surface is interpreted here as dorsal, the free surface as ventral, although there are no true umbilical structures in early spiral growth stages.

So far, three species are distinguished in the Gulf of Aqaba. The shell of *Planorbulina mediterranensis* d'Orbigny (1826) consists of a single layer of coarsely perforated orbitoidal chamberlets which are often of unequal size in one chamber ring (Fig. G.32a). The bilocular megalospheric apparatus is followed by five to eight spiral chambers. Adult orbitoidal chamberlets have two large interiomarginal foramina with imperforate lips. Occasionally supplementary sutural openings give rise to scattered additional chamberlets at the free surface of the shell. *P. mediterranensis* usually lives on plant surfaces and is found, therefore, frequently as detached shells in the sediment deposited in the euphotic zone.

In *Planorbulina acervalis* (Brady 1884) the chamberlets are arranged in more than one layer (Fig. G.32b). The megalospheric embryonic apparatus is trilocular and followed by two to seven spiral chambers. The latter have supplementary passages forming vertical stolons as access to a first layer of lateral chamberlets filling the umbilical region. During a single growth step in the adult, one ring of marginal chamberlets is formed simultaneously with one discontinuous layer of lateral chamberlets. There are short spines between the coarse pores on the ventral shell surface.

P. acervalis is a pioneer on short-lived hard substrate in shallow water, colonizing clean stone pebbles, macroids and all kinds of plastic refuse where surfaces are flush. Neighboring shells of the same generation may touch each other or may partially overlap having reached adult size. They may form more or less continuous crusts. Such population densities were observed only in larger encrusting foraminiferans, never in populations of free-living benthics.

Acervulina inhaerens Schulze (1854) possesses early growth stages similar to *P. acervalis,* but smaller than in the latter. They are difficult to locate in the huge adult shells covering surfaces of over 50 cm^2. Adult chambers from a polygonal network of anastomosing chamber cavities rather than true chamberlets (Fig. G.32c). Their perforation is comparatively fine. With each growth step the whole free surface is covered by a new network layer. Where the animal is in competition for space with another attached organism, the network margin may loose its connection to the substrate and grow freely into space in order to overgrow the attached margin of its neighbour (Fig. G.33a). The marginal apertures may be radially extended by lips or peristomes (Fig. G.34a). There is no definite number of marginal apertures per marginal chamberlet unit. Other openings are unevenly distributed along the sutures of each network layer all over the free surface. Feeding cysts, however, are deposited only in front of the marginal apertures, indicating that the ventral supplementary openings might be used only for the extrusion of protoplasm during the next growth step. While the outermost two or three layers of chamberlet network contain only very thin films of protoplasm along the shell walls, protoplasm of ectoplasmic structure with numerous food particles fills the chamber cavities below the seemingly empty layers. Vacuolated endoplasm is confined to the inner parts of the shell.

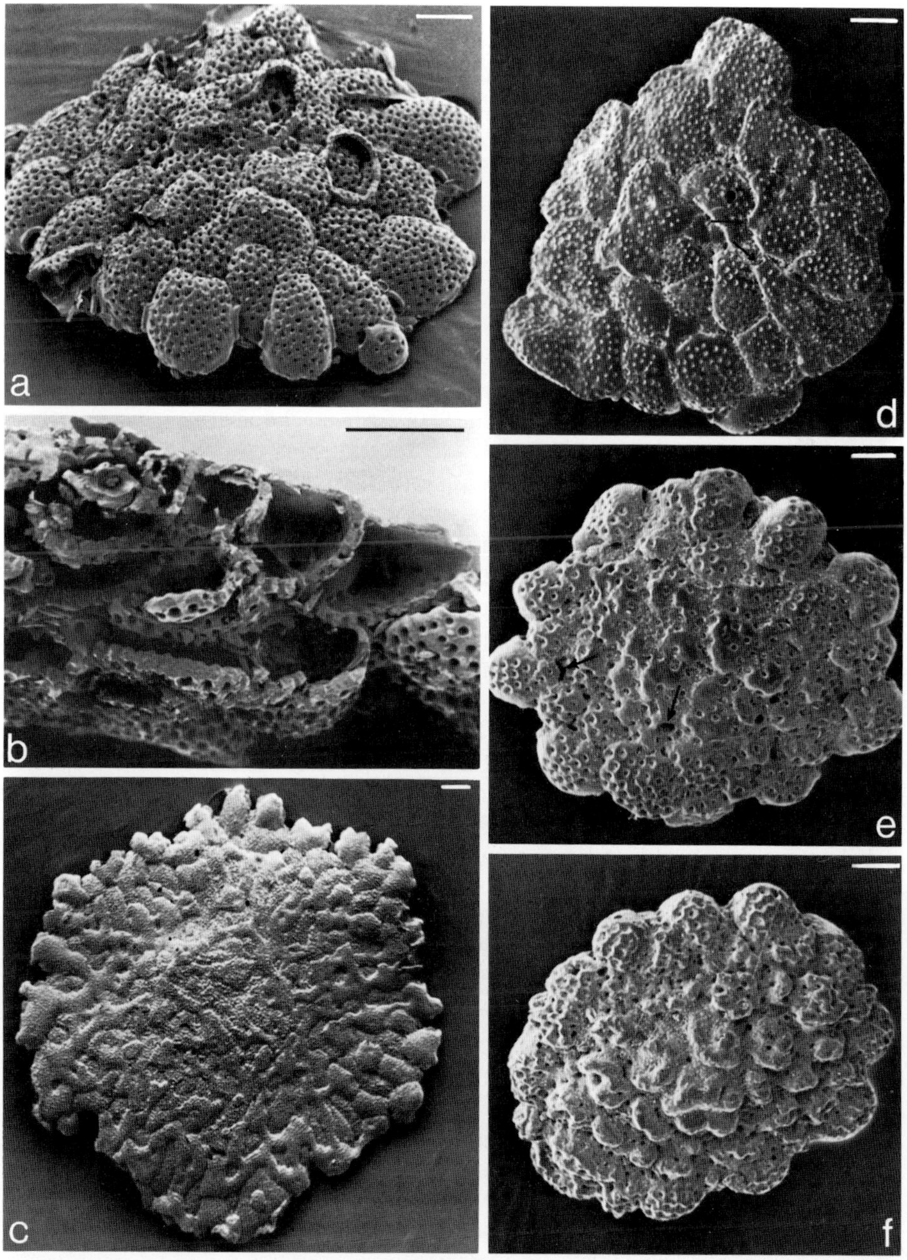

Fig. G.32 a–f. Perforated Foraminifera with orbitoidal or analogous growth patterns, SEM, scale bars 0.1 mm. **a** *Planorbulina acervalis,* oblique ventral view. **b** *P. acervalis,* broken specimen, dorsal, attached side upwards, showing superposition of chambers on venter. **c** *Acervulina inhaerens,* free, ventral side, early growth stage (adults reach diameters of 6–8 cm). **d** *Planorbulina mediterranensis,* doral surface with supplementary apertures *(arrow)*. **e** *Planorbulinella larvata,* young specimen, dorsal view, showing supplementary apertures *(arrow)*. **f** *P. larvata,* ventral view. (Adapted from Levinson 1976)

A. inhaerens is the most proliferous sessile foraminifer in the Gulf of Aqaba contributing extensively to the coating of macroids and to the cementation of coral or other coarse debris below the reef front and in the coral carpets. To at least 60 m depth it grows on the light-exposed surfaces, while coralline red algae and/or bryozoans cover the shaded side of the substrate. In shallow water, *A. inhaerens* may also take up cryptic habits living, for instance, in abandoned pholad bore holes. Living specimens of the same generation may touch or overgrow each other. Successive generations form continuous, thick but layered crusts. This species belongs to the second generation of pioneers competing for space with encrusting bryozoans or coralline algae. Its marginal growth mechanisms by detachment of the margin and radial acceleration of growth reflect this strategy, while the margins of arborescent perforate Foraminifera (see below) are firmly fixed to the substrate. As first-comers on new substrate, the arborescent forms have no competition for space, but need a firm mechanical support for their stem.

In the feeding cysts of *A. inhaerens,* investigated on specimens from 40 m depth, diatom frustules of different shapes and sizes, as well as accumulations of coccoliths were observed (Fig. G.33 b). The latter might be produced by fecal pellets falling from the pelagic realm on the foraminiferal crust and being used there as food. However, as the depth distribution of *A. inhaerens* in the Gulf of Aqaba is strictly limited to the euphotic zone (130 m), it is presumed that the main food source is benthic diatoms. The latter may also be responsible for the pale greenish color of the living acervulinid crusts.

Sphaerogypsina globulus (Reuss 1848) is another, globular representative of the acervulinid family present in the Gulf of Aqaba. There might be two species respectively exhibiting at the surface of spherical shells of equal size coarser, voluminous or distinctly smaller chamberlets of polygonal to quadrangular outline. Specimens with surfaces of attachment are a minority. The hemispherical to conical shells composed of much coarser chamberlets belonging to the species *Gypsina vesicularis* Parker and Jones, found frequently in association with *Marginopora* and *Sorites* on inner reef platforms of the Maldive Islands atolls, are not present in the Gulf of Aqaba, possibly because of bioprovincial distribution patterns. *S. globulus* contributes up to about 10% to the assemblages of dead shells in coarse coral sands of the euphotic zone. Living shells were not observed on their original substrate. Probably, *S. globulus* lives mainly on ephemerous algae present only during short periods of the year, in mid-winter.

The foramina in *S. globulus* are small, without lips and irregularly distributed along the sutures of each chamberlet on the previously formed shell (Fig. G.34 a). After the formation of a new chamberlet layer they are transformed into stolons connecting the chamberlets of two successive growth stages in almost tangential

Fig. G.33 a–f. Competition for space and nutrition of encrusting *Acervulina inhaerens*. **a** Ultimate and penultimate peripheral chambers forming imbricated network adapted to quick radial growth. **b** Detail of ultimate chamber showing multiple apertures on shell periphery *(arrows)*. **c** *A. inhaerens* overgrowing coralline red algae by forming bridge with last few chambers. **d** Ultimate chamber overgrowing bryozoan colony. Scale bars 0.2 mm. **e** Feeding cyst around periphery of *A. inhaerens* in front of marginal aperture *(arrow)*, scale bar 0.01 mm. **f** Detail of feeding cyst showing concentration of coccoliths by digestion of fecal pellets, scale bar 0.002 mm

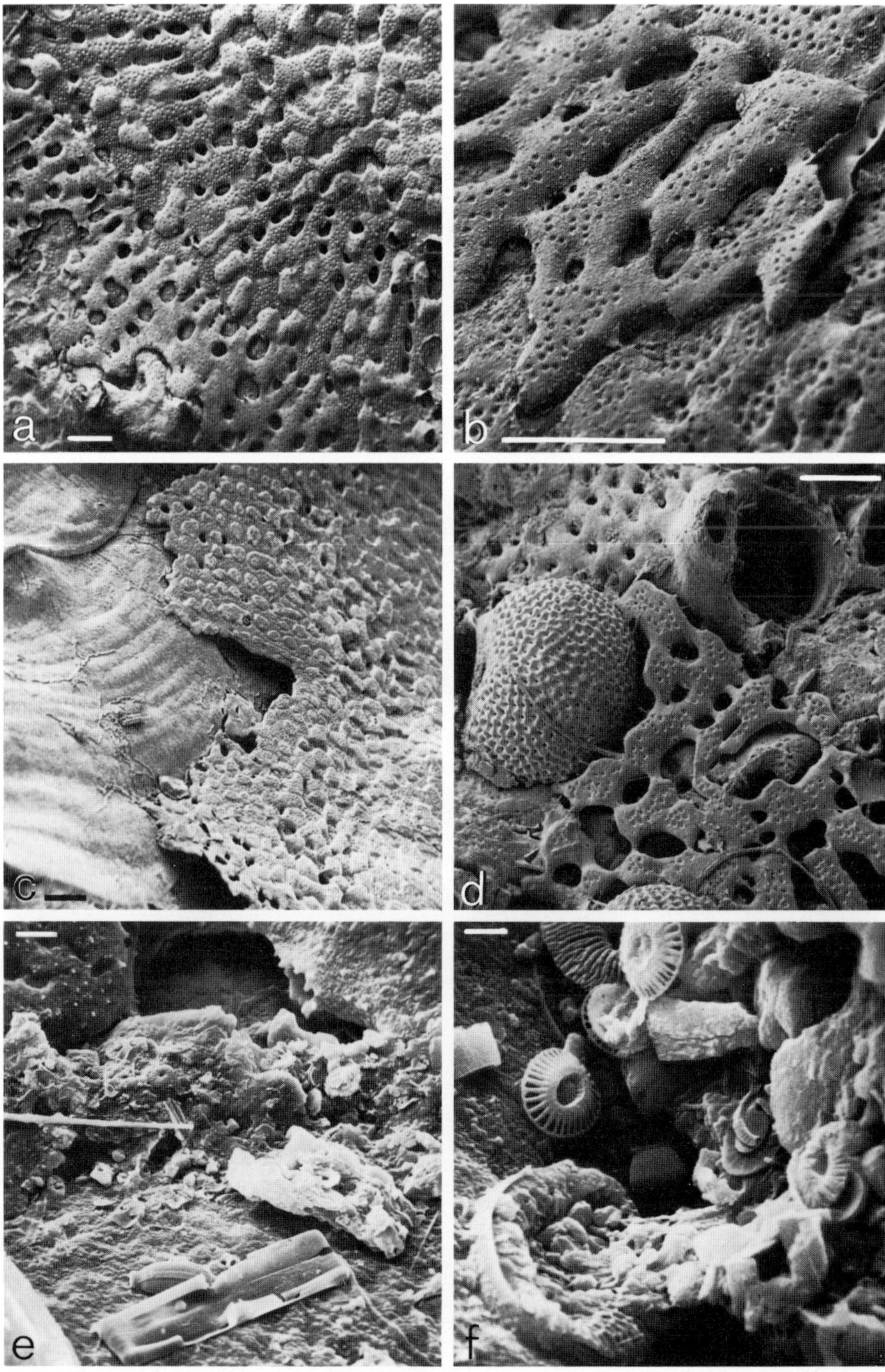

Fig. G.33 a–f

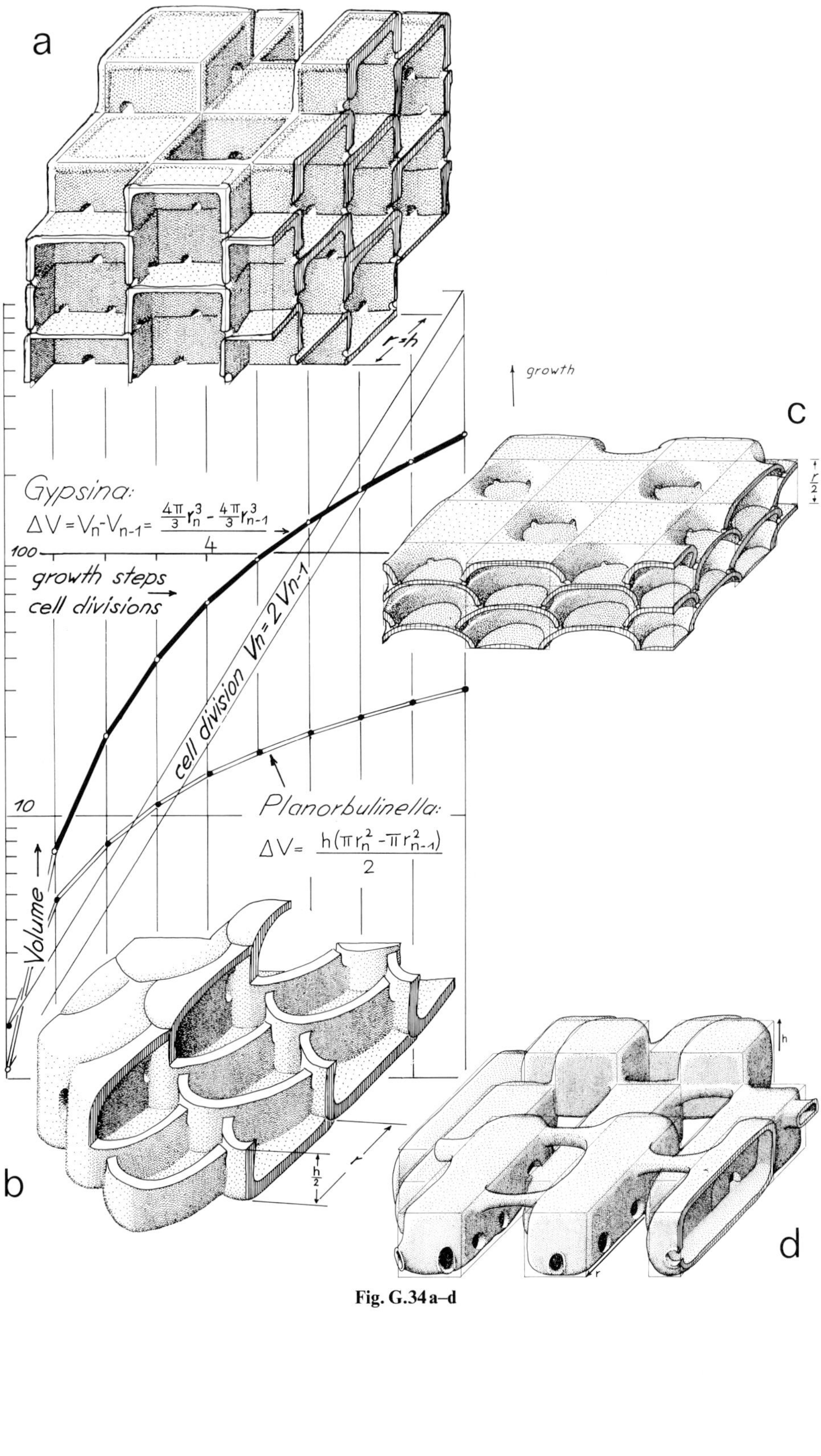
a
r=h
growth
c
r/2
Gypsina:
ΔV = Vn - Vn-1 = (4π/3 rn³ - 4π/3 rn-1³) / 4
100
growth steps
cell divisions
cell division Vn = 2Vn-1
Planorbulinella:
ΔV = h(πrn² - πrn-1²) / 2
10
Volume
b
h/2
r
h
d
r

Fig. G.34 a–d

direction in respect to the globular shell surface. The disposition of the chamberlets may be described as "chessboard pattern" where the free room of half a chamberlet's radial height between a network of "white" chamberlets is filled up by the succeeding network of "black" chamberlets during the next growth step. This mode of growth is similar to the one observed in other acervulinids, but forms a more regular pattern, where most of the chamberlets of each second growth step are superposed in regular radial stacks. Their disposition represents an orbitoidal growth pattern modified to three-dimensional, spheroconcentric shell growth, as exhibited, e.g., by the spherical orbitoid *Torreina* or, in combination with regular orbitoidal growth in the lateral layers, by the flaring median layer of *Omphalocyclus*. Including also the funnel-like invaginations of the lateral shell wall in *Miniacina* and particularly in their attachment discs (Fig. G.34 b), these patterns reflect a common strategy of growth consisting in a minimization of the amount of new protoplasmic volume necessary to realize complete rings or layers with chamberlets of a specific volume. This is to keep rates as linear as possible during growth of large-sized shells (Fig. G.34 c).

Arborescent perforate Foraminifera are represented in the Gulf of Aqaba by all three currently known genera: *Miniacina* is defined by cylindrical invaginations of the lateral wall of large chambers, covering large parts or all the surface of the previous shell (Fig. G.34 b). The invaginations form hollow "pillars" when covered by the next chamber. Similar structures in *Homotrema* are imperforate and solid. The shell of *Sporadotrema* is an arborescent, branching build-up of very numerous alternating chambers with a peculiar, coarse perforation where usually four original pores of the primary wall unite to form a large composite pore after deposition of secondary lamellae.

Origin and significance of the red color in the mineralized shell are unknown. Colorless shells of *Miniacina* are found occasionally, but it is not known if this is due to post-mortem alteration of the shell or some kind of physiological defect. In spite of their common and exclusive coloration, their close taxonomic relation within the subfamily Victoriellinae seems questionable because the mode of perforation in the chamber walls, the construction of the megalospheric embryo, the chamber arrangement, the intercameral communication system and the apertural features are different in the three genera. The general arborescent shape of the shell must reflect mechanisms of filter feeding and is doubtlessly an analogy. No symbionts are known in this group.

In shallow waters of the Gulf of Aqaba, the arborescent perforates, mainly *Miniacina* sp., are comparatively rare, living in cracks of plant-covered rock, in hollows below living reef corals or on coarse coral debris buried below several layers of rubble. In the Mediterranean, *Miniacina miniacea* is particularly frequent on shaded rizomes of dense, old *Posidonia* fields, while shallow-water

Fig. G.34 a–d. Chamber arrangement and growth rates of chamber volume according to chessboard **(a, c)** orbitoidal **(b)** or modified chessboard **(d)** pattern as adaptation to K-strategy of life compared to regular cell division (e.g., diatoms) of r-strategists. Growth steps in K-strategists compared to reproduction by cell division in r-strategists. Volume of chamberlet = 1 = volume of cell in r-strategists. *Stereographs* of chamber arrangement, schematic, not to scale: **a** *Sphaerogypsina;* **b** *Planorbulinella;* **c** *Miniacina;* **e** margin of *Acervulina inhaerens*. Note different mode of construction in **a** and **c.** **(a, b** after Hottinger 1982 a)

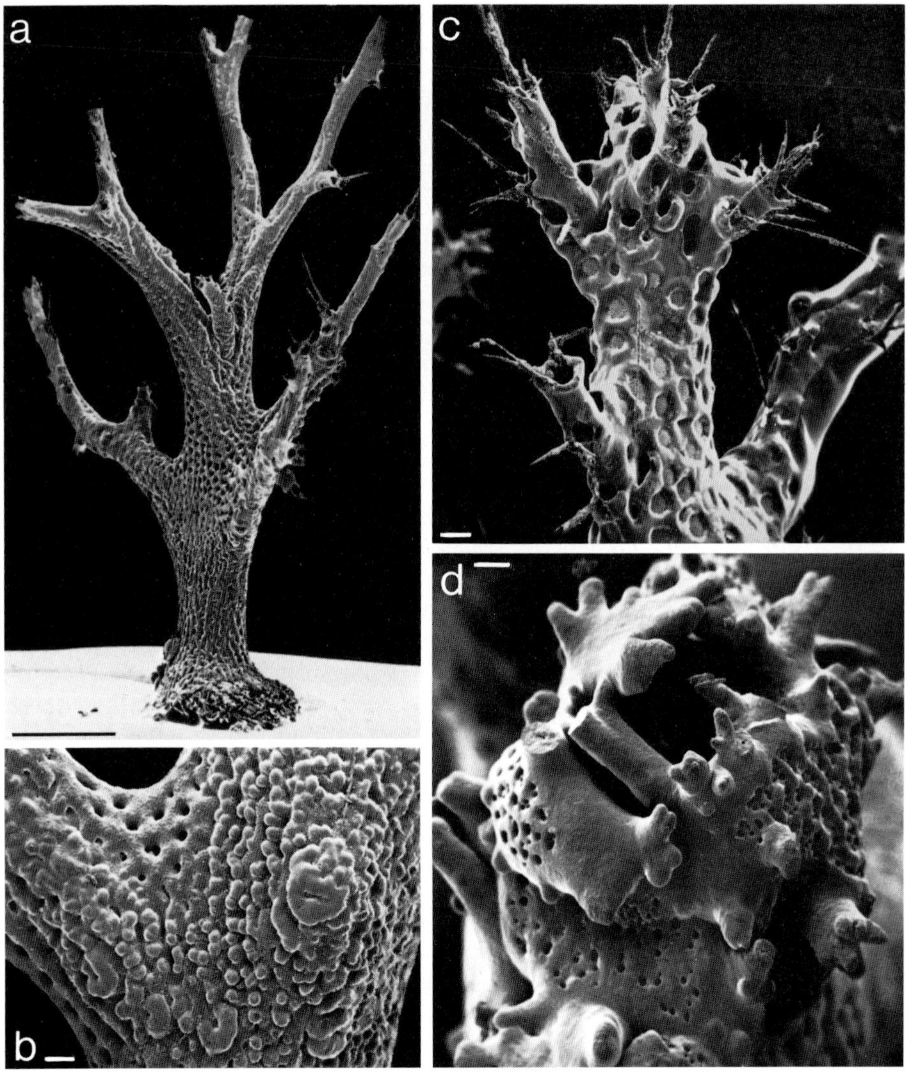

Fig. G.35 a–d. Arborescent perforated Foraminifera, from driftwood off MBL, Elat, 180 m (see Fig. F.10c). SEM. **a** *Miniacina* sp., scale bar 1 mm. **b** Detail with particular ornamentation of deep specimens. **c** *Homotrema* sp., tubular peristomes with glued sponge spicules functioning as fishing rods. **d** *Sporadotrema* sp., peristome (note particular perforation). Scale bars **b–d** 0.1 mm

Sporadotrema sp. from the Maldive Islands in the open Indian Ocean grows from the lower side of dead coral or coral debris downward into coarse sand or rubble layers and is feeding obviously on organisms in the interstitial water. Arborescent perforates occurring in shallow water are usually short, with broad bases and short peristomes. *Homotrema* sp. from shallow Caribbean waters may lose the peristomes, forming flattened crusts or hemispheres with flush multiple apertures.

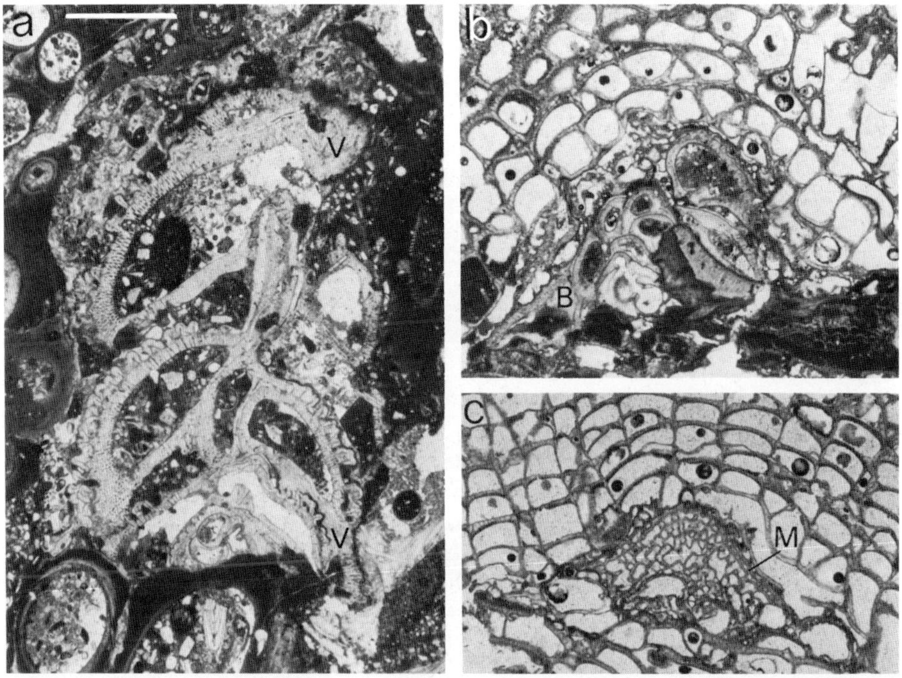

Fig. G.36 a–c. Pioneering, sessile Foraminifera overgrown by subsequent generation of settlers on solid substrate. Thin sections of macroids and bryozoan builds-up. **a** *Victoriella* sp. (*V*), 115 m; **b** *Carpenteria balaniformis* (*B*), 165 m; **c** *Miniacina* sp. (*M*), 165 m. Scale bar 1 mm for **a–c**

In the Gulf of Aqaba, below 90 m in depth, representatives of all three arborescent genera are extremely frequent on hard substrate. At these depths, cryptic life is abandoned and large, high and slender shells with small bases are produced. On hard substrate these arborescent Foraminifera may occupy almost all available space. *Miniacina* sp. and *Homotrema* sp. develop long, tubular peristomes carrying linear sponge spicules glued to the inner rim of the peristome as support for the extruding rhizopods like a fishing rod (Fig. G.35 a). Deep-water *Miniacina* sp. develop particular ornaments on their stem (Fig. G.35 b). It is not known if shallow-water and deep-water representatives of an arborescent genus belong to the same species.

The arborescent perforates are the most conspicuous sessile pioneer community on hard substrate, settling on fresh or unoccupied hard surfaces. They seem to have short life spans in the order of some months in spite of their large size, because they are regularly overgrown by organisms of the first successor community, particularly bryozoans or – if there is still some light available between 90 and 130 m – encrusting acervulinid Foraminifera and coralline algae. The life span of the first successor community must be in the order of 1 yr, being linked to the seasonal cycle as inferred from macroid generation.

Thus, the arborescent perforates are an important frame-builder in or on hard substrates. The frequency of their red fragments in the sediment samples is a very convenient indicator for hard substrate, emerging at a short distance from the

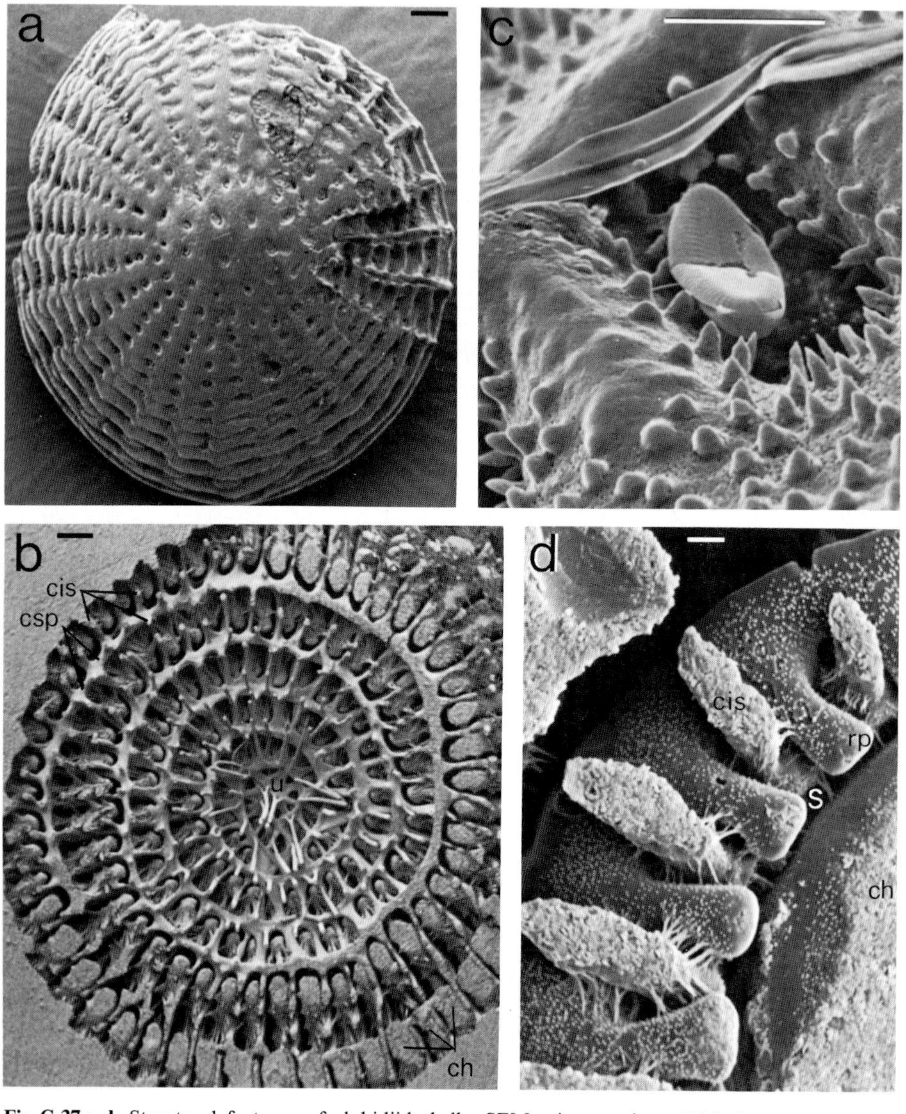

Fig. G.37 a–d. Structural features of elphidiid shells. SEM micrographs. **a** *Elphidium craticulatum*, megalospheric, lateral view showing canal mouths of umbilical and intraseptal canals. **b** Araldite cast of shell cavities, scale bars 0.1 mm. **c** Intraseptal canal mouth in *Elphidium* sp. Spikes keep diatom shell out of canal system. **d** Araldite cast of single chamber cavity, lateral part, in *Elphidium crispum*, scale bars 0.01 mm. *cis* intraseptal canal; *ch* chamber; *csp* spiral canal; *rp* retral process of chamber; *s* septum; *u* center of umbilicus sampling site, when samples have to be taken from surface vessels without visual control. Other representatives of the Victoriellinae, *Carpenteria balaniformis* (Gray 1858), *Victoriella* sp. and/or *Rupertia* sp. have been identified so far only in occasional thin sections of hard substrate from depths below 100 m (Fig. G.36) and seem to play a very minor part as pioneering frame-builders.

The **elphidiids**, represented in the Gulf of Aqaba by at least the species *E. advenum* (Cushman 1922), *E. complanatum* (d'Orb. 1839), *E. craticulatum* (Fichtel and Moll 1798), *E. crispum* (Linné 1758), *E. macellum* (Fichtel and Moll 1798), and *E. reticulosum* (Cushman 1933) produce significant components of the benthic foraminiferal assemblages only in the euphotic zone. The highest frequencies of dead shells in the sediment were observed between 20 and 50 m depth. Many species live in dense mats of short, ramified algal strands of *Cystoseira* habitus.

Elphidiids are defined by a double spiral umbilical canal and retral chamber processes generated by infoldings of the primary wall along the septum (Fig. G.37 a, b). The intercameral foramina arranged as a row of stolons at the base of the chamber are temporarily plugged by organic material assembled in the opening by particular cell organelles. The chamber plasma has a particularly thin, discontinuous organic lining and houses isolated chloroplasts of different types derived from diatoms or from some chrysophycean species. There are no eggholder structures ("pore pits") at the proximal end of the very fine pores in the lateral chamber walls as in all other perforated symbiont-bearing Foraminifera investigated so far.

The restriction of the elphidiids to the photic zone and details of canal anatomy and protoplasm ultrastructure suggest that the chloroplasts are stripped of their cellular envelope outside the chamber cavity by digestive processes and transported through the canal system into the chamber cavity, where they may continue to assimilate. They might be digested at a final stage, possibly prior to reproduction.

Locomotion studies on *Elphidium crispum* in Petri dishes show a peculiar alternation of the animal's position on the substrate when climbing over grits or mounting steps at the bottom of the dish. Such topographic obstacles are mastered by shifting the site of pseudopod extrusion from one side of the shell to the other. With the pull of the pseudopods the shell is turned over its periphery to the opposite side. This might be the first clue as to the significance of large sutural apertures of the canal system in involute shells of bilateral symmetry where the whole shell surface may act as "apertural face." The unusually large openings of the canal system in elphidiids are obviously protected by the large spines inserted radially in the canal walls (Fig. G.37 c) from obstructions caused by larger foreign particles, particularly by the empty frustules of diatoms serving most probably as food. Some rotaliid Foraminifera and all calcarinids have similar spikes in their canal system.

References: Billmann et al. (1980), Frenkel (1974), Halicz and Reiss (1979), Hansen and Dalberg (1979), Hansen and Lykke-Andersen (1976), Hansen and Reiss (1971), Hottinger (1981 a, 1982 a, 1983 a), Hottinger and Leutenegger (1980), Hottinger and Levinson (1978), Leutenegger (1984), Levinson (1976), Noy (1982), Perelis (1974), Perelis and Reiss (1975), Reiss (1959, 1977), Reiss et al. (1961), Said (1949, 1950 a, b), Thomas (1977), Zweig-Strykowski (1973), Zweig-Strykowski and Reiss (1975).

G.3 Significance of Shell Morphogenesis

Overall **shell shape** in benthic Foraminifera such as spheres, cones, discs, lenticular bodies, etc. are generated in different groups by different growth strategies. Patterns of chamber arrangement are to be combined with appropriate chamber shape and appropriate, usually repeatable location of communications between shell cavities and the ambient environment. The chamber cavities represent a **compartmentation** of the protoplasmic body. For physiological reasons the volume of each compartment has to be constant within specific ranges of variation throughout the ontogenetic process. There is a general strategy to keep **growth rates** as linear as possible where constance of shell shape during ontogeny (gnomonic growth) by simple "blow-up" processes would demand exponential growth rates. The growth strategy must be adapted further to dispose the openings for the extrusion of pseudopods from the shell in functional patterns throughout ontogeny. For example, orbitoid alternating chamberlets added concentrically to a discoidal shell and operculinoid spiral chambers with backward inclinations produce both discoidal shells. The disposition of the intercameral foramina repeated in each growth step provides orbitoids with short **internal communications** between shell cavities of successive growth stages and a series of evenly spaced marginal apertures for the **extrusion of the pseudopods,** while in the operculinoid shell the foramina provide only intercameral connections in a linear spiral. Short internal communications and the marginal openings for the pseudopods are produced by an additional cavity system. The latter can be produced only by shells of perforate-lamellar nature where the independent play of inner lining and outer lamella provide the morphogenetic potential to transform intrasutural spaces to canal systems (Fig. G.38). Intrasutural canals serve as recipient for permanently **differentiated pseudopodial protoplasm,** provide "back door" short communications between early and late growth stages and permit a distribution of openings for pseudopods on the shell surface in patterns independent of chamber arrangement. The ecological significance of this morphogenetic potential may be illustrated by the following example. In all Foraminifera, the chamber protoplasm retracts to the interior of the shell when irritated by chemical or physical perturbations in the ambient environment. The protoplasm in the external chambers is replaced by liquids (provided by the vacuoles) which constitute some kind of buffer or barrier between the ambient environment and the living animal (Fig. G.39). In single-layered shells the sites of pseudopod extrusion communicate with the cavities of the external chambers. When these are emptied, no protoplasm can extrude from the shell and the animal is immobilized. In canaliferous animals, the intrasutural space provides a "back door" communication to each chamber, connecting it directly with the ambient environment. Such animals keep up their motility during retraction of the chamber protoplasm and move away from the site of irritation. Incidentally, larger porcelaneous benthics live on plants or other substrates washed clean by the movements in the open water, while most canaliferous Foraminifera live close to the bottom where the activity of microorganisms may produce areas of unhealthy microclimate.

The geometry of the canal system reflects thus a specific function. Fossil shells of very different modes of shell construction, e.g., trocholinids, rotaliids, elphidi-

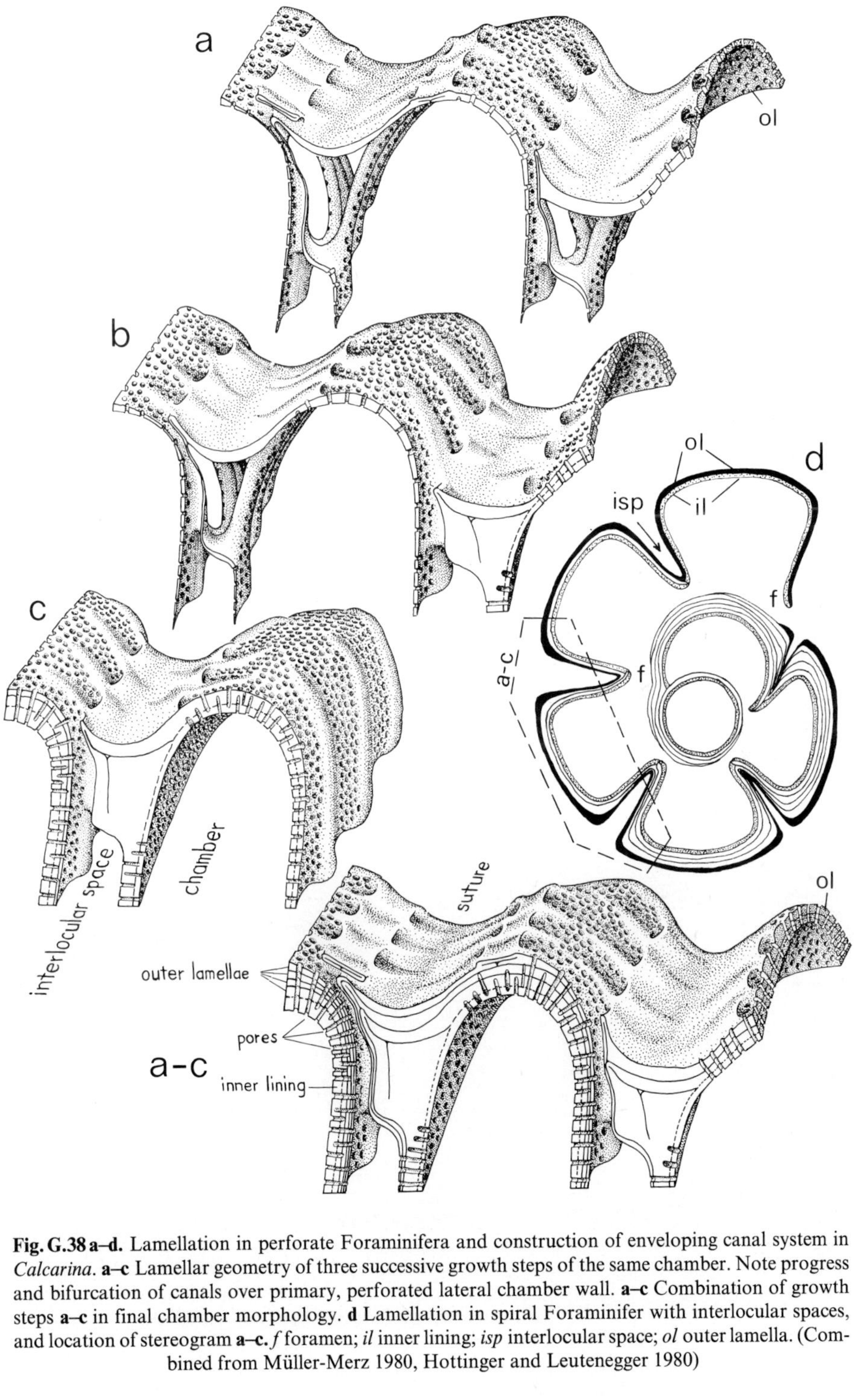

Fig. G.38 a–d. Lamellation in perforate Foraminifera and construction of enveloping canal system in *Calcarina*. **a–c** Lamellar geometry of three successive growth steps of the same chamber. Note progress and bifurcation of canals over primary, perforated lateral chamber wall. **a–c** Combination of growth steps **a–c** in final chamber morphology. **d** Lamellation in spiral Foraminifer with interlocular spaces, and location of stereogram **a–c.** *f* foramen; *il* inner lining; *isp* interlocular space; *ol* outer lamella. (Combined from Müller-Merz 1980, Hottinger and Leutenegger 1980)

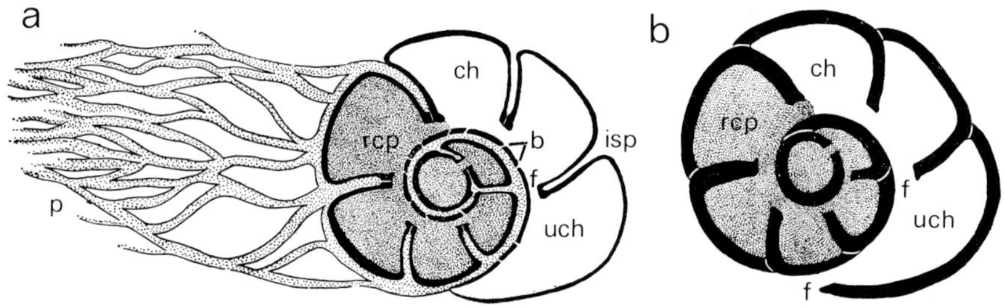

Fig. G.39 a, b. Motility in canaliferous Foraminifera during periods of retracted protoplasm after irritation. **a** Motile form with a canal system; **b** simple shell without canal system. Schematic, not to scale (Hottinger 1977 b). *b* Passages connecting chamber lumen with canal system; *f* foramen (direct communication between successive chamber lumina); *isp* interlocular space transformed to canal system; *p* pseudopods; *rcp* retracted chamber plasma; *uch* ultimate chamber

ids, nummulitids, calcarinids and siderolitids, were tested as to the geometric pattern of their canal system. In all cases, the geometric patterns meet the functional requirement of "back door" connection between all chamber cavities and the ambient environment (Fig. G.40).

Thus, the morphology of the shell has to satisfy "internal" demands of optimal growth rates, optimal compartment volumes and short communications. Shell shape combined with distribution pattern, location or shape of the functional openings for pseudopod extrusion must satisfy "external" demands of locomotion, adaptation to substrate, food uptake, light capture or other activities of autecological relevance. The larger the animal the more important are the

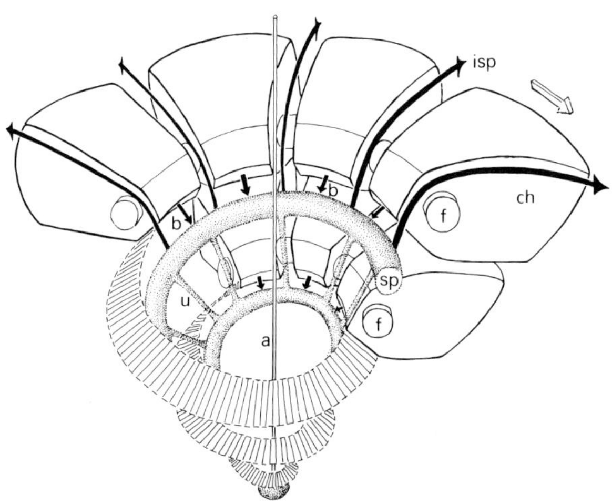

Fig. G.40. Geometric requirements to fulfil canal system functions in a trochospiral Foraminifer, schematic, not to scale (Hottinger 1978 a). *a* Coiling axis; *b* umbilical connexion chamber lumen-canal system; *ch* chamber lumen; *f* foramen; *isp* interlocular space; *sp* spiral canal; *u* umbilical subsutural canal; *white arrow* direction of growth; *black arrows* direction of protoplasmic flow through interlocular space; *hatched spiral surface* idealized cover of spiral suture by subsequent shell whorl transforming suture into spiral canal adaptations of the shell to external demands. On the other hand, nature of the shell walls, chamber arrangement and peculiarities of the apertural face are predetermined by the genetically controlled morphogenetic process and reflected by taxonomic units. Chamber shape, ornamentation and wall thickness may be subject to intraspecific variation. Genetically controlled morphology reflects environmental conditions by the presence or absence of a taxonomic unit in a particular environment, while intraspecific variation may reflect ecologic gradients in the environment by physiological mechanisms as, for instance, biocalcification processes.

References: Hallock and Hansen (1979), Hansen and Reiss (1971), Hottinger (1977b, 1978a, 1981b, 1982a, 1983a, b), Hottinger and Dreher (1974), Hottinger and Leutenegger (1980).

G.4 Significance of Symbiosis

Symbiosis in Foraminifera is widely distributed among the porcelaneous-imperforate and the lamellar-perforate taxonomic units. Virtually nothing is known about symbiosis of agglutinated Foraminifera. Symbiosis is neither restricted to large-sized animals, nor to benthic or tropical forms. While symbiontic relationships have been studied to various degrees of detail in larger benthic und planktic (see Chap. E) hosts, the presence or absence of symbionts in many smaller benthics, particularly in smaller miliolids, is not yet known. There is no secure criterium available at present to detect symbiosis by specific morphological elements in the empty shell.

Many **different types of symbionts** occur in foraminiferal hosts (Fig. G.41). These types are defined by ultrastructural peculiarities of the chloroplast and its pyrenoid as seen under the TEM in situ. Usually, a singular type of symbiont was found in all specimens of one particular species irrespective of season, depth or region characterizing the host's provenance. The same type of symbiont may be found in more than one, often closely related species, but there is no strict correlation between the taxonomic system of the hosts and the structural types of the symbionts. Some symbionts have been specifically identified within the taxonomic system of unicellular algae by extracting the naked symbiont cells from the host for separate culturing where they produce taxonomically relevant shells. Taxonomic identification of the symbiont does not improve correlations with the host's taxonomic position. Pigment analysis has shown little difference between different structural types of symbionts. This seems to be reflected by rather indistinct differential depth distribution of the hosts according to different symbiont types (Fig. G.42). Moreover, the upper limits of the host's depth distribution subjected to photoinhibition might reflect behavioral patterns such as negative phototaxis, rather than light requirements of the symbionts. Of course, the symbiont bearers are restricted to the euphotic zone (130 m in the Gulf of Aqaba). No species having symbionts in the photic zone and lacking symbionts below the photic zone is known either in the Gulf or elsewhere.

Suborder	Superfamily	Family	Subfamily	Foraminifer species	Locality	Year of collection and depth at collecting site	Algal type see legend	Algal class
Rotaliina (perforated foraminifers)	Nonionaceae	Nonionidae		NONION germanicum	North Sea, Danmark	?:0-0.5m	Chl.	chloroplasts
				NONIONELLA stella	Pacific, California	1978:29m	〃	
	Rotaliaceae	Elphidiidae		ELPHIDIUM excavatum	North Sea, Danmark	?: 6m	〃	
				E. williamsoni	North Sea, Danmark	?: 0-0.5m	〃	
				E. crispum	Mediterranean Sea	1976:1-5m	〃	
				E. macellum aculeatum	Mediterranean Sea	1976:1-5m	〃	
				Elphidium sp.	Red Sea	1978:5m	〃	
				E. craticulatum	Pacific, New Caledonia	1981:18m	〃	
		Calcarinidae		CALCARINA calcar	Red Sea	1978:5m	B1	Diatoms
				C. spengleri	Pacific, Japan	1979:shallow	〃	
				BACULOGYPSINA sphaerulata	Pacific, Japan	1979:shallow	〃	
		Nummulitidae	Cycloclypeinae	OPERCULINA ammonoides	Red Sea	1971:30-50m	B3 d	
				HETEROSTEGINA depressa	Red Sea	1971,73,74:30-90m	〃	
					Indic, Maldives	1977,80:?,50m	〃	
					Pacific, Hawaii	?	〃	
				HETEROCYCLINA tuberculata	Red Sea	1973:90m	〃	
				CYCLOCLYPEUS carpenteri	Indic, Maldives	1977:70m	〃	
			Numm.	NUMMULITES cumingii	Indic, Maldives	1977:?	B3 e	
	Discorbaceae	Amphisteginidae		AMPHISTEGINA lobifera	Mediterranean Sea	1974:38m	B2 a	
					Red Sea	1973,74,78:1-10m	〃	
				A. lessonii	Red Sea	1971,78:3-45m	〃	
					Pacific, New Caledonia	1981:20,55m	〃	
					Pacific, Hawaii	?:10m	〃	
				A. bicirculata	Red Sea	1978:20-68m	〃	
					Indic, Maldives	1977,80:?,50m	〃	
				A. papillosa	Red Sea	1972,76,78:60-68m	B2 b	
				A. radiata	Indic, Maldives	1977,80:?,50m	B4	
					Pacific, New Caledonia	1981:18m	〃	
				unidentified amphisteg.	Pacific, New Caledonia	1981:18m	B3 c	
Miliolina (imperforated foraminifers)	Miliolaceae	Alveolinidae		BORELIS schlumbergeri	Red Sea	1978:10-35m	B3 a	
				ALVEOLINELLA quoii	Indic, Maldives	1977:45m	B3 b	
		Soritidae	Soritinae	SORITES orbiculus	Mediterranean Sea	1974:20-40m	D1	Dinophyceans
					Red Sea	1973,74:30,10m	〃	
				S. "orbitolitoides"	Red Sea	1973,74:10,30m	〃	
				S. marginalis	Atlantic, Florida	1974:shallow	D1 ?	
				AMPHISORUS hemprichii	Red Sea	1973,74,78,80:5-40m	D1	
					Indic, Maldives	1977:5,45m	〃	
					Pacific, Hawaii	?	(D)	
				MARGINOPORA vertebralis	Pacific, Australia	1970:8-15m	〃	
			Archaiasinae	CYCLORBICULINA compressa	Atlantic, Florida	1974,?:shallow	C1	Chlorophyc.
				ARCHAIAS angulatus	Atlantic, Florida	1974,?:shallow	C2	
			Peneroplinae	PENEROPLIS proteus	Mediterranean Sea	1976:3m	C2	
				P. planatus	Mediterranean Sea	1974:38m	R1	Rhodophyceans
					Red Sea	1974:10m	〃	
				P. pertusus	Mediterranean Sea	1976:3m	〃	
					Pacific, Hawaii	?	(R)	
				SPIROLINA arietina	Pacific, Hawaii	?	(R)	

Fig. G.41. Symbiosis in Foraminifera, critical review (Leutenegger 1984)

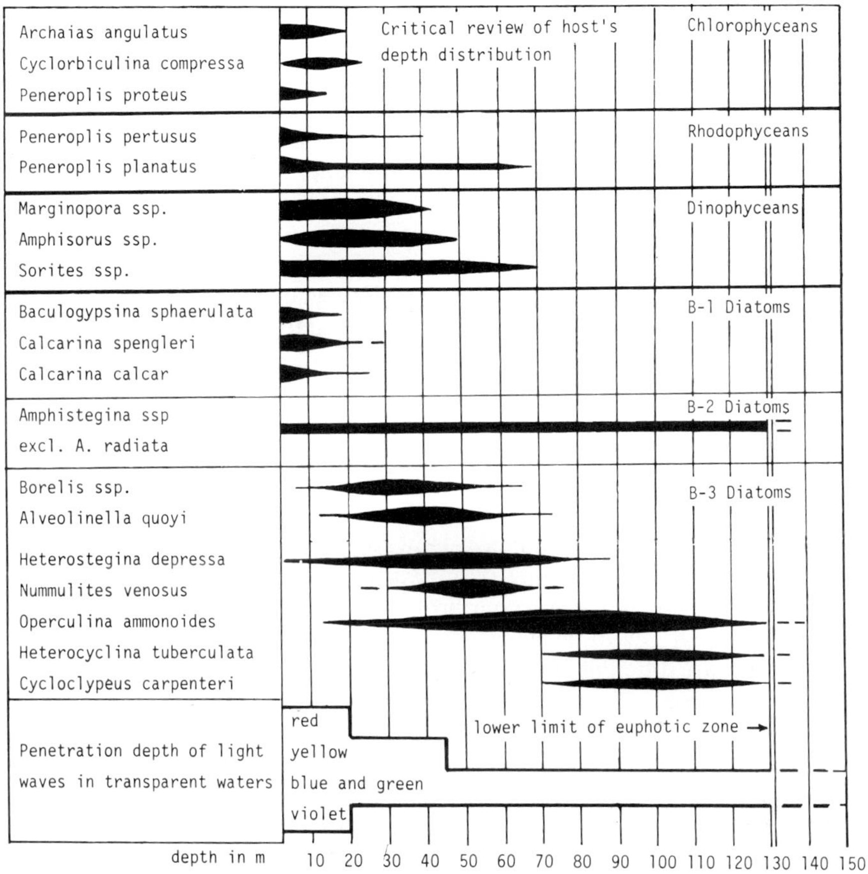

Fig. G.42. Depth distribution of selected larger Foraminifera grouped according to ultrastructural type of chloroplast in symbionts and compared to penetration depth of light with different wave lengths. Note behavior of hosts of diatom symbionts in shallow water living in shade of cavities *(Heterostegina)* or below sediment–water interface (alveolinids, calcarinids)

As a **nutritional factor,** symbiosis seems to play different roles in different hosts. *Heterostegina depressa* was reported to live to a large extent from symbiosis, as do many corals closely associated with this particular species, while amphisteginids and soritines are known to take up algal food and to produce large feeding cysts in spite of their harboring many symbionts. In *Peneroplis planatus* the rhodophycean symbionts produce starch grains occurring as free bodies in the host's chamber protoplasm. It is not known if the free starch grains are available to the host as storage product or if they are waste left over from the digestion of the symbiont by the host. It is not known either if the lipid droplets observed in most host protoplasts and interpreted as storage products are produced by the symbionts, or if and to what extent their eventual generation by the host depends on symbiosis.

In elphidiids, the chamber plasma houses **free chloroplasts** stripped from their cellular envelope by partial digestion outside of the shell. The significance of this particular kind of symbiosis is not understood at present, as the light-dependent depth distribution of the host may reflect either the need of light for continuing photosynthetic activity of the chloroplasts or the host's dependence on a specific food source needing light to grow in the ambient environment. The absence of egg-holder structures below the pores of extremely small diameter and the even distribution of the chromatophores in the chamber protoplasm suggest a relative independence of the chloroplasts from gas exchange. This may point again in opposite directions: either the chloroplast has stopped or reduced respiration before being digested by the host or its photosynthetic activity continues and is totally integrated in the biochemical processes of the host protoplasm as in its proper cell.

Thus, most of the different roles played by symbiotic processes in Foraminifera involve feeding processes reflected by $\delta^{13}C$ values in the carbonate deposited in their shell (see below). A direct connection between **biomineralization rates** and **photosynthetic rates** seems questionable, as calcification continues in light when photosynthesis is artificially stopped. However, as symbiosis is a nutritional factor for the Foraminifera, rising photosynthetic rates must enhance metabolic rates in general and produce higher growth rates, which implies the quickening of calcification. It has to be expected that the quantities produced by the symbiontic factor are different for each particular nutritional pattern, while the qualitative dependence on light might be finally similar in all Foraminifera bearing symbionts.

Larger Foraminifera with symbionts proliferate in tropical zones of low productivity all over the world, where they are associated with coral reefs representing a strategy of life designed to recycle nutrients in the presence of a large energy source from light under conditions of high water temperatures. Muscatine and Porter (1977) point out the role of coral symbionts in **nitrogen recycling** by summarizing the available evidence for uptake and retention of the host's excretory nitrogen by the zooxanthellae and for the translocation of nitrogen-bearing compounds (e.g., alanine and leucine) from the symbiont to the host. Although there is no direct proof for similar relationships in the symbiosis between autotrophic algae and heterotrophic Foraminifera, the circumstantial evidence provided mainly by distribution patterns of the hosts is strong enough to admit such relationships as a working hypothesis, pointing to by far the most important ecological significance of symbiosis in larger Foraminifera, i.e., as an adaptation to nutrient-poor environments.

References: Hansen and Lykke-Andersen (1976), Hottinger (1982a), Kremer et al. (1980), Lee (1980), Lee and Bock (1976), Lee (1983), Lee et al. (1979a, b, 1980a, b, c), Leutenegger (1977a, c, 1983, 1984), Muscatine and Porter (1977), Schmaljohann and Röttger (1978).

G.5 Stable Isotopes and Related Problems

The stable isotope composition of skeletal carbonate, especially of Foraminifera, has become a major tool in paleoceanographic and paleoecological studies. Oxygen isotope ratios have been widely studied and successfully used in the interpretation of the thermal history of the oceans, of past fluctuations in sea level and ice volume, evaporation precipitation rates and density stratification. Carbon isotope ratios in skeletal material, although less well studied, are used in interpreting the organic fraction of the total CO_2 in the ambient water; hence the latter's fertility and, implicitly, dissolved oxygen levels. Problems arise in interpreting both oxygen and carbon isotope ratios because of incorporation into the foraminiferal shell of metabolic CO_2 derived from photosynthetic activity of endosymbionts hosted by many benthic and planktic species and/or from food. Furthermore, skeletal carbon isotope ratios are influenced by the nature (marine or terrestrial) and diagenetic stage of the organic material oxidized and incorporated through the bicarbonate of the ambient water. Additional problems arise because of changes in oxygen and carbon isotope disequilibrium values in skeletal carbonate due to growth rates, metabolic activity and seasonal changes of the environment during ontogeny, as well as to specific departures from equilibrium values ("vital effect") explained in part by chemical composition of the skeletal carbonate (e.g., magnesium content), but actually not well understood. [For recent reviews of the progress and problems in isotope research see Berger et al. (1981), Savin and Yeh (1981), Vincent and Berger (1981)].

The unique setting in terms of temperature, salinity, seasonality and light penetration make the Gulf of Aqaba an ideal environment for the study of stable isotopes and related aspects in Foraminifera.

From the very early phases of the ecological micropaleontological research in the Gulf of Aqaba, attention was attracted to the symbiont-bearing Foraminifera, particularly through laboratory studies on photo- and thermotactic responses of *Amphistegina,* attributed at least in part to endosymbionts. Subsequent depth-distribution studies in the Gulf of different species of *Amphistegina* (and other symbiont-bearers) seemed to support the assumption that light requirements by symbionts determine this distribution. Biometrical studies on inter- and intraspecific variations in the morphology of *Amphistegina* species pointed to adaptations of thickness/diameter ratios (flatness) to light levels. The degree of involution of *Operculina* was equally regarded as correlatable with depth. Detailed measurements of lamellar thickness in *Amphistegina* revealed correlation (thinning) with the depth gradient, accounting in part for the morphological variation of the shell and raising the problem of calcification rates as possibly related to symbiosis and light gradient.

The apparent correlation between depth distribution of morphotypes and the thickness of their individual shell lamellae, on the one hand, with light gradient, hence implicitly with photosynthetic activity of endosymbionts on the other, led to a first attempt to investigate **stable isotope composition** in the foraminiferal shells in the Gulf of Aqaba. Symbiont-bearing species of *Amphistegina, Operculina* and *Heterostegina,* as well as symbiont-free *Bolivina* and *Planorbulinella,* sampled together with molluscan shells over the upper 150-m range, were exam-

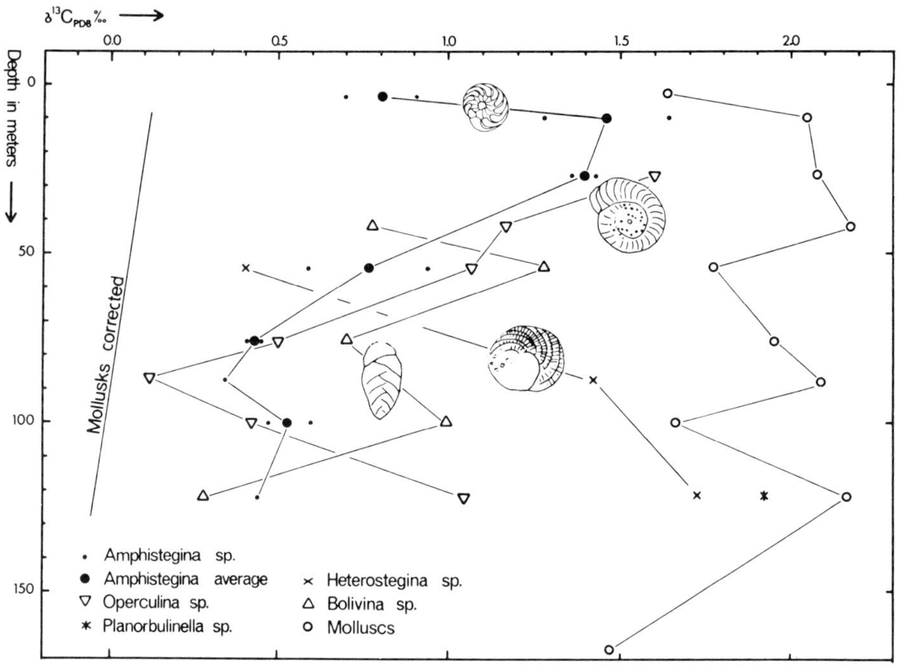

Fig. G.43. Corrected $\delta^{18}O$ variation of shell carbonate with depth. Analytical precision better than ±0.1‰ (Buchardt and Hansen 1977)

ined (Fig. G.43). $\delta^{18}O$ values in molluscan shells were close to calculated equilibrium and did not vary significantly with depth. Correcting for temperature effects on carbonate $\delta^{18}O$, the values point to a nearly constant isotopic composition of the sea water and dissolved bicarbonate in the upper 150 m (cf. also the direct water measurements carried out later; Chap. D). Compared to the molluscan values the symbiont-free Foraminifera were shown to be depleted in ^{18}O by about 0.6‰ (vital effect) with very little variation over the whole depth range. Symbiont-bearing species exhibited an additional depletion of up to 1.5‰ (for a total of about 2.0‰), the magnitude of depletion decreasing with depth and differing from genus to genus. This strongly suggested the incorporation of metabolic CO_2 of the algae into the shell carbonate and a correlation between light-dependent photosynthetic rates and oxygen isotopic composition. Although the carbon isotopic composition of the foraminiferal shells indicated disequilibrium with the dissolved bicarbonate over a wide range, no significant differences between symbiont-bearing and symbiont-free species was observed. The data suggested even a general negative correlation of $\delta^{13}C$ with depth (Fig. G.44).

It should, however, be emphasized that more recent measurements of the total dissolved CO_2 and the carbon isotopes in the waters of the Gulf (Shemesh et al. 1981), as compared with calculated equilibrium values (Emrich et al. 1970), indicate that molluscs are not in carbon isotopic equilibrium; this does not affect the trends mentioned above.

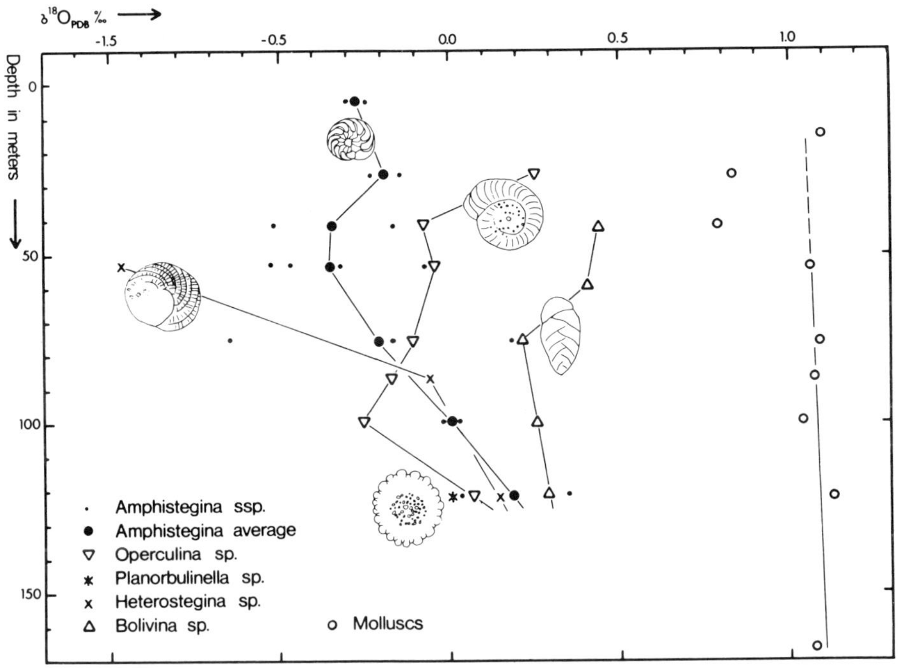

Fig. G.44. $\delta^{13}C$ variation of shell carbonate with depth. Analytical precision better than ± 0.1‰ (Buchardt and Hansen 1977)

A study on shells of living epiphytic *Amphisorus hemprichii* collected in the Gulf of Aqaba at 4 m depth over a whole year for a population dynamics study (see above) and examined for oxygen and carbon isotopes is of interest in this respect (Fig. G.45 a–d). For comparison, the isotopic composition of the sea water and of the dissolved bicarbonate were analyzed with respect to the same machine reference as used for the carbonate percentage. $MgCO_3$ analysis was carried out on the shells by atomic absorption spectrophotometry in order to determine equilibrium composition with regard to water and bicarbonate. A model curve for oxygen isotopic composition was used, to which the data were compared. The plot of the data shows a remarkably good fit with the exception of juveniles which are in disequilibrium. Part of the deviations from the model curve can be explained by nonuniform growth rates, expressed also by the growth cycle of the populations (there is very good correlation between diameter and weight, i.e., calcium carbonate addition; see previous subchapters).

It follows that the symbiont-bearing, porcelaneous *Amphisorus* deposits its shell in oxygen isotopic equilibrium with the sea water.

Further studies at different localities at 10 m depth in the Gulf of Aqaba confirmed the observation that porcelaneous symbiont-bearing genera (*Amphisorus, Sorites, Peneroplis*) deposit their shells in oxygen isotopic equilibrium (showing the temperature effect), while *Amphistegina* shells are in disequilibrium by 0.7–1.0‰ $\delta^{18}O$, with slight seasonal variation as a function of temperature.

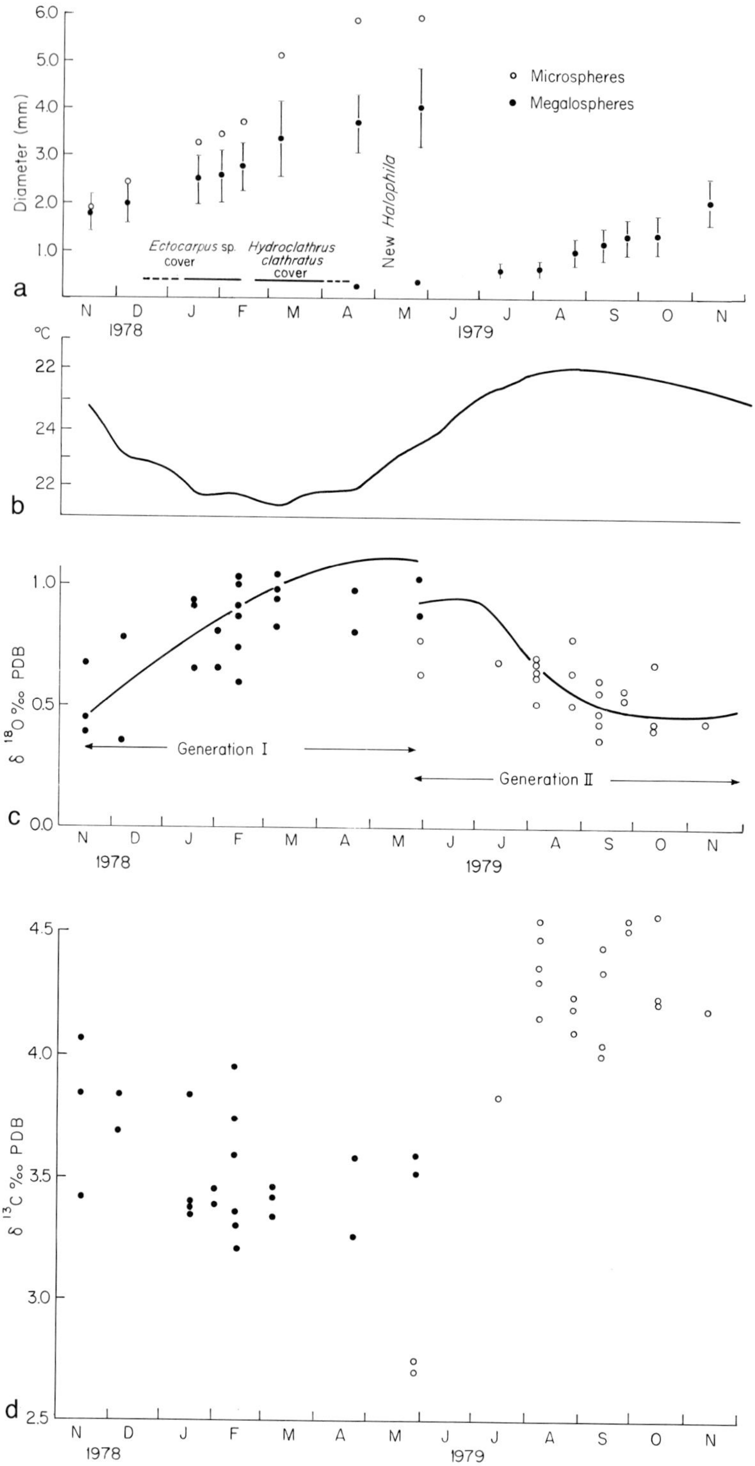
Diameter (mm)
Microspheres
Megalospheres
New Halophila
Ectocarpus sp. cover
Hydroclathrus clathratus cover
°C
δ ¹⁸O ‰ PDB
Generation I
Generation II
δ ¹³C ‰ PDB
N D J F M A M J J A S O N
1978
1979
a
b
c
d

Fig. G.45 a–d

Most of the results of the carbon-isotope analyses of *Amphisorus* fall within the expected equilibrium range or above it. Again, the only exception is embryonic stages which are about 4‰ lighter than equilibrium calcite. $\delta^{13}C$ values of the nepionic-neanic-ephebic stage show lower values during the colder season and in older specimens.

This trend is comparable to that of *Cyclorbiculina* from Bermuda. *Marginopora vertebralis* from the Philippines shows an opposite trend, with lower $\delta^{13}C$ in summer, but not toward the end of the ephebic stage (Wefer and Berger 1980), thus indicating that the variations depend on the seasonal change in environmental conditions and less upon the ontogenetic stage. It seems that changes in the microenvironment control part of the $\delta^{13}C$ variations. During spring and summer the substrate *Halophila* in the Gulf of Aqaba is blooming and ^{13}C is enriched in the dissolved HCO_3^- near the leaves due to preferential consumption of ^{12}C for photosynthesis. This may explain the higher than equilibrium values recorded at that period. In contrast, $\delta^{13}C$ is within equilibrium range in winter, when filamentous algal mats cover the *Halophila* meadow and light penetration is restricted. At the same time growth rates also slow down. Another possibility is the availability of food and the seasonal changes in $\delta^{13}C$ of the latter (e.g., diatoms are very abundant in winter). Thus, rate of metabolism, as well as the $\delta^{13}C$ of metabolic CO_2 incorporated in the shell, may change seasonally and affect the isotopic composition of the shell.

The role of microenvironment is shown by studies on seasonal changes in isotopic composition of *Amphisorus, Sorites,* and *Peneroplis* at different localities in the Gulf of Aqaba and at the same depth (4 m); specimens of these genera are enriched in ^{13}C when living attached to *Halophila* leaves, but depleted (up to 1‰) when attached to stems. Again, either photosynthetic uptake or different food sources on the two substrates may be responsible for these differences.

A stable oxygen and carbon isotope study on the symbiont-bearing *Operculina ammonoides* in the Gulf of Aqaba was carried out in conjunction with examination of specimen frequency and biometry of protoconch (Fig. G.46). It is noteworthy that, while both involute and evolute specimens occur over a depth range from 10 to 150 m, the involute type is more frequent on *Halophila*/coral substrates, while the evolute type is more frequent on muddy bottom (see also Chap. F). *Operculina* is shown to be grossly out of equilibrium, the $\delta^{18}O$ being

Fig. G.45. a Annual growth cycle of megalospheric and microspheric *A. hemprichii* generations, expressed by temporal changes in mean diameter (standard deviations of microspheric forms' mean diameter are not included, because of their low frequency in the population). Seasonal algal cover over the *Halophila* meadow is indicated (Zohary et al. 1980). **b** Water temperature during the sampling period and $\delta^{18}O$ of *A. hemprichii*. *Closed circles* generation I; *open circles* generation II. **c** *Solid curve* model-derived equilibrium-$\delta^{18}O$, i.e., $\delta_t = \frac{\delta_{t'} \times W_{t'} + \delta_T(W_t - W_{t'})}{W_t}$, where δ_t is the integrated equilibrium $= \delta^{18}O$ t days after start of the generation; δ_T is the $\delta^{18}O$ of equilibrium calcite formed at temperature T °C calculated with the generation: $\delta^{18}O_{cq}$ (‰PDB) $= \delta^{\Gamma\Psi\theta}{}_{water}$ (‰SMOW) + 15.93-(130.95 + 7.69 T) + 0.06 m (M = $MgCo_3$ content in mole %); W_t and $W_{t'}$ are estimates of individual weight, t and t′ days after start of the generation, where t′ corresponds to the immediately preceding sampling period (Luz et al. 1983). **d** $\delta^{13}C$ of *A. hemprichii*. *Closed circles* generation I; *open circles* generation II. The equilibrium range is 3.2–4.1‰ (Luz et al. 1983)

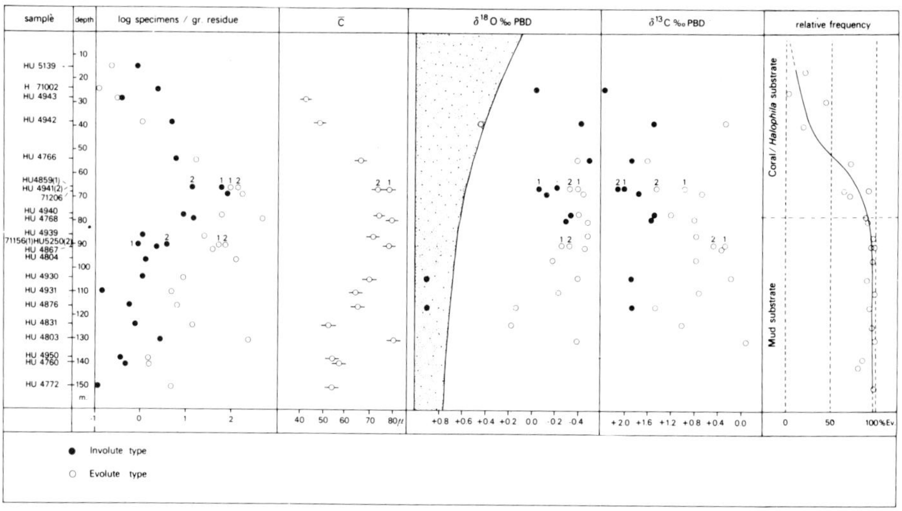

Fig. G.46. Relations with depth of the various parameters of *Operculina* in the Geziret Fara'oûn area. *Solid circles* involute type; *open circles* evolute type; *left to right* log of the number of specimens per gram residue, diameter of the protoconch of the evolute type ± 1 SE, $\delta^{18}O$ values, $\delta^{13}C$ values, and relative frequencies of both types of *Operculina*. *Shaded area* in the column represents the interval between winter and summer $\delta^{18}O$ equilibrium values of the carbonate [equation of Epstein et al. (1953), $\delta^{18}O$ of seawater of $+2\%$, Nissenbaum in: Erez (1978)]. *1* and *2* refer to different samples from the same depth (Fermont et al. 1983)

on the average by 1‰ too low. There is a modest negative correlation between $\delta^{18}O$ and number of specimens per sample on one hand and a very distinct correlation between $\delta^{18}O$ and protoconch size. Since both the number of specimens and the size of the protoconch probably depend on the productivity of *Operculina* (with a maximum at about 60–70 m) and since maximum productivity was assumed to be related to maximum photosynthetic activity, it was concluded that the strongest depletion in ^{18}O reflects this latter activity. The $\delta^{13}C$ values did not correlate with any of the investigated parameters, but fell into two groups, involute *Operculina* having a more positive $\delta^{13}C$ than evolute ones (Fig. G.47). These differences were explained most probably as a result of different food requirements, or different foraging tactics of the two forms, or of differences in the microflora serving as food on coral/*Halophila* and mud substrates respectively.

The influence of microenvironment and apparently of shell-building pattern on stable isotope composition is also shown in studies on symbiont-free benthic Foraminifera from Quaternary cores in the Gulf of Aqaba (see Chap. H). As far as oxygen isotopes are concerned, the porcelaneous (miliolid) species are at or close to equilibrium, while perforate-lamellar ones (buliminids, cibicidids) are in disequilibrium. Carbon isotopes seem to reflect mainly habitat (epi- or in-faunal) and possibly food quality, as well as the organic CO_2 in the deep and in the interstitial waters respectively, corresponding to dissolved oxygen levels.

Radioactive tracer (^{14}C and ^{45}Ca) in situ experiments on *Amphistegina* living at a depth between 5 and 30 m in the Gulf of Aqaba indicated correlation between photosynthetic rates of the symbionts and calcification rates of the host, with

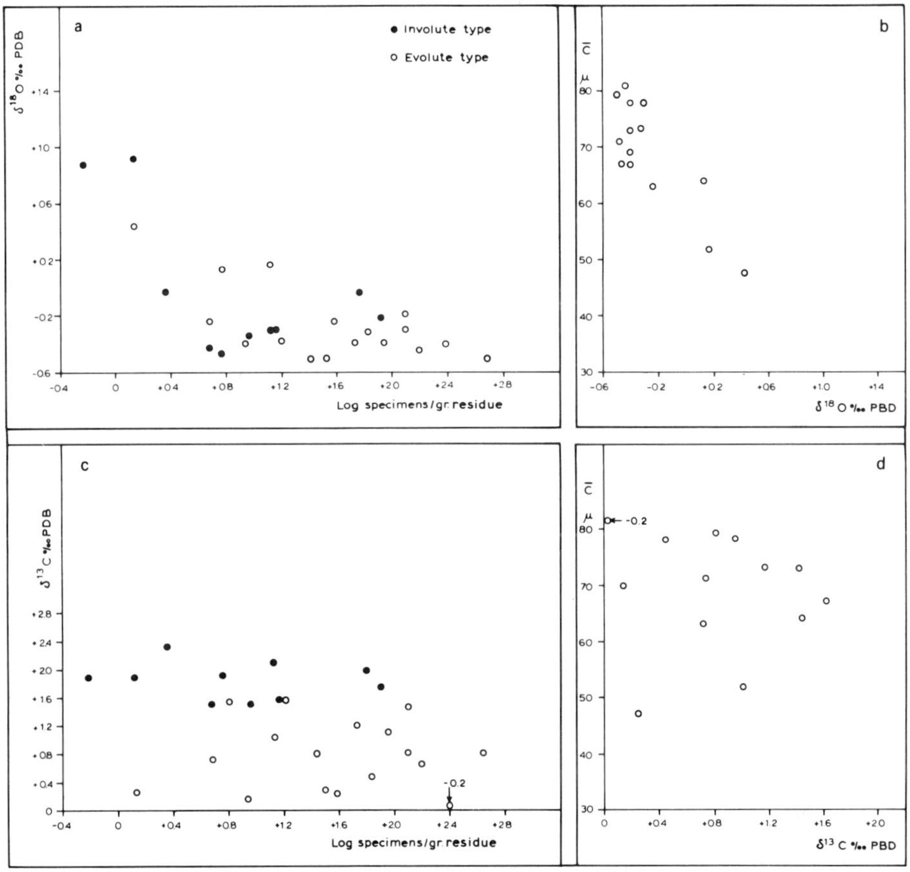

Fig. G.47 a–d. Scatter diagram of **a** log of the number of specimens per gram residue against $\delta^{18}O$ values; **b** $\delta^{18}O$ values of the evolute type of *Operculina* against $\bar{c}$, the mean of the diameter of the protoconch; **c** log of the number of specimens per gram residue against $\delta^{13}C$ values; **d** $\delta^{13}C$ values of the evolute type versus $\bar{c}$, the mean diameter of the protoconch (Fermont et al. 1983)

mid-water maxima at about 25 m and surface water minima, the latter attributed to photoinhibition of symbionts in the strongly illuminated waters (Fig. G.48). Radiotracer experiments on *Amphistegina, Heterostegina, Borelis,* and *Amphisorus* living at the same (10 m) depth indicated a general correlation between photosynthetic rates of symbionts and ^{18}O depletion of the shell, suggesting that at higher photosynthetic rates more algal metabolic CO_2 is incorporated in the shell and the isotopic composition becomes lighter (Fig. G.49). On the other hand, while correlation between calcification and photosynthetic rates with the depth gradient over the upper 30 m could be shown in *Amphistegina lobifera,* no good correlation between depth or calcification rates with stable isotope composition could be demonstrated in the in situ experiments in this depth range. (The maximum ^{18}O depletion has, indeed, been previously observed to occur at a depth between 40 and 50 m; Fig. G.1.) On the other hand, the genera examined

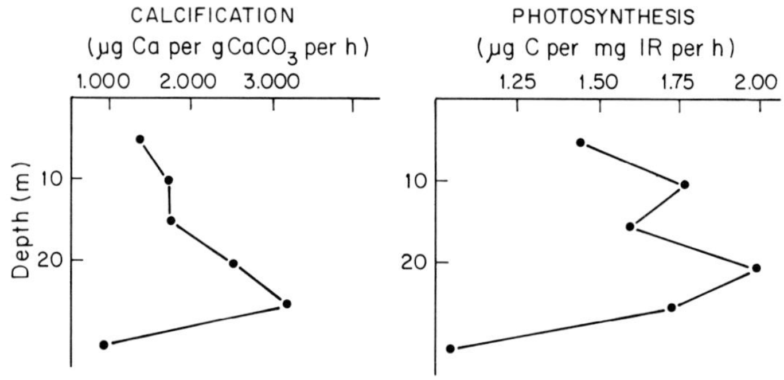

Fig. G.48. Calcification and photosynthesis depth profiles for *Amphistegina lobifera* (in situ radiotracer experiments). *IR* insoluble residue after dissolution of the skeleton (Erez 1978)

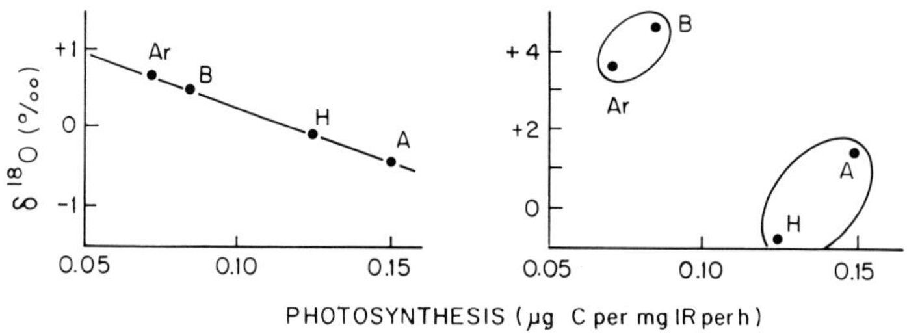

Fig. G.49. Stable oxygen isotopic composition of shells versus photosynthetic rates of symbionts in *Amphistegina (A), Heterostegina (H), Borelis (B),* and *Amphisorus (Ar)* at 10 m depth, Gulf of Aqaba (Erez 1978)

from the same depth fell into two groups, i.e., (a) the **porcelaneous** *Borelis* and *Amphisorus* with low photosynthetic rates of symbionts and stable isotope (oxygen and carbon) composition near or at equilibrium and (b) the **perforate-lamellar** *Amphistegina* and *Heterostegina* with high photosynthetic rates of symbionts and the stable isotope composition in strong disequilibrium (lighter).

These results raised a number of interesting problems, involving among others the relationship between calcification rates of the host and photosynthetic rates of symbionts on one hand, with oxygen isotope composition on the other; role of host–symbiont specificity at generic and perhaps at higher taxonomic levels of Foraminifera (dinoflagellates and rhodophytes are present in porcelaneous forms in addition to diatoms, but absent in benthic perforate-lamellar ones; see also above); significance of the biomineralization pattern which seems to be fundamentally different in porcelaneous and perforate-lamellar ones (Berthold 1976); the importance of CO_2 uptake method for symbiosis (in addition to that provided by recycling in the host), i.e., through pores in the lamellar Foraminifera and directly through thin areas of the lateral walls in porcelaneous ones, as shown by ^{14}C laboratory experiments on species from the Gulf of Aqaba.

In laboratory experiments on *Amphistegina* considerable differences in calcification rates were determined between specimens cultured in the dark and in

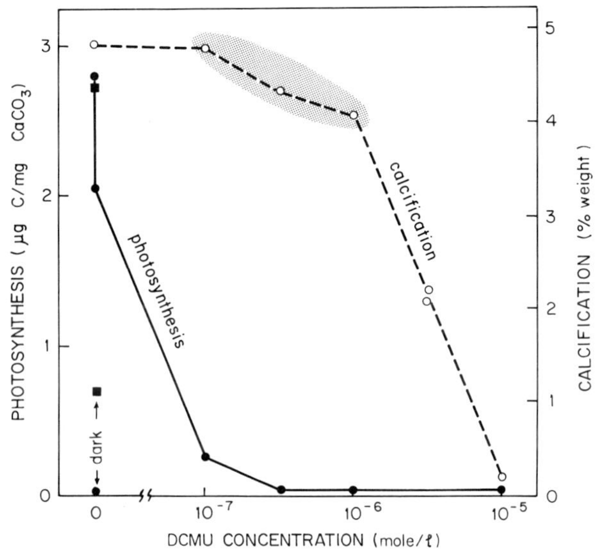

Fig. G.50. Relationship between photosynthesis rates of symbionts, calcification rates and DCMU concentrations in *Globigerinoides sacculifer.* (Original J. Erez)

light, differences attributed to the influence of photosynthesis on calcification. Later nonradioactive, long-term, in situ (cage) studies on growth rates (size parameters and addition of calcium carbonate) in *Amphistegina lobifera* and *Amphisorus hemprichii* in the Gulf of Aqaba confirmed the results of the radiotracer experiments with regard to depth optima of calcification and concerning the light/dark differences. However, although both species grew slower in the dark by a factor of roughly 10, it was shown that feeding influences considerably the amount of $CaCO_3$ added per day: specimens fed in the dark added 3 times as much $CaCO_3$ as those starved, while in the light, fed specimens added 1.5 times more $CaCO_3$ than those not fed. In addition, it was shown in the laboratory that gentle shaking about once a day of the culture containers of *Amphistegina* led to an increase in thickness/diameter ratios in the growing specimens.

These data suggested that light, as well as food and water energy influence rates of calcification in Foraminifera. However, in the dark/light experiments the effects of photosynthesis and light were not separated.

Earlier work on hermatypic corals using high concentrations of DCMU as inhibitor of photosynthesis of their symbionts seemed to suggest that photosynthesis is the main factor influencing calcification. Radiotracer experiments on the symbiont-bearing planktic foraminifer *Globigerinoides sacculifer* from the Gulf of Aqaba indicated that high concentrations of DCMU indeed inhibit both photosynthesis and calcification (Fig. G.50). However, at concentrations of 10^{-6} to 10^{-7}, i.e., 2–3 orders of magnitude lower than those used in earlier coral experiments, it was possible to demonstrate that, in the light, when photosynthesis is completely inhibited by DCMU, calcification proceeds at very high rates. Thus, light-enhanced calcification in Foraminifera is attributable largely to a direct

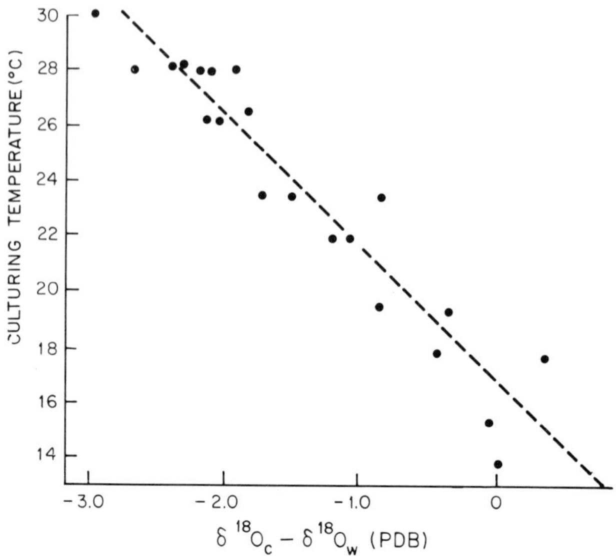

Fig. G.51. Culturing temperature of *Globigerinoides sacculifer* vs. the differences $\delta^{18}O_c$–$\delta^{18}O_w$ (i.e., the carbonate minus the water oxygen isotopic composition). The line is the best fit for the data and yields the following paleotemperature equation: $t = 17.00–4.52\,(\delta^{18}O_c–\delta^{18}O_w) + 0.03(d^{18}O_c–d^{18}O_w)^2$ with r = 0.95 (Erez and Luz 1983)

photobiological effect and only perhaps in part to photosynthesis of the symbionts. This has considerable bearing on problems of incorporation of metabolic CO_2 (from host and symbionts) in the foraminiferal shell and on stable isotope fractionation.

It may be noteworthy in this connection that *Artemia*-fed *Globigerinoides sacculifer,* which hosts dinoflagellate symbionts (like various porcelaneous species) but has very high rates of calcification (in contrast to the latter) was shown in controlled laboratory cultures to deposit its shell in oxygen isotopic equilibrium with the ambient water (Fig. G.51). (Disequilibrium values found in living specimens collected from the plankton were explained as a result of presence of uncombusted cytoplasmic matter in the shells examined.)

The studies on Foraminifera from the Gulf of Aqaba have yielded interesting results with regard to stable isotopes in shell material and their apparent relationship with symbiosis, productivity, life cycles, substrate, in- or epifaunal habitat, feeding patterns, light, water energy, growth and calcification, shell morphology and structure and possibly even biomineralization patterns.

It appears that while light influences photosynthesis of symbionts and directly the **host's calcification rates,** the latter **are not always dependent upon photosynthesis** and **are not correlatable** in a simple manner **with stable isotope composition of the shell.** Symbionts do not necessarily lead to oxygen isotopic disequilibrium, although in various forms they influence such a disequilibrium, depending perhaps upon the nature of the symbionts. The light gradient not only affects symbiontic activity, but also the **food quality** and the latter seams to have considerable influence on stable carbon isotopic composition of the shell. Food, as well as **feeding**

patterns, is apparently of considerable importance in this respect. Since in the case of benthic Foraminifera these latter factors are intimately connected with the **nature of the substrate,** there seems to be a circumstantial correlation between substrate and isotopic disequilibria. Optimal conditions for **growth and reproduction** are different from species to species and involve among others also optimal conditions for food. High **metabolic activity** of the foraminifer itself seems to influence isotopic fractionation, the latter being modulated in various symbiont-bearers by incorporation in the shell of photosynthetic CO_2. The possibility is not excluded that **shell-formation patterns** influence isotopic composition of the shell.

Despite intensive biological work on the foraminiferal symbionts from the Gulf of Aqaba (for references see Lee et al. 1979a, Lee 1980, McEnery and Lee 1981, Leutenegger 1983, 1984), many problems still remain to be solved with regard to the influence of photosynthetic activity on isotopic fractionation of the host shell. Extremely little is known as yet on the physiology and biochemistry of the Foraminifera themselves, especially with regard to pathways and processes of carbon and oxygen incorporation in their shells; work on these topics will be crucial to a better understanding of oxygen isotope composition of foraminiferal shells.

References: Berger et al. (1981), Buchardt and Hansen (1977), Emrich et al. (1970), Erez (1978, 1983), Erez and Luz (1983), Fermont et al. (1983), Hallock and Hansen (1979), Hansen and Buchardt (1977), Hansen and Dalberg (1979), Lee (1980), Lee et al. (1979a, b), Lee et al. (1980a, b, c, d), Leutenegger (1983, 1984), Leutenegger and Hansen (1977a), Luz and Reiss (1983), Luz et al. (1983), McEnery and Lee (1981), Noy (1982), Savin and Yeh (1981), ter Kuile and Erez (1984), Vincent and Berger (1981), Zmiri et al. (1972, 1974).

G.6 Thanatocoenoses

The composition of the assemblages of dead foraminiferan shells in the sediment reflect simultaneously the multiple ecological processes in the water column and on the bottom, because no large-scale transport takes place across the slope. However complex and overlapping these processes may be, there is a significant change with depth in the cumulative percentage distribution of foraminiferan genera and groups of genera in the sediment between shore and the shelf edge, but hardly any change on the shelf slope and below (Fig. G.52). The increase of the plankton percentage at depth levels of 40–50 m to 10–20% of the total assemblage corresponds to the upper depth limit to zooplankton layers during the day, while that at 130–150 m to 20–30% reflects to lower limit of the euphotic zone, where light-dependent benthos disappears.

The assemblages of benthic Foraminifera are dominated by porcelaneous forms in shallow depths of less than 50 m and by perforated-lamellar forms between 50 and 130 m depth. The porcelaneous forms are epiphytic and bottom-dwelling species of which all larger and many smaller-sized species have symbionts. Other species lacking symbionts may indirectly depend on light by their nutritional habits. Even smaller porcelaneous forms living in the pore space of the

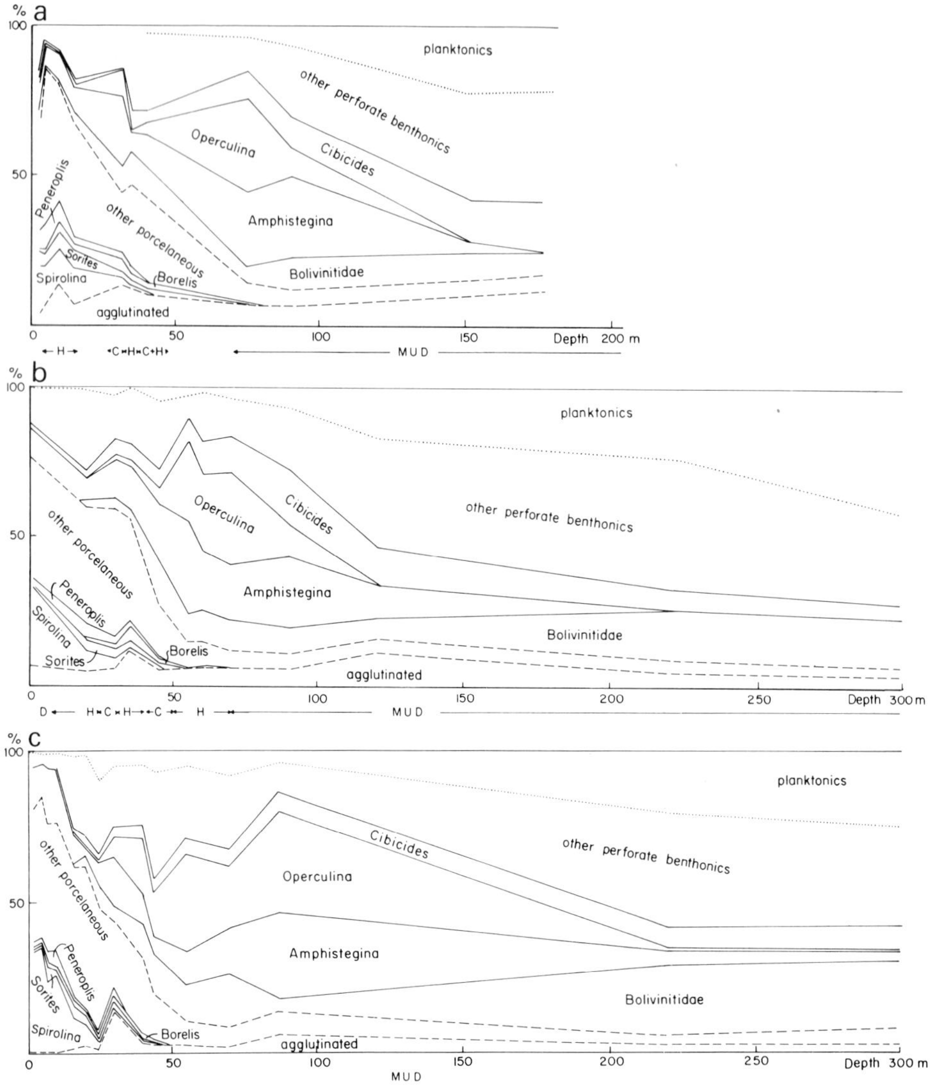

Fig. G.52 a–c. Cumulative percentage distribution of foraminiferal genera and groups of genera in Gulf of Aqaba sediments. **a, b** W-E traverses, Geziret Fara' oûn area. **c** N-S traverse between northern end of gulf (Sea Star Hotel, Elat) and off Wadi Taba. *Substrates: D Diplanthera; H Halophila; c coral carpet* (Reiss 1977)

sediment may feed on diatoms retreating in the pore space of the sediment or on plant debris buried in the sediment by burrowing and may thus indirectly depend upon light. At depths of more than 50 m, nummulitids and amphisteginids contribute over 50% to the total assemblages. Their abundance must also be light-controlled by their symbiosis with diatoms and/or by their food. Their high contribution to the assemblages may obscure other changes taking place in the com-

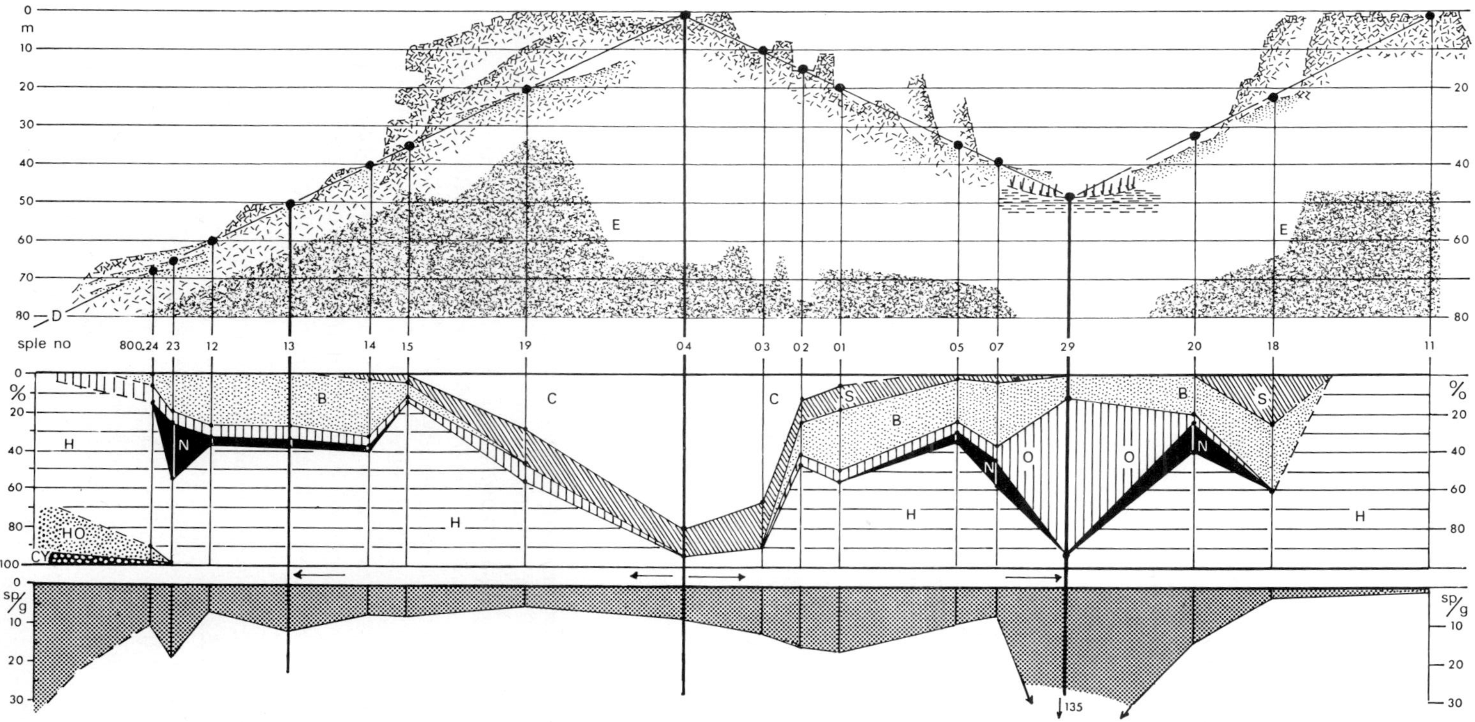

Fig. G.53. Schematic section through reef barrier and sheltered patch reef of South Male Atoll, Maldivas (open Indian Ocean) and cumulative percentage distribution of selected larger Foraminifera (species or groups of species) compared to water energy and shell frequency. *D* Depth gradient drawn symmetrically in respect of shallowest sample 04. *E* Estimation of water energy relative to lowest value on muddy lagoon bottom. Samples collected in March by scuba diving. Foraminifera: *B* Alveolinids: *Borelis schlumbergeri, B.* cf. *pulchra* and *Alveolinella quoii. C Calcarina calcar. CY Cycloclypeus carpenteri. H Heterostegina depressa. HO Heterostegina operculinoides. N Nummulites cummingii (=venosus). O Operculina ammonoides, O.* n. sp.?. *S* Soritinae: *Sorites orbiculus, S. orbitolitoides, Amphisorus* n. sp.?, *Marginopora* aff. *vertebralis. sp/g* Number of specimens of above mentioned species per g dry sediment. Note asymmetry of distribution pattern in depth gradient corresponding to levels of water energy (Hottinger 1980)

position of the thanatocoenoses of smaller benthic Foraminifera. Thus, at the level of groups of genera, the depth-dependent changes in cumulative percentage distribution of foraminiferan shells in the sediment reflects mainly light dependence as an overruling ecological relation.

At higher levels of taxonomic resolution, particularly on species level in larger Foraminifera, the assemblages of shells in the sediment may be distinguished by at least four depth-dependent groups: (1) the "soritine assemblage" between 0 and 20 m, where *Sorites orbiculus* and *Amphisorus hemprichii* dominate the assemblages of larger Foraminifera; (2) the "alveolinid assemblage" between 20 and 40 m, where *Borelis schlumbergeri* is comparatively frequent; (3) the "nummulitid assemblage" between 40 and 80 m, where *Operculina ammonoides* is most abundant and (4) the "cycloclypeid assemblage" between 80 and 130 m, corresponding to the depth distribution of *Heterocyclina tuberculata*. Below the euphotic zone, *Planorbulinella elatensis* is the only benthic form of larger size. The species composition of the assemblages changes, however, in the same depth zone according to the characteristics of substrate and/or water energy. In shallow water, *A. hemprichii* is replaced by *Calcarina calcar* in high water energy environments, while in deeper waters *O. ammonoides* on soft substrate is replaced by *Heterostegina depressa* on hard bottoms as a dominant faunistic element. The same succession of species with depth and the same dependence on substrate types was observed in the open Indian Ocean on the Maldive Islands, where species diversity is, however, much higher. In particular, the presence of *Nummulites cummingii, Alveolinella quoii* and *Marginopora vertebralis* does not change the depth sequence in abundance maxima of the other species of larger Foraminifera common to the Maldive Islands and the Gulf of Aqaba. Moreover, the sequence of larger foraminiferan species is the same on exposed, high water energy slopes or sheltered, low water energy environments of coral reefs on the Maldive Islands, as well as on the shores in the Gulf of Aqaba, from the sheltered Elat region to the exposed Ras Muhammed section. Different levels of water energy thus do not change the **sequence of assemblages in the depth gradient,** but may shift the depth limits between successive assemblages, as reflected in the asymmetry of percentage distribution of species on Maldivan exposed and sheltered reef slopes (Fig. G.53).

The depth-dependent sequence of marine environments and of their populations of larger Foraminifera is recorded by the sediments distributed along the depth gradient in a depth-related sequence. The latter may be compared to fossil sequences recorded in transgressive or regressive sedimentary cycles reflecting respectively the deepening or shallowing of the environment of deposition. The microfossil content in such fossil sedimentary series may be compared to recent ones if the time-span of the sedimentary cycle is significantly shorter than the repetitive and often synchronous evolutionary cycles in larger Foraminifera. The sequences of assemblage observed in Late Cretaceous and Tertiary sedimentary cycles (Fig. G.54) are comparable to the recent assemblage sequence by an analogous succession of shells with similar morphological characteristics, but belonging to other, extinct taxa. Some types of larger fossil Foraminifera, namely the conical-agglutinated and the orbitoid type, are altogether extinct. Their niche might be occupied today by an extension of the habitat of other morphological types or simply remain empty.

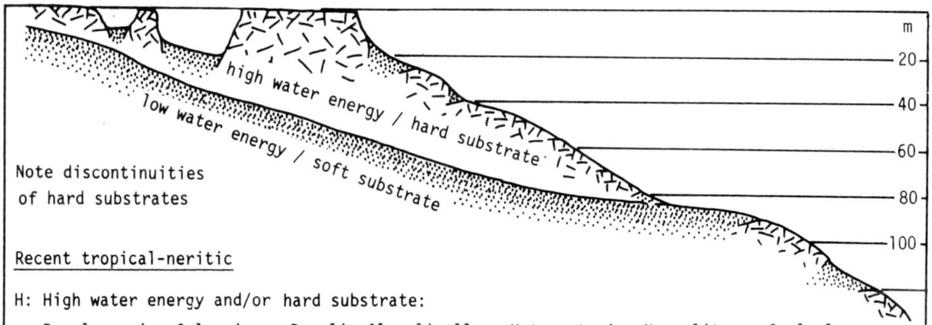

Recent tropical-neritic

H: High water energy and/or hard substrate:
Baculogypsina,Calcarina → Borelis,Alveolinella → Heterostegina,Nummulites → Cycloclypeus

L: Low water energy and/or soft substrate. In the deepest section, the H/L difference may be inconspicuous or irrelevant.
Sorites,Amphisorus,Marginopora → Borelis,Alveolinella → Operculina → Heterocyclina

Miocene

H: Baculogypsina → Heterostegina (inv.) → Flosculinella → Nummulites → Cycloclypeus

L: "Archaias",Sorites → Borelis,Flosculinella → Operculina,Planoperculina

Oligocene

H: "Pararotalia" viennoti → Borelis → Nummulites → Heterostegina (inv.) → Spiroclypeus

L: Praerhapydionina → Borelis,Bullalveolina → Nummulites → Heterostegina (ev.) → Operculina

Late Eocene

H: Baculogypsinoides,Fabiania → Borelis → Nummulites,Heterostegina (inv.) → Spiroclypeus
Pellatispira (inv.)

L: Fallotella?,Orbitolites → Borelis → Nummulites → Operculina gr. gomezi & gr. alpina
Biplanispira,Pellatispira(ev.) → Bullalveolina

(inv.)=involute
(ev.) =evolute

Middle Eocene

H: Calcarina?,Fabiania,Dictyoconoides → Alveolina → Nummulites

L: Coskinolina,Dictyoconus → Orbitolites,Fabularia → Alveolina → Nummulites → Operculina s.l.

Early Eocene

H: Lockhartia,Dictyoconoides → Alveolina → Nummulites

L: Coskinolina,Dictyoconus → Orbitolites,Somalina,Fabularia → Alveolina → Nummulites → Operc.

Middle-Late Paleocene

H: Miscellanea, Daviesina (ornate),Kathina → Alveolina → Nummulites

L: Fallotella,Coskinon,Glomalveolina → Alveolina → Nummulites,Ranikothalia → Operculina
Broeckinella,Saudia Daviesina (compressed)

Upper Cretaceous

H: Siderolites,Daviesina,Sulcoperculina → Lacazina compressa → Pseudosiderolites

L: Calveziconus,Montsechiana → Periloculina,Subalveolina → Lacazina elongata - Meandropsinids
Ilerdorbis,Dicyclina

Fig. G.54. Sequence of selected larger Foraminifera in the depth gradient according to substrate and/or water energy levels in extant tropical seas compared to Tertiary and Late Cretaceous faunal sequences. Ranges widely overlapping. Amphisteginid and orbitoid species form similar parallel sequences (Hottinger 1983 b)

In the shallow water "soritine assemblage" the dominant species have imperforate planispiral-flabelliform to discoidal shells with at least mature evolute growth stages tending to become cyclical. The multiple apertures are distributed radially or in crosswise oblique direction along the shell margin. Within the assemblage, thicker forms occur in shallower, thin and successively more evolute forms in deeper water. In shallow high water energy environments, perforate forms of thick-lenticular to subglobular shell shape appear. They are characterized by an enveloping canal system with openings disseminated all over the shell surface and often canaliculate spines. The "alveolinid assemblage" is dominated by imperforate planispiral-involute or concentric, globular to fusiform shells. The multiple apertures are either restricted to one of the poles (larger miliolids) or arranged in a single band reaching form pole to pole. Within the assemblage, globular forms are generally characteristic of shallower, elongate forms of deeper waters. The "nummulitic" assemblage is characterized by perforate planispiral shells with a marginal cord, distributing the pseudopodial protoplasm in horizontal furrows around the shell margin. This complex structure represents a grade of organization realized for the first time in the earliest Tertiary. Most probably there is no earlier occupant of this particular niche. Within the nummulitid assemblage, involute, thick shells were living in shallower, involute or evolute, thin shells in deeper environments. The "cycloclypeid assemblage" is characterized by the presence of extremly flattened nummulitids with late annular growth stages. They appear in the Oligocene and may have eliminated the largest and most flattened species of the orbitoids. These types of shell morphology, distributed in sequential order according to the depth gradient, are repeated in the course of time as an adaptive response of a foraminiferal K-strategist to the challenge of environmental conditions changing with depth according to the different shell construction potential of imperforate nonlamellar and perforate-lamellar shell walls. A more detailed comparison of depth-dependent sequences of morphotypes repeated in the course of time is essential for identification of analogous, adaptive morphological features in fossil and recent Foraminifera leading to a better understanding of the significance of the morphology recorded in the fossils through geologic history.

References: Hottinger (1978a, 1983b), Hottinger and Drobne (1980), Reiss (1977).

H. 150,000 Years Gulf of Aqaba

H.1 Deep Sea Cores

Intensive studies of numerous deep-sea cores from the **Red Sea proper** have elucidated its paleoceanographic history during the Late Quaternary and have thereby significantly contributed to the reconstruction of regional Pleistocene-Holocene paleoclimates in this arid-semi-arid area (Degens and Ross 1969, Deuser et al. 1976, Herman 1968, Rosenberg-Herman 1965). Indeed, the Red Sea has been regarded as a "natural laboratory" for investigating changes in partly restricted marine environments (Berggren 1969).

However, at the time of the deep-sea core studies in the Red Sea relatively little was known on the latter's present-day oceanographic characteristics, especially with regard to the distribution of those organisms which provide the major tools for paleoceanographic investigations (see Morcos 1970). For this reason, paleoecological interpretations had to be based nearly exclusively on interpolations from normal marine, mostly open-ocean environments. Moreover, most of the deep-sea cores from the Red Sea proper were raised in its southern and middle parts which differ in many respects from the northern one.

Although a large body of information on the oceanographic setting and on the ecology of paleoceanographically important shell-bearing planktic and benthic organisms from the **Gulf of Aqaba** and **northernmost Red Sea** became available through the DCPE project (see previous chapters), the application of this knowledge to the reconstruction of Quaternary environmental changes in the Gulf was prevented by the lack of material: until 1977, only one short deep-sea core, Vema V 14–126, drilled in the Dakar Deep and penetrating turbidites (Emery 1963, Rosenberg-Herman 1965) was available from the Gulf of Aqaba.

In 1977, The Woods Hole Oceanographic Institution research vessel, *Atlantis* II, raised six deep-sea cores within the Gulf of Aqaba and two cores immediately outside the Straits of Tiran, in the Red Sea proper.

The study of these cores (project CATE – **C**ores **A**tlantis **T**wo **E**lat, see Reiss et al. 1980) has facilitated the reconstruction of the Late Quaternary paleoceanographic history of the Gulf of Aqaba and has provided further clues to the paleoceanography of the Red Sea proper. This study has also contributed to a better understanding of the ecology of various marine, calcareous shell-bearing organisms, widely used in paleoecological studies.

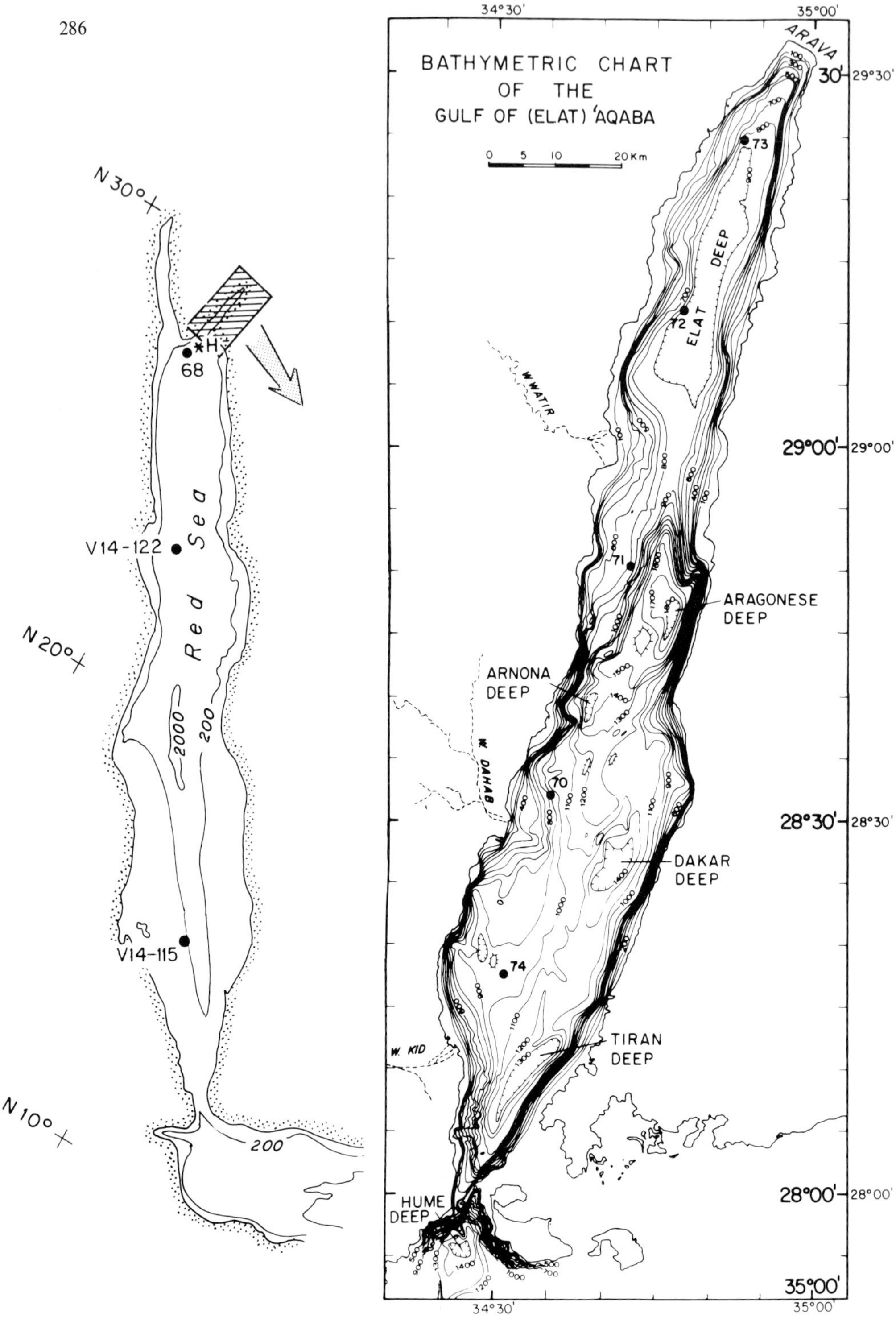

Fig. H.1. Location map of *Atlantis* II cores (● 74) and Vema (V 14) examined within the framework of Project CATE

Table H.1. Gulf of Aqaba (Elat) and Red Sea cores Atlantis II – Leg 19 Expedition 93

Station	Core no. PG – gravity PC – Piston	Lat. N	Lon. E	Depth (m)	Length (cm)
94	PC 86	27°38.4′	34°28.1′	1,087	827
94	PG 68	–	–	–	90
96	PC 70	28°32.1′	34°35.0′	963	745
96	PG 70	–	–	–	90
97	PC 71	28°50.1′	34°42.4′	735	519
97	PG 71	–	–	–	116
98	PC 72[a]	29°11.5′	34°47.6′	550	673
99	PC 73[a]	29°24.0′	34°53.2′	858	683
99	PG 73[a]	–	–	–	64
101	PC 74	28°18.2′	34°31.4′	884	729
101	PG 74	–	–	–	61

[a] Turbidity sequences

The locations of the *Atlantis* II drilling sites were chosen in the vicinity of the DCPE-stations (see Fig. E.5) and at a water depth considered adequate to avoid the turbidite-filled deeps (Fig. H.1, Table H.1, see also Chap. C).

For comparison, samples from two Vema cores from the Red Sea proper, provided by the Lamont-Doherty Geological Observatory, were examined in detail, viz. V14–122 situated at N.Lat. 23°55′ E, Long. 36°28′ at 1,490 m depth and V14–115 situated at N.Lat. 16°04′ E, Long. 41°28′ at 1,195 m depth. *Atlantis* II cores 70, 71, 72 and 73 were obtained from the slope edge, while core 74 was drilled in a relatively flat part of the sea bottom. Nevertheless, cores 72 and 73 represent largely turbiditic sequences.

The normal sequences cored consist of two main types of sediment, i.e. (1) gray-green, partly silty, calcareous mud and (2) silty-clayey sands, mostly quartzose and occasionally arkosic. A thick arkosic sequence occurs in the lower part of core 70 drilled at the edge of the Dahab delta. (For lithology of cores 70, 71, 74 and 68, see Halicz and Reiss 1981.)

Recovery of the cores was generally very good. However, the piston-core tops were lost and for this reason the records of the trigger-weight and piston cores were combined.

Considering the high sedimentation rates known from the Red Sea proper, the study on the *Atlantis* II cores in the Gulf of Aqaba and northernmost Red Sea was carried out on samples taken at 10–20 cm intervals. (For methods and techniques, see Almogi-Labin 1982, Halicz and Reiss 1981, Reiss et al. 1980, Winter 1982 a).

H.2 Microfossil Assemblages from Cores

The stratigraphic and paleoenvironmental interpretation of the core sequences are based on (a) presence/absence and relative abundance patterns of planktic Foraminifera, Coccolithophorida and Pteropoda, as well as of benthic Foraminifera, and (b) stable oxygen and carbon isotope analysis of planktic and benthic Foraminifera and Pteropoda.

The following planktic taxa were recognized in the core samples examined (species marked " × " occur only in cores and not in living plankton or in Recent sediments; see Chap. E):

Coccolithophorida

× *Braarudosphaera bigelowi* (Gran and Braarud 1939)
Calcidiscus leptoporus (Murray and Blackman 1890)
× *Crenalithus* spp. (Roth 1973)
× *Cyclolithella subtilis* (Müller 1976)
Emiliania huxleyi (Lohmann 1902)
× *Gephyrocapsa caribbeanica* Boudreaux and Hay (1967)
G. ericsoni McIntyre and Bé (1967)
G. oceanica Kamptner (1943)
G. ornata Heimdal (1973)
G. protohuxleyi McIntyre (1970)
Oolithus fragilis (Lohmann) subsp. *cavum* Okada and McIntyre (1977)
Umbilicosphaera sibogae (Weber-van Bosse 1901)
× *U. sibogae* (Weber van Bosse) var. *foliosa* (Kamptner 1963)
Helicosphaera carteri (Wallich 1877)
× *H. pavimentum* Okada and McIntyre (1977)
× *Heliolithus crassus* Müller (1976)
Pontosphaera discopora Schiller (1925)
Scyphosphaera apsteinii Lohmann (1902)
Acanthoica aculeata Kamptner (1941)
Anthosphaera quatricornu (Schiller 1913)
Discosphaera tubifera (Murray and Blackman 1898)
× *Neosphaera coccolithomorpha* Lecal-Schlauder (1950)
Rhabdosphaera clavigera Murray and Blackman (1898)
Umbellosphaera irregularis Paasche (1955)
U. tenuis (Kamptner 1937)
Anoplosolenia brasiliensis (Lohmann 1919)
Calciopappus caudatus Gaarder and Ramsfjell (1954)
Halopappus adriaticus Schiller (1914)
Michaelsarsia sp. Borsetti and Cati (1976)
Syracosphaera corrugis Okada and McIntyre (1977)
S. elatensis Winter (1979)
S. halldali Gaarder and Hasle (1971)
S. histrica Kamptner (1941)
S. mediterranea Lohmann (1902)
S. pirus Halldal and Markali (1902)

S. pulchra Lohmann (1902)
S. variabilis (Halldal and Markali 1955)
Calyptrosphaera catillifera (Kamptner 1937)
C. oblonga Lohmann (1902)
Helladosphaera cornifera (Schiller 1913)
Homozygosphaera wettsteini (Kamptner 1937)

Foraminifera

Globigerina bulloides d'Orbigny (1826)
Globigerina calida Parker (1962)
Globigerina falconensis Blow (1959)
Globigerina quinqueloba Natland (1938) (recorded together with *T. clarkei; see Chap.* E_3)
Globigerinella siphonifera (d'Orbigny 1839)
× *Beella* sp.
× *Globigerinoides conglobatus* (Brady 1879)
Globigerinoides ruber (d'Orbigny 1839)
Globigerinoides elongatus (d'Orbigny 1826, recorded together with *G. ruber*)
Globigerinoides sacculifer (Brady 1877)
Globoturborotalita rubescens (Hofker 1956)
Globoturborotalita tenella (Parker 1958)
Orbulina universa d'Orbigny (1839)
Orbulina sp. ("small") Reiss et al. (1974)
Turborotalita clarkei (Rögl and Bolli 1973)
× *Turborotalita humilis* (Brady 1884)
Globigerinita glutinata (Egger 1893)
× *Globigerinita iota* Parker (1962)
Globigerinita uvula (Ehrenberg 1861)
× *Neogloboquadrina pachyderma* (Ehrenberg 1861)

Pteropoda

× *Cavolinia inflexa* (Lesueur 1813)
Cavolinia longirostris (de Blainville 1821)
Cavolinia uncinata (Rang 1829)
Clio convexa (Boas 1886)
Creseis acicula (Rang 1828)
× *Creseis chierchiae* (Boas 1886)
Creseis virgula virgula (Rang 1828)
Diacria quadridentata (de Blainville 1821)
× *Diacria trispinosa* (de Blainville 1821)
× *Hyalocylis striata* (Rang 1828)
× *Limacina bulimoides* (d'Orbigny 1836)
Limacina inflata d'Orbigny 1836)
Limacina trochiformis (d'Orbigny 1836)
× *Peraclis* sp.
Peraclis reticulata (d'Orbigny 1836)
× *Styliola subula* (Quoy and Gaimard 1827)

All taxa listed are illustrated and most of them discussed by Reiss et al. (1974), Halicz and Reiss (1981; Foraminifera), Almogi-Labin and Reiss (1977), Almogi-Labin (1982; Pteropoda), Winter et al. (1978, 1979) and Winter (1982a; Coccolithophorida).

Benthic Foraminifera from the cores were determined at species or species-group levels, but recorded at generic and higher taxonomic levels, viz.: Agglutinated genera, like *Rheophax, Textularia* and *Spiroplectammina*, Miliolidae including: *Quinqueloculina, Triloculina, Spiroloculina, Pyrgo, Massilina, Pateoris, Miliolinella, Pseudomassilina, Sigmoilina,* and *Schlumbergerina.* Opthalmidiidae including: *Cyclogyra,* Buliminacea including *Bulimina, Globobulimina,* and *Fursenkoina, Bolivina, Brizalina, Uvigerina, Siphouvigerina.* "Rotaliformes" including such genera as *Cibicides, Cibicidoides, Carribeanella, Heterolepa, Hanzawaia, Anomalina* (s.l.), *Siphonina, Rosalina, Conorboides, Discorbis, Stomatorbina.* Nodosariacea including *Nodosaria, Dentalina, Lenticulina, Robulus, Astacolus, Lagena, Amphicoryna, Globulina, Glandulina,* and *Fissurina. Cassidulina* (and *Islandiella*). *Gyroidina* (*Gyroidinoides*). *Bucella. Sphaeroidina. Hoeglundina.* Other genera occurring, but not recorded in the graphs below include: *Spirillina, Lamarckina, Robertina, Reussella, Pullenia, Nonion, Melonis, Astrononion, Elphidium, Elphidiononion, Cancris, Baggina, Cymbaloporella, Cymbaloporetta, Tretomphalus,* etc.

H.3 Stratigraphy

The assemblages analyzed from the core sequences show numerous prominent quantitative shifts in response to changing environmental conditions through time (Figs. H.2 to H.4). At present many of the components of these assemblages live in the Gulf of Aqaba and in the northern Red Sea at the edge of their ecological tolerance, under hypersaline and oligotrophic conditions. Thus, even small environmental changes to more extreme conditions in the past have strongly affected these taxa. Since different groups are not influenced equally by the same ecological factors, their quantitative responses are at least in part not synchronous. In this manner, a sequence of ecostratigraphic events of the different groups can be determined in the cores and recognized throughout the Gulf of Aqaba and outside it, in the Red Sea proper.

The various quantitative shifts can be intercalibrated with the stable oxygen isotope curve obtained from planktic Foraminifera and from Pteropoda (Figs. H.2 and H.4). The shape of this curve is similar to that recorded previously from Red Sea cores and corresponds to the generalized global isotopic curve, thus facilitating the recognition of isotope stages 1 to 6 (Fig. H.5), which correspond to the Holocene, last Glacial, last Interglacial and penultimate Glacial periods. Although no radiometric ages are available from the *Atlantis* II cores, the well-documented ages of at least certain points of the generalized isotopic curve (even beyond the radiocarbon range) can be attributed to some of the major events of the sequence, i.e., 11,000 yr *BP* to the boundary between isotope stages 1 and 2 (Termination I), 75,000 yr *BP* to the boundary between stages 4 and 5 and 127,000 yr *BP* to that between stages 5 and 6 (Termination II). The highest values

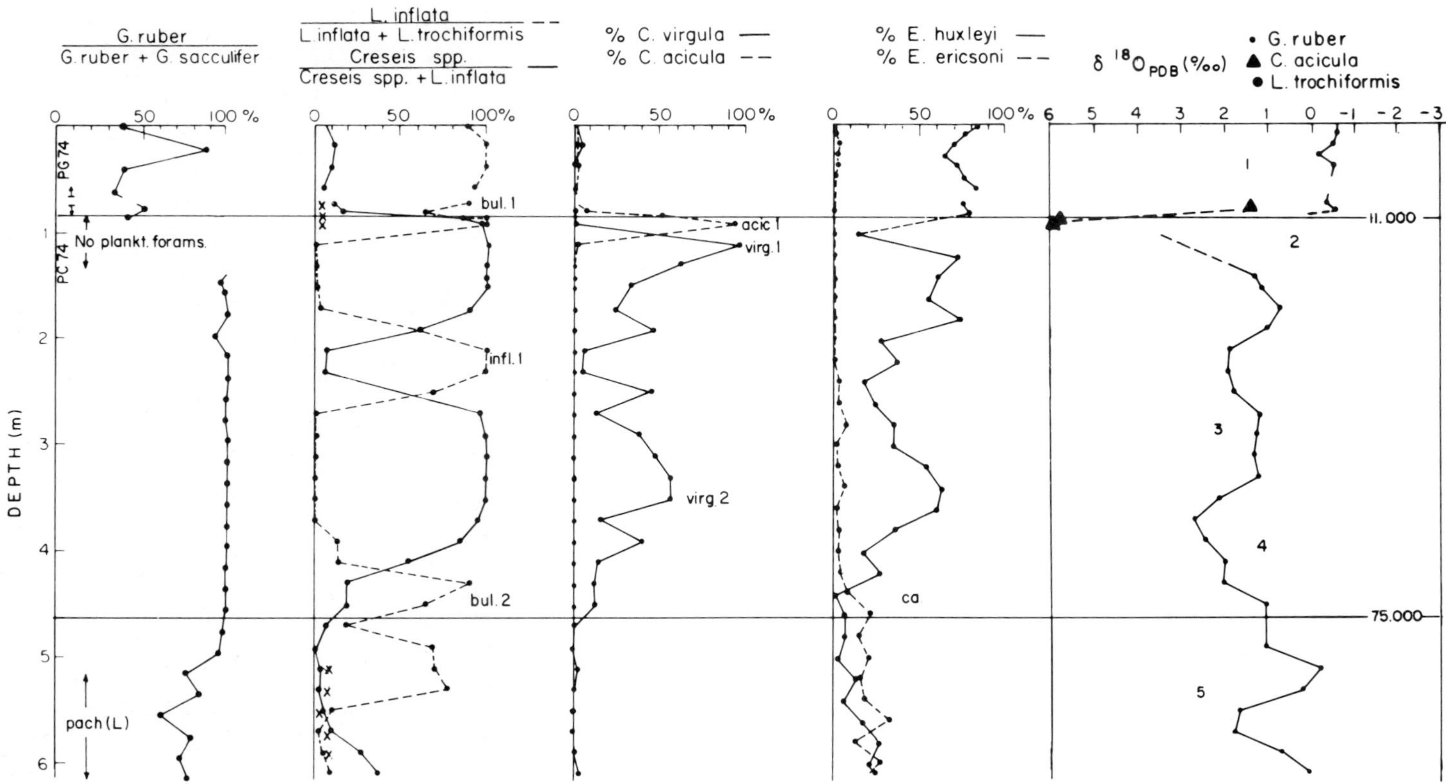

Fig. H.2

Fig. H.2–H.4. Selected faunal, floral, and isotope curves, characteristic species peaks, and isotope stratigraphy of Gulf of Aqaba and Red Sea cores. PG = trigger weight gravity core; PC = piston core; infl. = peak *L. inflata*; bul. = peak *L. bulimoides*; x = pteropod molds in part; ● acicular aragonite overgrowth on pteropod specimens in part; acic. = peak *C. acicula*; virg. = peak *C. virgula*; D = diatoms; Si = silicisponge spicules (Reiss et al. 1980)

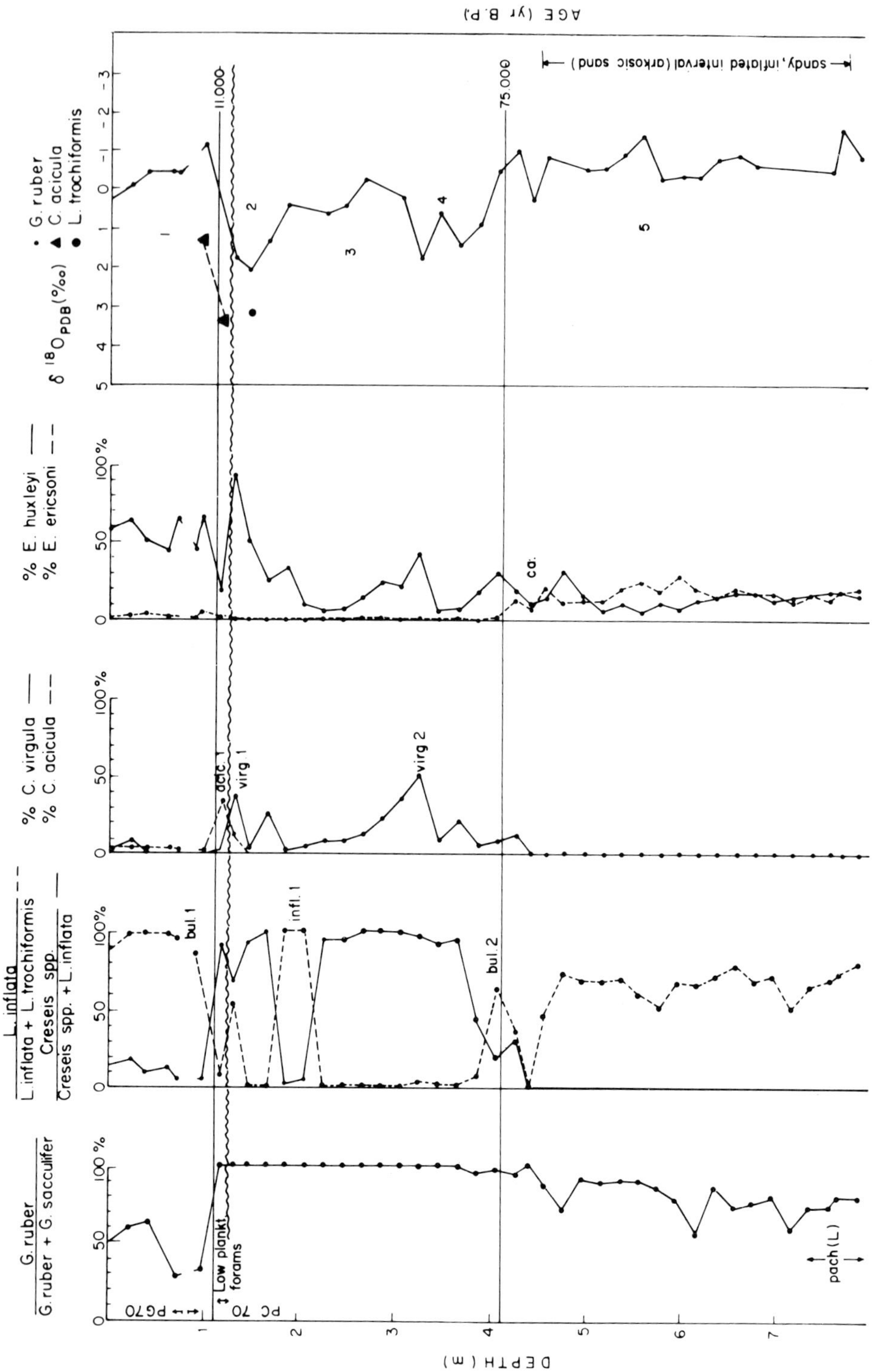

Fig. H.3 (legend see p. 291)

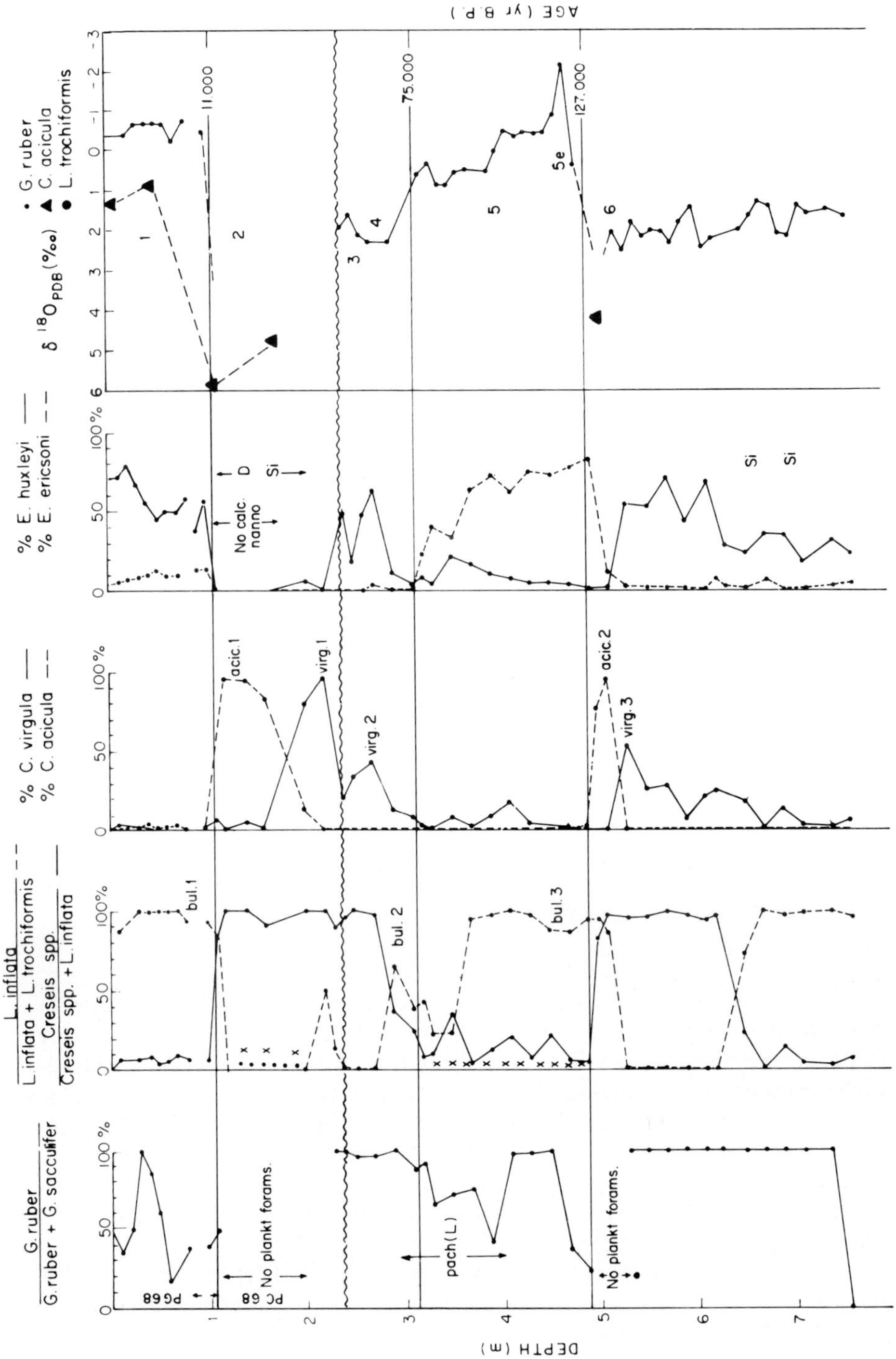

Fig. H.4 (legend see p. 291)

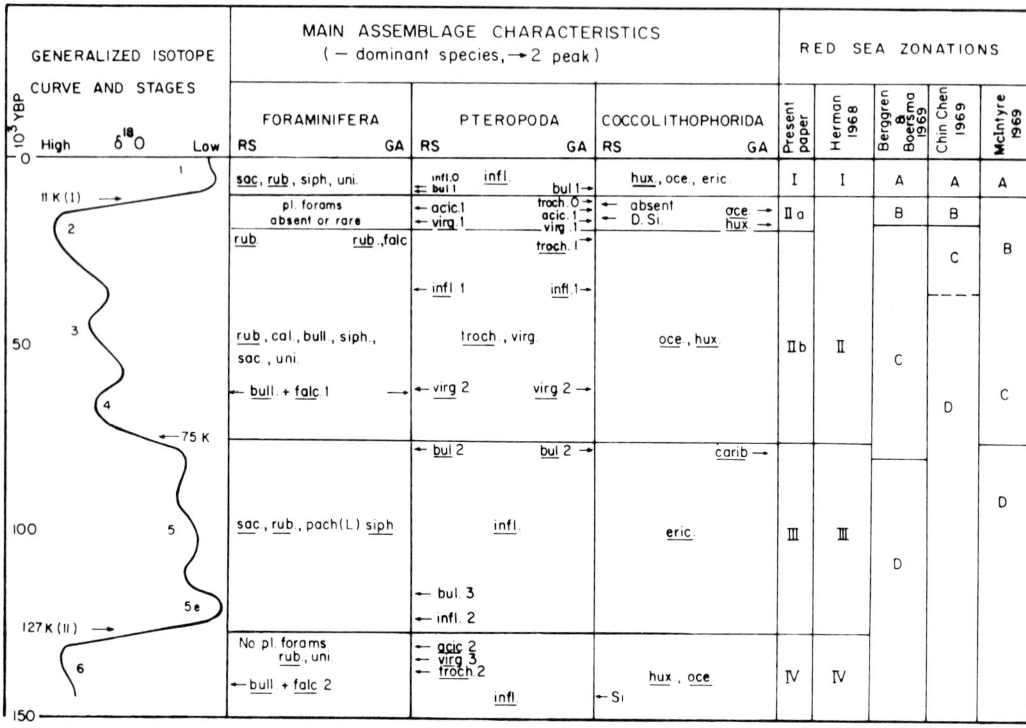

Fig. H.5. Biostratigraphic subdivision of Gulf of Aqaba (*GA*) and northernmost Red Sea (*RS*) cores and comparison with other Red Sea zonations with reference to the generalized oxygen-isotope curve. sac. = *G. sacculifer*; rub. = *G. ruber*; siph. = *G. siphonifera*; uni. = *O. universa*; falc. = *G. falconensis*; cal. = *G. calida*; bull. = *G. bulloides*; pach. (L). = *N. pachyderma* (left-coiling; < 63 µm); infl. = *L. inflata*; troch. = *L. trochiformis*; bul. = *L. bulimoides*; acic. = *C. acicula*; virg. = *C. virgula*; hux. = *E. huxleyi*; oce. = *G. oceanica*; eric. = *G. ericsoni*; caribb. = *G. caribbeanica*; D. = diatoms; Si. = Silicispongiae spicules. Roman numerals in isotope curve: terminations after Broecker and Van Donk (1970). Arabic numerals in curve: isotope stages after Emiliani (1966) and Shackleton and Opdyke (1973). Time scale after Broecker and Van Donk (1970) and Shackleton and Opdyke (1973) (Reiss et al. 1980)

of the $\delta^{18}O$ record below stage 1 indicate the last maximum Glacial, dated at about 18,000 yr *BP*. The heaviest values below the stages 5–6 boundary indicate the penultimate Glacial maximum. The intercalibration of the **major ecostratigraphic events** with the dated stable isotope curve, as well as with biostratigraphic zonations recognized earlier in the Red Sea proper is also shown in Fig. H.5. Reliable radiocarbon dates (less than about 30,000 yr), obtained from the biostratigraphic zones in Red Sea cores, support the age assignments of the various events in the *Atlantis* II core sequences. (For references, see Luz 1982, Reiss et al. 1980.)

As can be seen from Figs. H.2 to H.4, unconformities are of varying extent and occur within the last Glacial period, although not at the same time and not necessarily during the Glacial maximum. This seems to indicate that the unconformities are due to submarine sliding or erosion and that no dessication of the Gulf of Aqaba or the Red Sea took place during the Glacial maximum; the Red

Table H.2. Comparison of Red Sea zonations (Winter 1982a)

RED SEA ZONATIONS					
1	2	3	4	5	6
I	A	I	I$_{n}$	I	I up
					I mid
					I low
II	B	II$_{a}$	II$_{n_0}$	II$_{a}$	IIa up
			II$_{n_1}$		IIa low
			II$_{n_2}$	II$_{b_1}$	IIb up
				II$_{b_2}$	
		II$_{b}$	II$_{n_3}$	II$_{b_3}$	IIb mid
	C		II$_{n_4}$	II$_{b_4}$	IIb low
			II$_{n_5}$		
III	D	III	III$_{n_1}$	III	III up
			III$_{n_2}$		III mid$_1$
					III mid$_2$
					III low
IV		IV	III$_{n_3}$	IV$_{a}$	IV up
			IV$_{n_1}$		
			IV$_{n_2}$	IV$_{b_1}$	IV mid
				IV$_{b_2}$	
			IV$_{n_3}$	IV$_{b_3}$	IV low

1 Red Sea, based on Foraminifera and Pteropoda (Herman 1968)
2 Red sea, based on Coccolithophores (McIntyre 1969)
3 Gulf of Aqaba and Red Sea, based on Coccolithophores, Foraminifera and Pteropoda (Reiss et al. 1980)
4 Gulf of Aqaba, based on Coccolithophores (Winter 1982a)
5 Gulf of Aqaba, based on planktic and benthic Foraminifera (Halicz and Reiss 1981)
6 Gulf of Aqaba, based on Pteropoda (Almogi-Labin 1982)

Sea was apparently continuously connected with the Gulf of Aden through the Straits of Bab-el-Mandeb.

Recently developed, more refined bio(eco)stratigraphic zonations of the core sequences, based on coccolithophorids, foraminiferans and pteropods (Table H.2) provide, together with the stable oxygen isotope record, a **high-resolution multiple stratigraphic scheme** for the Late Quaternary of the Gulf of Aqaba and northern Red Sea. The major features of these zonations are shown for coccolithophorids and foraminiferans in the combined record of the most complete sequences [core 74 (stages 1–4) and core 68 (stages 5 and 6)] in Figs. H.6 to H.8 respectively, and for pteropods in the various cores in Figs. H.9 and H.10. This latter figure also indicates the high resolution correlation of the combined pteropod and stable isotope records, as well as the extent of the various unconformities in the cores. (For detailed descriptions of the zonations and complete core records, see Almogi-Labin 1982, Halicz and Reiss 1981, Winter 1982a).

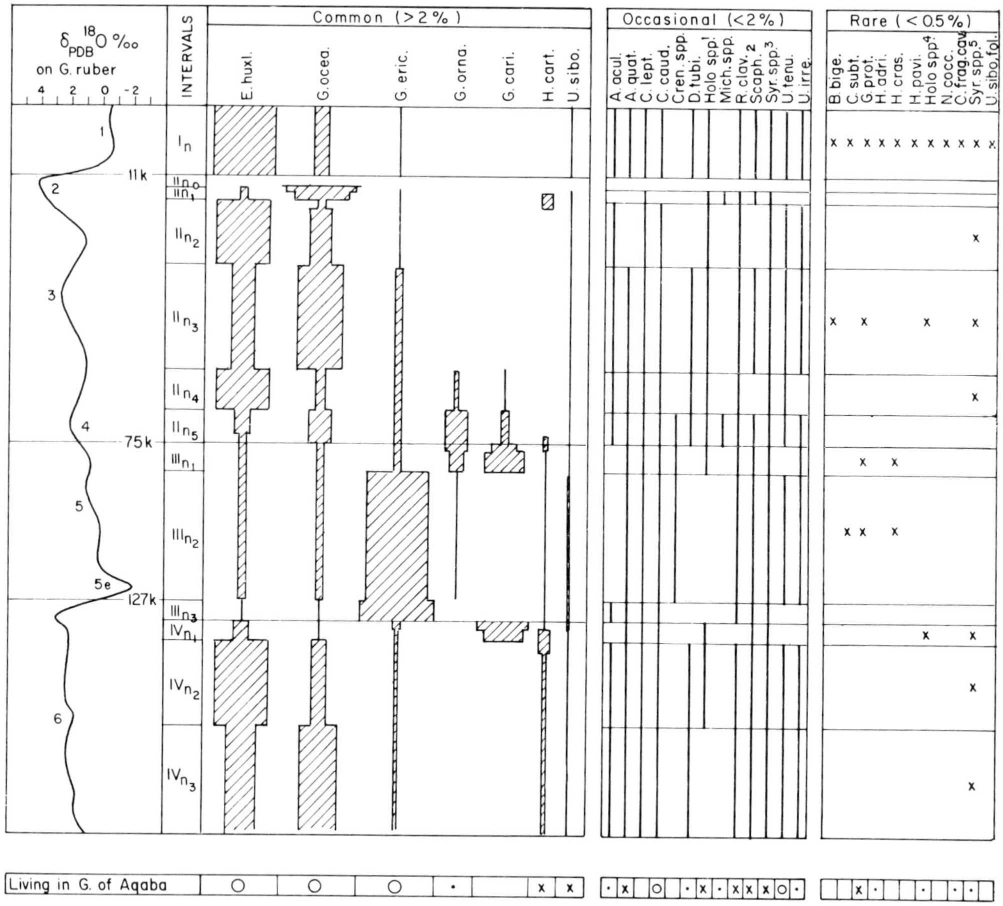

Fig. H.6. Generalized distribution of coccolithophorid species in Gulf of Aqaba and northern Red Sea cores. *1 C. oblonga + H. wettsteini*; *2* all scapholites identified have been placed in *A. brasiliensis*; *3 S. mediterranea + S. pulchra*; *4 C. catillifera + H. cornifera*; *5* all other *Syracosphaera* spp. O-abundant (50%); X-frequent (10%); *-present (Winter 1982a)

Considering the thickness of sediment accumulated between dated levels in the cores, accumulation rates can be calculated for the lutitic sediments in the Gulf of Aqaba. Generally, this rate is on the order of several centimeters per 1,000 years. Red Sea core 68 has similar rates of accumulation. The rate within the inflated section of arkosic sands in core 70 (410–790 cm) is more than 40 cm/1,000 yr. This accumulation rate in the lower part of core 70 and the relatively significant unconformity in the upper part of core 68 are the reasons for the presence of older strata (lower stage 5 and upper 6) in core 68, despite the same approximate length of both cores. High accumulation rates are also indicated in cores 72 and 73, which are composed of several meters of alternating silty muds and quartz sands of turbidite nature and have a fossil assemblage of Holocene aspect. Calculated accumulation rates in the Gulf of Aqaba and in the northern-

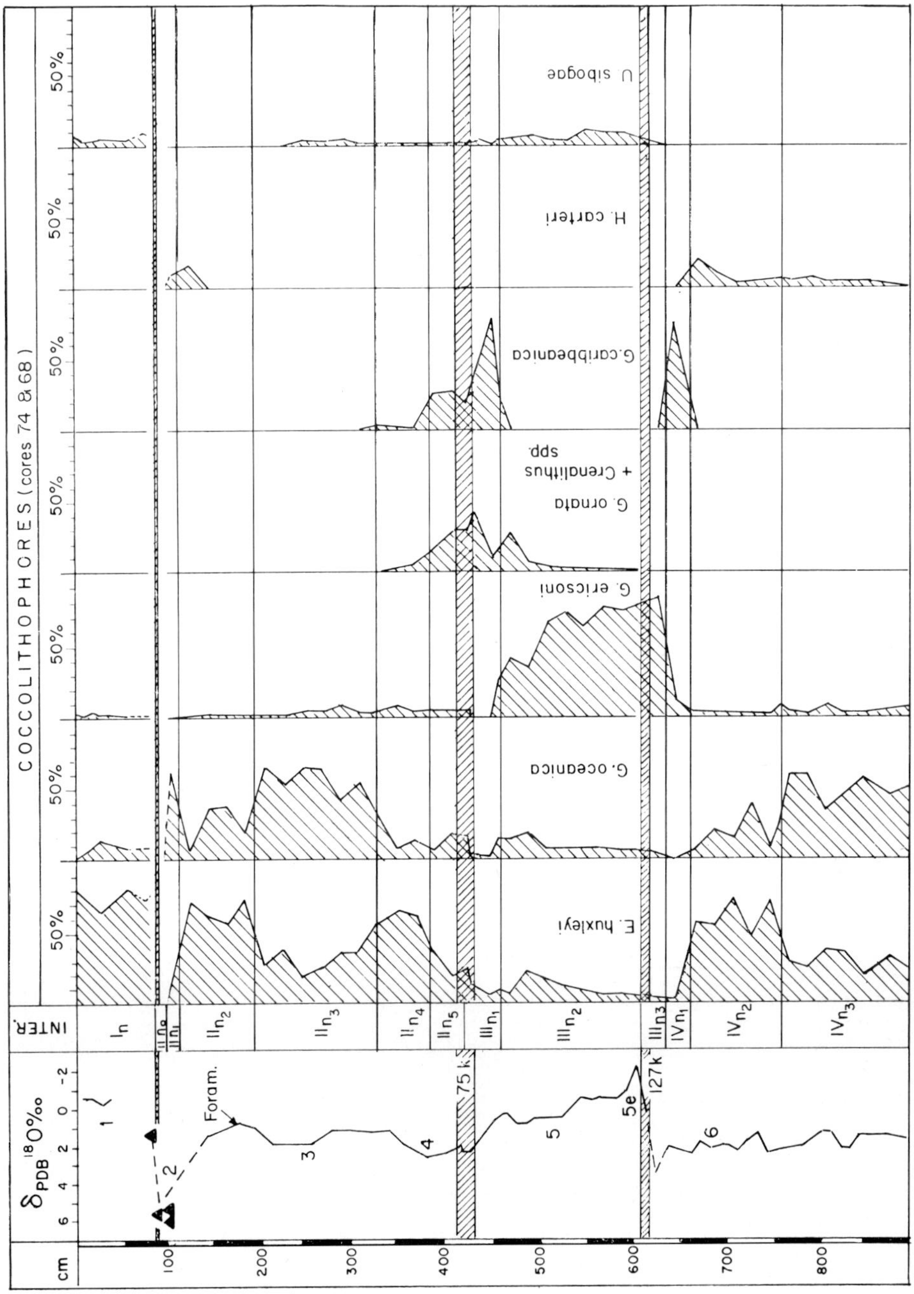

Fig. H.7. Composite selected core record of Coccolithophorids. Based on core 74 for intervals I and II and on core 68 for intervals III and IV. Isotope curve from Luz and Wolf in Reiss et al. (1980) (Winter 1982a)

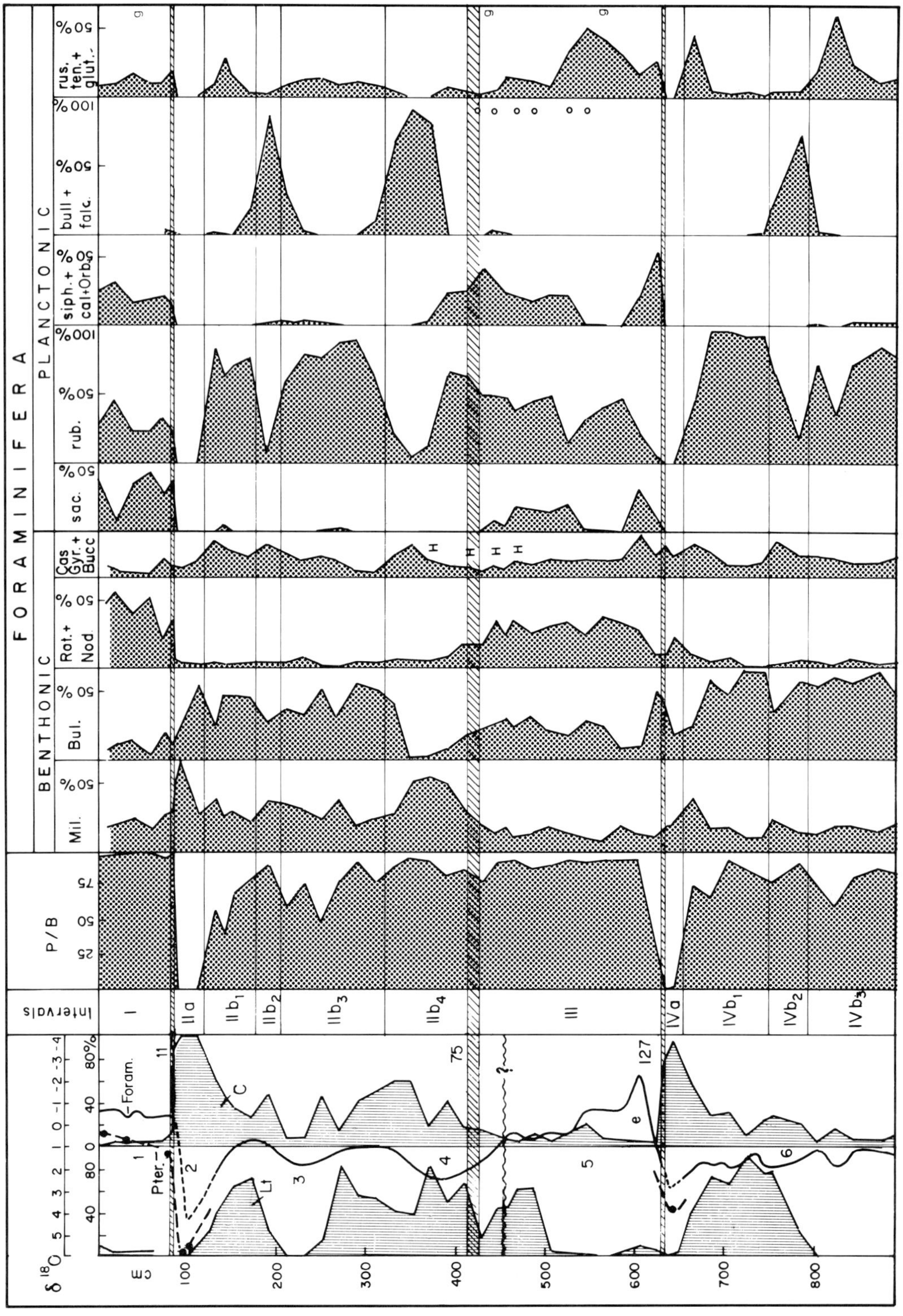

FORAMINIFERA
BENTHONIC
PLANCTONIC
Mil.
Bul.
Rot.+ Nod
Cas Gyr.+ Bucc
sac.
rub.
siph.+ cal+Orb.
bull + falc.
rus. ten.+ glut.
P/B
Intervals
I
IIa
IIb1
IIb2
IIb3
IIb4
III
IVa
IVb1
IVb2
IVb3
δ18O
Foram.
Pter.
C
Lt
e
11
75
127
cm

Fig. H.8

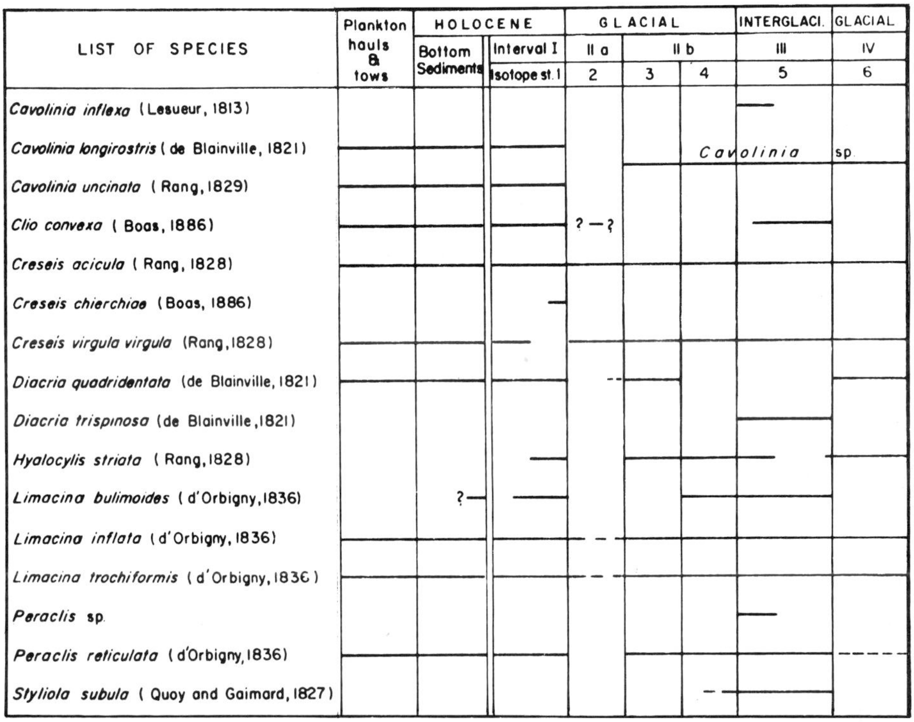

Fig. H.9. Distribution of pteropod species occurring in the water column, bottom sediments and cores of the Gulf of Aqaba and northernmost Red Sea (Almogi-Labin 1982)

most Red Sea during the Holocene (about 10 cm/1,000 yr) are at least twice as high as they were in the Late Pleistocene. This is probably due to compaction effects and possibly also to changes in abiogenic sediment transport into the sea. The fact that the lost piston-core intervals are represented in the trigger-weight cores may suggest that the Holocene sediments are softer than those of Pleistocene age.

◄

Fig. H.8. Composite generalized core record of Foraminifera (and selected Pteropoda). Based for intervals I to IIb4 on core 74 and for intervals III-IVb3 on core 68. Lt-precentage plot of *Limacina trochiformis*; C-percentage plot of *Creseis virgula* and *C. acicula*. Foram. = $\delta^{18}O_{PDB}$ curve derived from planktic Foraminifera (*G. ruber*); Pter. = $\delta^{18}O_{PDB}$ curve derived from Pteropods (*L. trochiformis* and *C. acicula*) (see also Figs. 3.6). 11,75,127-years BP as based on calibrated oxygen isotope curve (Kukla 1977). Mil. = Miliolacea (Miliolidae and Ophthalmidiidae); Bul. = Buliminacea (Buliminidae, Bolivinidae and Uvigerinidae); Rot. = "rotaliformes" (see text); Nod. = Nodosariacea; Cas. = *Cassidulina* and *Islandiella*; Gyr. = *Gyroidina-Gyroidinoides*; Bucc. = *Buccella*; sac. = *Globigerinoides sacculifer*; rub. = *G. ruber*; siph. = *Globigerinella siphonifera*; cal. = *Globigerina calida*; Orb. = *Orbulina universa* and sp.; bull. = *Globigerina bulloides*, falc. = *G. falconensis*; rus. = *Globoturborotalita rubescens*; ten. = *Globoturborotalita tenella*; glut. = *Globigerinita glutinata*; H = *Hoeglundina*; o = *Neogloboquadrina pachyderma* (leftcoiling), smaller than 149 μ; g = *Globigerinoides conglobatus* (Halicz and Reiss 1981)

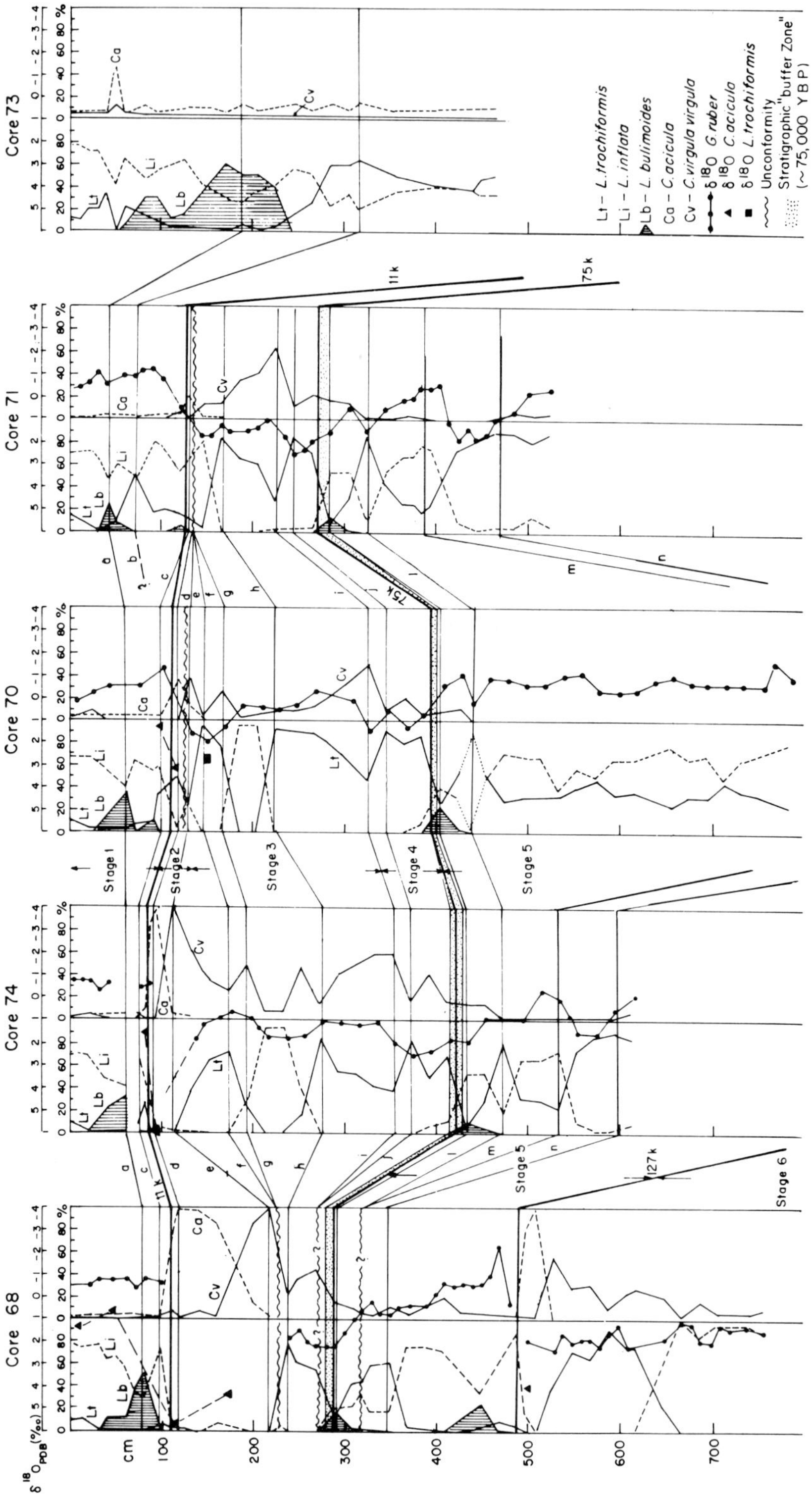

Fig. H.10. Correlation of cores examined by means of quantitative distribution of five species of Pteropods. Oxygen isotopes from B. Luz and M. Wolf (in Reiss et al. 1980) (Almogi-Labin 1982)

H.4 Paleoenvironments

The two independent lines of evidence – micropaleontology and stable isotopes – applied in stratigraphic analysis of the *Atlantis* II cores were also used in **paleoenvironmental interpretation.**

The **presence/absence pattern** of the different taxa recognized in the core-sequences can be seen from Figs. H.6, H.8 and H.9 (see also list of species). The **frequency distribution** of the most important taxa is shown in Figs. H.2 to H.4, H.7, H.8 and H.10. A **composite core record** summarizing the most significant features of the Late Quaternary sequence represented in the *Altantis* II cores is shown in Fig. H.11.

As pointed out above, the **shape** of the stable oxygen isotope curve obtained in cores from the Gulf of Aqaba and in the Red Sea is similar to that of the generalized curve determined from the open ocean. However, the **values of $\delta^{18}O$** obtained on planktic Foraminifera from Glacial intervals are much higher in the Red Sea–Gulf of Aqaba than in the open ocean. The same is true of values obtained from Pteropoda, although only well-preserved specimens were examined from the *Atlantis* II cores, while speciemens examined from Red Sea cores were mostly encrusted by aragonite. Thus, the heavy isotopic composition of pteropods during Glacial maxima must be attributed to their life environment and not to diagenetic effects. Since the difference in $\delta^{18}O$ between planktic Foraminifera and Pteropoda is fairly constant throughout the core sequences, the latters $\delta^{18}O$ values were used in the core study wherever planktic Foraminifera were absent (Glacial maxima, stages 2 and upper 6; see below).

The overall similarity of the oxygen isotope curves in the Red Sea and Gulf of Aqaba to those from the open ocean indicates isotopic composition of the sea water as the major factor determining them. The global isotopic signal is amplified in the Red Sea at times of lowered sea level (Glacials) and much reduced water exchange with the Indian Ocean through the Straits of Bab-el-Mandeb. The reduction of water exchange by itself, even without changes in precipitation/evaporation ratios, may explain the extremely high $\delta^{18}O$ values during Glacial maxima (see Assaf 1977, Deuser et al. 1976, Reiss et al. 1980).

The differences in $\delta^{18}O$ in planktic foraminiferal and pteropod shells between the last Glacial maximum, about 18,000 yr *BP* and the present is about 1.6‰ in the open ocean, less than 2‰ in the Gulf of Aden and in excess of 5‰ in the Red Sea and in the Gulf of Aqaba (see Fig. H.11). This difference does not take into account global and temperature effects and expresses the Glacial/Postglacial change in sea water composition alone. It therefore does not provide a direct clue to changes in temperature and/or salinity.

However, throughout the deep-sea cores studied from the Gulf of Aqaba displaced shells of shallow-water, benthic, tropical-subtropical and symbiont-bearing Foraminifera occur, among them *Amphisorus*, whose present-day minimum winter isotherm is about 17 °C (see Chap. G). This indicates that (a) throughout Glacial times minimum winter temperature of the **upper waters** in the Gulf were **never lower than 17 °C**, hence 4 °C (corresponding to 1‰ $\delta^{18}O$) below that at present, and (b) that the $\delta^{18}O$ values reflect a **rise in salinity** of about 10‰ (cor-

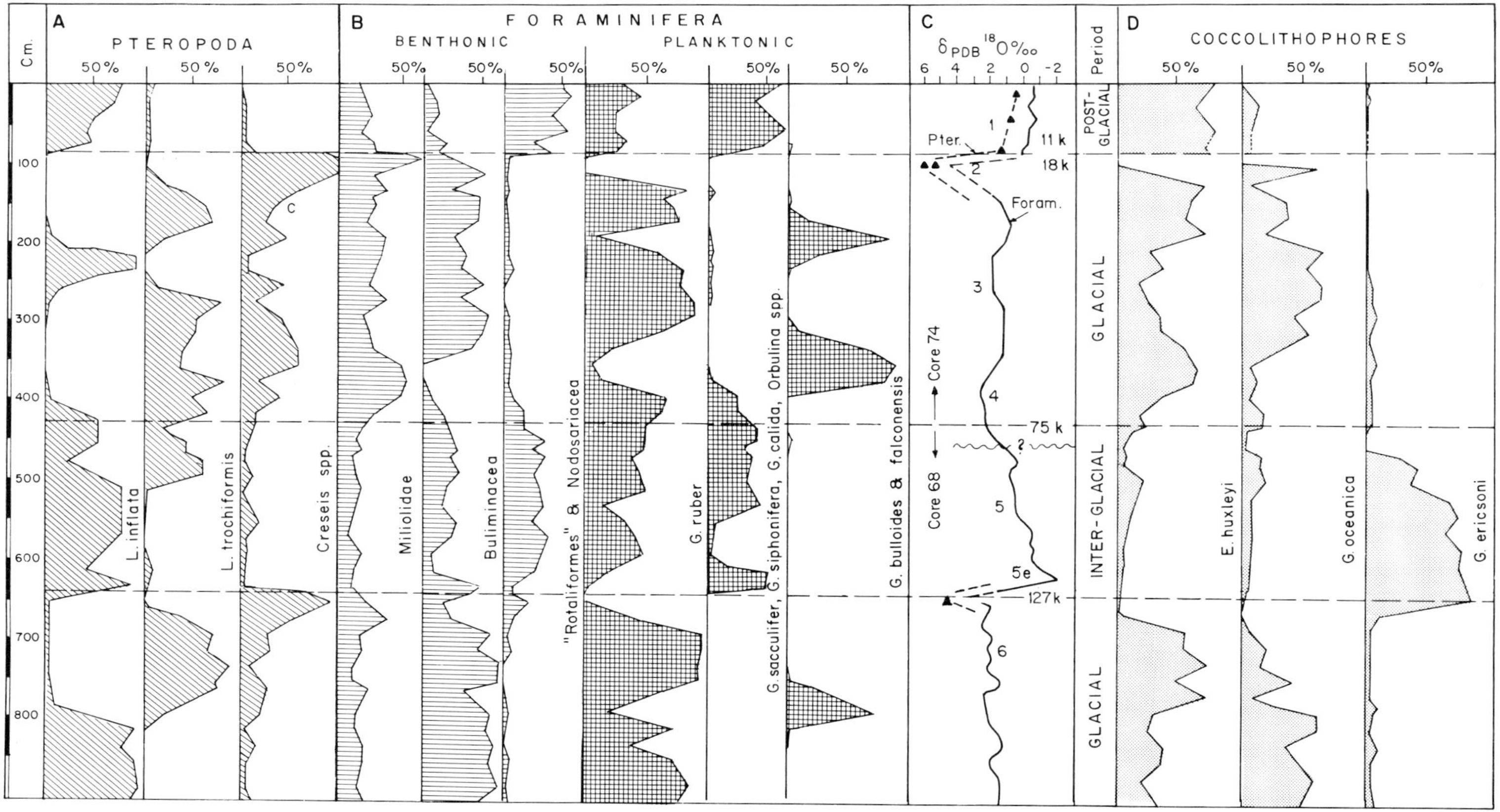

Fig. H.11. Composite generalized core-record of Pteropoda, benthic and planktic Foraminifera, Coccolithophorida and stable oxygen isotopes (based on cores 74 and 68). Stable isotope curve based on *G. ruber* (*Foram*) and *L. trochiformis* and *C. acicula* (*Pter.*). Percentage plots of Foraminifera and Pteropoda based on counts of about 400 specimens larger than 149 μ on the average per sample and on 300 coccoliths per sample (counted under the SEM) (Reiss et al. 1983)

responding to about 4‰ $\delta^{18}O$) as compared to the present, to **more than 50‰** during the last Glacial maximum.

This extremely high salinity seems to be the reason for the total **absence of planktic Foraminifera** during Glacial maxima: indeed, laboratory experiments on living planktic Foraminifera species from the Gulf of Aqaba (Erez, unpubl., Winter 1982a, b) point to a salinity of about 50‰ as the upper tolerance limit for planktic Foraminifera. As proven by the presence of abundant and well-preserved aragonitic pteropod shells (mainly the euryhaline *Creseis acicula*) in strata without planktic Foraminifera, dissolution cannot be invoked to explain the absence of the latter's low-Mg calcitic shells. Furthermore, the benthic foraminiferal assemblages in these intervals (Figs. H.8 and H.11) indicate that deposition took place at depths much greater than those in which planktic Foraminifera are extremly rare to absent in the Gulf of Aqaba (30–50 m) (see Chap. E.3).

Prominent **changes in frequency-pattern** of the group *Globigerinoides sacculifer, Globigerinella siphonifera, Globigerina calida*, and *Orbulina* spp. occur in the sequence (Fig. H.11). It is abundant in Holocene and in Interglacial strata and is nearly absent in the last Glacial interval (except in lower stage 4) and completely so in the penultimate Glacial. Especially important is *Globigerinoides sacculifer* which is also absent in lower stage 4 (Fig. H.8). The absence of *G. sacculifer* (and dominance of *G. ruber*) in the Red Sea in strata deposited during Glacial times was judged by various authors to reflect increase in salinity; this interpretation was based on present-day distribution of this species in the Atlantic Ocean and especially of dominance of *G. ruber* and absence of *G. sacculifer* in the Eastern Mediterranean, where salinities exceed 39‰. However, at present *G. sacculifer* dominates in the Gulf of Aqaba with 41‰ salinity (see Chap. E.3) and is shown in the laboratory to survive salinities of about 48‰. Present-day densities in the Gulf of Aqaba are also far above those suggested as restricting the distribution of *G. sacculifer* in the present-day oceans. On the other hand, considering the surface temperature ranges given for *G. sacculifer*, especially in the Indian Ocean (Fig. H.12), it seems probable that winter surface temperatures below 17 °C prevent this species from surviving. This may be particularly so if water **temperature at depth of reproduction** is considered (see Chap. E.3), especially in strongly illuminated, biologically poor regiones. This may also explain the absence of *G. sacculifer* for the last 1,500 years in the Eastern Levantine Basin, where minimum winter temperature at about 100 m, is 17 °C and fertility of the sea is very low. As will be seen below, it is also reasonable to assume that in the Gulf of Aqaba a stronger thermocline was in existence during Glacial times than at present. *Globigerinella siphonifera, Globigerina calida*, and *Orbulina* have apparently lower temperature limits and shallower reproduction depths than *Globigerinoides sacculifer* (Fig. H.12) and for this reason occur in lower Glacial stage 4. *G. ruber* with a lower surface temperature limit of about 14 °C is the most epipelagic species (Chap. E.3) and survives in Glacial times, except at their maxima when **salinity** rises to more than 50%.

In Glacial intervals where *G. sacculifer* is absent, dramatic peaks of *Globigerina bulloides-falconensis* are observable, compensating mostly for *Globigerinoides ruber* (Figs. H.8 and H.11).

PLANKTONIC FORAMINIFERA SPECIES	SURFACE WATER T° C Range (0 5 10 15 20 25)	Depth Prefer. m and Diel migr.	Up. waters PO_4–P Ind. Oc. optimum µg at/l	Dissolution Resistance	Ind. Oc. Sed. Assemb. S_1	S_2	S_3	S_4	S_5	G. Aqaba 800 m % Pktn. >210 µm	Sed. >63 µm	G. Aqaba seasonality Plankton Northern basin	Southern basin	Red Sea X G. Aqaba O Cores H	GI	IG1
G. pachyderma	(1) 4.8±4.7 (2)	>100	1.7±0.5	H_{17}	X	·	·	·	·	–	–			– / –	X / –	– / O
G. quinqueloba (incl. *G. clarkei*)	9.9±5.4	50 I	1.2±0.6	L_9	X	·	·	·	·	–	54.0			X / O	X / O	X / O
G. bulloides (incl. *G. falconensis*)	13.4±7.8	50–100 H	1.0±0.6	L_8	X	·	·	·	X	–	0.1			X / O	X / O	X / O
G. glutinata	22.2±7.5	50–100 L	0.2±1.2	L_{10}	·	X	X	·	·	–	8.0			X / O	X / O	X / O
O. universa	21.7±2.9	50–100 H	0.3±0.2	L_2	·	·	(X)	·	·	0.2	0.4			X / O	X / (O)	X / O
G. siphonifera	23.5±4.5	50–100 I	0.5	L_3	·	·	X	X	·	27	5.4	W 23%	W 16%	X / O	– / (O)	X / O
G. ruber	24.2±4.0	50 L	0.5	L_1	·	·	X	X	(X)	20	7.6	S 11%	T 7%	X / O	X / O	X / O
G. sacculifer	25±3.2	50 L	0.5	L_5	·	·	X	X	·	51	8.3	S–W 64%	S–W 78%	X / O	– / –	X / O
G. conglobatus	23.9±2.3	50 L	0.5	L_7	·	·	(X)	·	·	–	–			(X) / O	X / –	X / O
G. rubescens	? 23±4.2	50	0.2	L_4	·	·	(X)	(X)	·	–	13.0			X / O	– / O	X / O
G. tenellus	?	50–100 I		L_6	·	·	X	·	·	–	0.9			X / O	X / O	X / O
G. calida	?	50–100 H								0.5	0.3			O	(O)	O
G. uvula	? 7.5±5.2	H	1.4±0.4		·	·	·	·	·					O	O	O
G. bradyi	? 7.5±5.0	H	1.4											O	O	O
G. humilis				H_{22}										–	O	O
G. iota														O	–	O
Beella														–	–	O

Fig. H.12. Distribution and selected ecological ranges and optima of planktic Foraminifera occurring in *Atlantis* II cores from the Gulf of Aqaba and northernmost Red Sea. Generic names on this figure must be revised according to the list given p. 289. *1* Total temperature range, *2* Indian Ocean temperature range and optimum acc. to Bé (1977) and Bé and Hutson (1977). *Stippled:* range in Gulf of Aqaba. Depth preference, diel migration and fertility of water mass (PO_4-P) acc. to Bé (1977). Dissolution resistance ranking acc. to Berger (1970). Indian Ocean sediment assemblages acc. to Bé and Hutson (1977) (x-dominant, o-present). Distribution in living plankton and in sediment faunas of the Gulf of Aqaba according to Almogi-Labin (unpublished). Red Sea core according to Rosenberg-Herman (1965) (Halicz and Reiss 1981)

While *G. ruber* is a well-known low-fertility-associated species, *G. bulloides-falconensis* is a group with a wide temperature range and associated with high nutrient content of the sea, especially abundant in areas of upwelling. It is also increasingly frequent along the stretch Gulf of Aqaba – Gulf of Aden, following the gradient of increasing productivity (see references in Halicz and Reiss 1981, Chap. E.3 and Fig. H.12.) **Higher fertility of the photic zone** in the Gulf of Aqaba and northern Red Sea during Glacial times is, therefore, suggested by the planktic Foraminifera.

Coscinodiscid diatoms occur in abundance in strata of the last maximum Glacial in *Atlantis* II core 68 in the northernmost Red Sea (see Figs. H.4 and H.5), indicating increased fertility of the upper waters and possibly strong upwelling outside the Straits of Tiran.

The frequency distribution in the *Atlantis* II cores of the major constituents of the **coccolithophore assemblages** is shown in Figs. H.4 to H.7 and H.11. The most conspicuous quantitative feature of the core record is the interplay of *Emiliania huxleyi* and *Gephyrocapsa oceanica*, which together dominate the core assemblages, except during the last Glacial maximum and during Interglacial stage 5, when *G. ericsoni* is the major species.

Considering present-day **temperature** ranges of the major species (Fig. H.13), temperatures during Glacials in the Gulf of Aqaba would have had to deviate greatly from present-day ones in order to affect the coccolithophore community. However, as explained above, minimum winter temperature of the upper waters apparently did not drop by more than 4 °C and could not have significantly influenced the coccolithophore floras. The small present-day temperature variation along the stretch Gulf of Aden – Red Sea – Gulf of Aqaba, well within the optimum range of both *E. huxleyi* and *G. oceanica*, cannot be responsible for the prominent change in frequency of these species along this stretch, from a *G. oceanica*-dominated to an *E. huxleyi*-dominanted assemblage and neither can temperature be invoked to account for the dramatic *huxleyi/oceanica* ratio changes in the cores (see also Chap. E.2 and below).

Temperature may have influenced the dominance of *G. ericsoni* during most of Interglacial stage 5, where it replaces *E. huxleyi*. *G. ericsoni* is generally a warm water indicator, occasional to common in the waters of the Equatorial to Transitional Zones of the Pacific and Atlantic Oceans, its optimum temperature range being, however, lower than that of *G. oceanica*. The fact that the $\delta^{18}O$ of stage 5 (especially its lower part, where *G. ericsoni* dominates) are generally lighter than those of the Holocene–Recent may be evidence that *G. ericsoni* prefers generally warm, although cooler waters than at present in the Gulf of Aqaba. It may also be noteworthy that *G. ericsoni* dominates in the penultimate Glacial maximum (uppermost stage 6) where planktic Foraminifera are absent, but where $\delta^{18}O$ values of pteropod shells are lighter than those in the last Glacial maximum (Figs. H.4 and H.11).

The dominance of *G. ericsoni* in isotope stage 5 seems to be a widely distributed phenomenon and reported among others from the Red Sea proper and from the North Atlantic. However, the absence of this species in other Pleistocene cycles in the N. Atlantic Ocean indicates that temperature alone does not explain its distribution. Judging from living plankton and number of coccoliths in surface

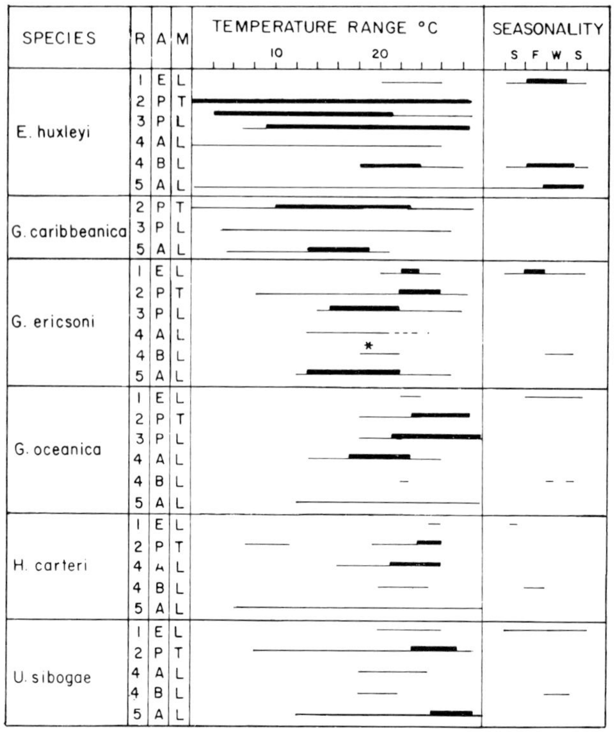

Fig. H.13. Temperature ranges and seasonality of common coccolithophore species discussed in text. R = References: *1* Winter et al. (1979), *2* Roth and Berger (1975), *3* McIntyre et al. (1970), *4* McIntyre and Bé (1967), *5* Okada and McIntyre (1979). A = Area: *E* Gulf of Aqaba, *P* North Pacific, *A* North Atlantic, *B* Bermuda. M = Material: *L* living, *T* core tops (Winter 1982 a)

sediments of the Gulf of Aqaba, *G. ericsoni* produces less coccoliths than (multiple-layered) *E. huxleyi* and low numbers of *ericsoni* coccoliths in sediments may thus be due to masking effects (as explained previously, no dissolution occurs in the Gulf). Possibly in the past *G. ericsoni* had a higher plate production which led – together with ecological influences (e.g., temperature) – to relatively greater proportional representation. Temperature may be the factor influencing the frequent occurrence of *G. caribbeanica* in the cores, at times of rapid temperature changes (Fig. H.6 and H.7). This species optimum lies between 13° and 19 °C, and in surface sediments of the Pacific Ocean it is found mainly in "transitional" regions (see also discussion of pteropods below).

E. huxleyi is the most ubiquitous, most eurythermal and most euryhaline coccolithophore species, ranging from the subarctic to the tropics and living in **salinities** between 10‰ (Sea of Azov) and 41‰ (Gulf of Aqaba). *G. oceanica* has been reported to have at present a more restricted range, preferring warm waters and marginal seas with normal salinity. The fact that, in the last Glacial maximum interval of the cores in the Gulf of Aqaba, where planktic Foraminifera are already absent, *G. oceanica* is the only coccolithophore present, *E. huxleyi* having disappeared before, indicates that – although euryhaline – **E. huxleyi is less tolerant of**

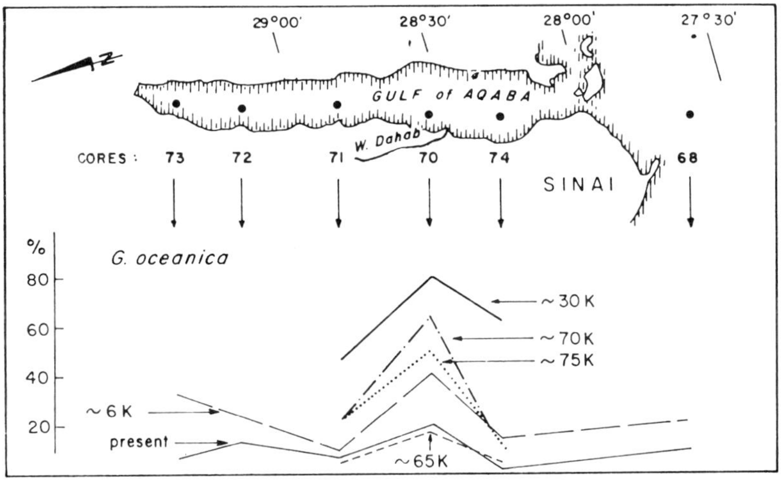

Fig. H.14. Percentage of *G. oceanica* in *Atlantis* II cores, Gulf of Aqaba at different Late Quaternary times (Winter 1982a)

extremely high salinities than G. oceanica (Winter et al. 1983). Indeed, laboratory experiments have shown 45‰ salinity to be the upper tolerance limit for *E. huxleyi*. The upper salinity limit of *G. oceanica* lies probably nearer to 50‰, considering isotope and planktic foraminiferal evidence. At even higher salinities, no coccolithophorids are present any longer in the Gulf of Aqaba cores. It is noteworthy that during stage 2 coccolithophorids disappear earlier in the northernmost Red Sea than in the Gulf of Aqaba; salinity differences may be the cause (see below). Salinity, however, does not explain either the low numbers of *G. oceanica* in the present-day Gulf of Aqaba as compared with the southern Red Sea or the Indian Ocean or, for that matter, the dominance of *G. oceanica* over *E. huxleyi-G. ericsoni* in Glacial intervals of the cores. As explained above on the basis of present-day distribution and seasonality (Chap. E.2), the ratio *huxleyi/oceanica* is strongly influenced by **fertility of the sea,** *E. huxleyi* dominating in oligotrophic environments and (the "nerithic") *G. oceanica* being abundant in fertile waters. The dominance of *G. oceanica* in Glacial intervals of the cores seems, therefore, to indicate higher **fertility of the photic zone**, in support of the conclusions drawn on the basis of planktic Foraminifera (see also discussion of pteropods below). It may be noteworthy that throughout the intervals representing the last 75,000 yr *BP* of core 70 drilled at the edge of the Dahab delta (and containing the arkosic sediments of stage 5), the percentage of *G. oceanica* is higher than in any other core in the Gulf of Aqaba (Fig. H.14). In the same core *Calcidiscus leptoporus* is more frequent than in other cores; this latter species shows the same north–south trend of increase from the Gulf of Aqaba to the Indian Ocean, paralleling that of the fertility-influenced *G. oceanica*.

Wadi Dahab carries at present considerable quantities of freshwater from Sinai into the Gulf of Aqaba during winter floods, thus contributing to higher nutrient content in the vicinity of its mouth. As indicated by the arkosic sequence in core 70, W. Dahab was active throughout the Late Quaternary and influenced the higher abundance of *G. oceanica* and *C. leptoporus* in core 70.

Although the relation between water **stratification** and coccolithophore distribution is not clear at present, some species are known to occupy preferentially certain parts of the water column (see Chap. E.2). *G. oceanica* is known to prefer the upper to middle euphotic layer. Interestingly enough, the abundance, of *G. oceanica* decreases somewhat (and that of *E. huxleyi* increases) in Glacial intervals of the cores deposited during times of stronger stratification of the water column, as interpreted from the difference in $\delta^{18}O$ between planktic and benthic Foraminifera (Fig. H.15 and see below). Possibly a **strong pycnocline** has adversely affected *G. oceanica* in the past. It may be noteworthy in this respect that living *G. oceanica* are absent during summer periods of (albeit slight) stratification in the Gulf of Aqaba.

The **pteropod assemblages** are of particular interest to the paleoenvironmental interpretation of the core sequences.

The stratigraphic distribution of the species is given in Fig. H.9 and the frequency distribution of the major components is shown in Figs. H.2 to H.4, H.8 and H.10, H.11 and H.15.

Prominent are the opposite trends of the curves for *Limacina inflata* and *L. trochiformis*, as well as the orderly succession of the frequency peaks *L. inflata* – *L. trochiformis* – *Creseis* spp. Among the *Creseis* species, *C. virgula* peaks are followed in the Glacial maxima by monospecific assemblages of *C. acicula*. *L. bulimoides* peaks occur at or near the boundaries between Glacial and Inter(Post-)Glacial stages. As can be seen from Fig. H.16, the trends of the *L. trochiformis* and the planktic-benthic foraminiferal $\triangle\,\delta^{18}O$ are parallel, while peaks of *Globigerina bulloides-falconensis* always occur within intervals of abundant *L. trochiformis*.

There is also a definite order of disappearance of pteropod species groups from Interglacial to Glacial stages and within the latter, viz. (1) *Clio convexa, Diacria trispinosa, Cavolinia inflexa* and *Peraclis moluccensis*; (2) *Styliola subula* and *Limacina bulimoides*; (3) *L. inflata, L. trochiformis* and *Hyalocylix striata*; (4) *Creseis virgula*; and (5) *C. acicula*. An explanation of this order of disappearance in terms of **temperature** is both difficult and internally inconsistent (see Fig. E.30). Indeed, if the total known surface temperature range of the species recognized is considered, one would conclude that temperature in the Gulf of Aqaba never fell below 18 °C at any time during the Late Quaternary, an estimate which closely supports the paleotemperature derived from stable oxygen isotopes and displaced benthic Foraminifera. If Indian Ocean temperature ranges of pteropods are considered, one would even be tempted to suggest minimum winter temperatures during the Late Quaternary of not less than 20 °C.

The abundance of *Limacina inflata* in Red Sea cores has been related to warmer periods (Herman 1968), but – although *L. inflata* is indeed abundant in stages 1 and 5 and absent in Glacial periods – the minimum water temperature during those times was above the lower temperature limit of this species. Moreover, prominent peaks of *L. inflata* occur also within Glacial stages 3 and 6. *L. trochiformis* occurs at present in the world oceans at temperatures slightly higher than *L. inflata*. In the Gulf of Aqaba this species is nearly absent during winter months when temperature (21 °C) is lower by about 6 °C than in summer, while *L. inflata* occurs both during summer and winter. On the other hand, *L. tro-*

chiformis is abundant in Glacial intervals, when temperature was certainly lower than at present.

The exclusive presence of *Creseis acicula* during Glacial maxima indicates that this species is able to withstand highly saline conditions.

Most of the species occurring in Glacial stages live at present under hypersaline conditions in the Gulf of Aqaba. *Styliola subula, Limacina bulimoides* and *Hyalocylix striata*, reported from the world oceans in **salinities** below 39‰, although absent in the present-day Gulf of Aqaba, are present and occasionally abundant in Glacial intervals of the cores with calculated salinities far above 41‰.

On the other hand, *Diacria trispinosa, Clio inflexa, Peraclis ?moluccensis* and *Styliola subula*, present in Interglacial stage 5, are absent in the Holocene when salinities were probably similar to those of stage 5. Thus, most of the presence/absence pattern of pteropods in the cores cannot be explained by salinity changes only. This is also true to the abundance pattern of various species. For example, the prominent peaks of *Limacina inflata* and *L. trochiformis* alternate within intervals (stages 3 and 6) with the same calculated temperature and salinity.

At present the pteropod assemblages living in the Gulf of Aqaba are of epipelagic and mesopelagic character, despite the depth of 1,830 m (see Chap. E.4). During Glacial stages the assemblages are solely epipelagic, with the exception of middle stage 3 and lower stage 6 (*L. inflata* peaks).

Although it would be tempting to interpret this change in terms of **depth**, i.e., of a considerable fall of sea level, the evidence from the *Atlantis* II core, drilled at 800–1,000 m water depth, [continuity of isotopic and fossil record, indicating connection throughout Glacial times of the Gulf of Aqaba and the Red Sea with the Gulf of Aden through the Straits of Tiran (252 m) and Bab-el-Mandeb (137 m), varying extent and position of unconformities, characteristics of authochtonous benthic foraminiferal assemblages; see above] does not support a conclusion that sea level in the Gulf of Aqaba fell below the global one (see also Chap. C).

Futhermore, an attempt to interpret the prominent and rapid changes in frequency of *Limacina inflata* and *L. trochiformis* in upper stage 5 in terms of depth (Chen 1969) would lead to the conclusion that depth changed drastically several times over a short time interval. The frequency of *L. bulimoides* in Red Sea cores has been suggested to indicate transition between moderately deep and deep seas; in the *Atlantis* II cores *L. bulimoides*-peaks occur indeed in lower stage 5 and at the base of stage 4; however, such a prominent peak occurs also **within** the Holocene, when considerable changes in depth of the sea certainly did not occur.

An alternative explanation for the exclusively epipelagic nature of the Glacial pteropod assemblages may be sought in **changes in circulation pattern**. Mesopelagic species, like *Limacina inflata, Styliola subula* and *Clio convexa*, are known to have pronounced **diel migration**. Strong **stratification** of the water column may hamper such migration. The difference in $\delta^{18}O$ between planktic and benthic Foraminifera in the core sequences (Fig. H.15) strongly suggests increased stratification during Glacial stages as compared with the Holocene and Recent or with the last Interglacial. Further evidence for stronger stratification is provided by

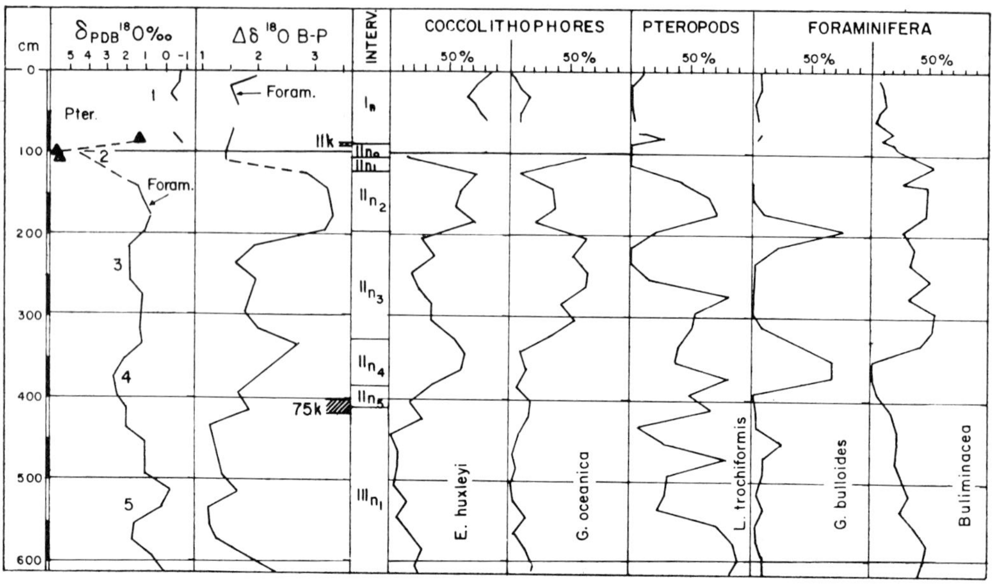

Fig. H.15. Abundance of selected faunal and floral species and stable isotope curves in core 74. Isotope data from Luz and Wolf (in Reiss et al. 1980) and from Luz (pers. com., 1981); Pteropoda curve from Almogi-Labin (1982); Foraminifera curves from Halicz and Reiss (1981) (Winter 1982a)

benthic foraminiferal assemblages and their stable carbon isotopes (see below), as well as by theoretical considerations of physical oceanography (Anati 1980b).

The observations in the present-day Gulf of Aqaba and analysis of distribution patterns of pteropods in the world oceans have drawn attention to an important ecological factor to which, until recently, little attention has been paid, viz. **fertility of the sea** (see Chap. E.4).

The most abundant species in the present-day oligotrophic Gulf of Aqaba is *Limacina inflata.* This species, although occurring frequently in the Somali current, is mostly characteristic of oligotrophic gyre centers (see Chap. D for comparison between the Gulf of Aqaba with oceanic gyre centers). *L. inflata* is also dominant in the Holocene and in the last Interglacial, where it is accompanied in parts of the sequence by *L. bulimoides*, as well as – in stage 5 – by frequent *Styliola subula. L. bulimoides* and *S. subula* are likewise gyre-center types, although *L. bulimoides* occurs also frequently in the Somali and Brazil currents. *L. inflata* and *S. subula* have strong diel migration. *L. trochiformis* is characteristic of highly productive areas, like boundary currents and upwelling regions and is abundant in the Gulf of Aqaba and northermost Red Sea during a short period in autumn, when primary production increases (Chap. E.4).

It seems, therefore, that the opposite trends of *L. inflata* and *L. trochiformis*, despite their similar temperature and salinity ranges, indicate changes in the fertility of the sea, coupled with changes in circulation pattern (stratification); abundance of *L. trochiformis* seems to indicate more productive and more stratified waters, while abundance of *L. inflata* seems to point to biologically poorer and

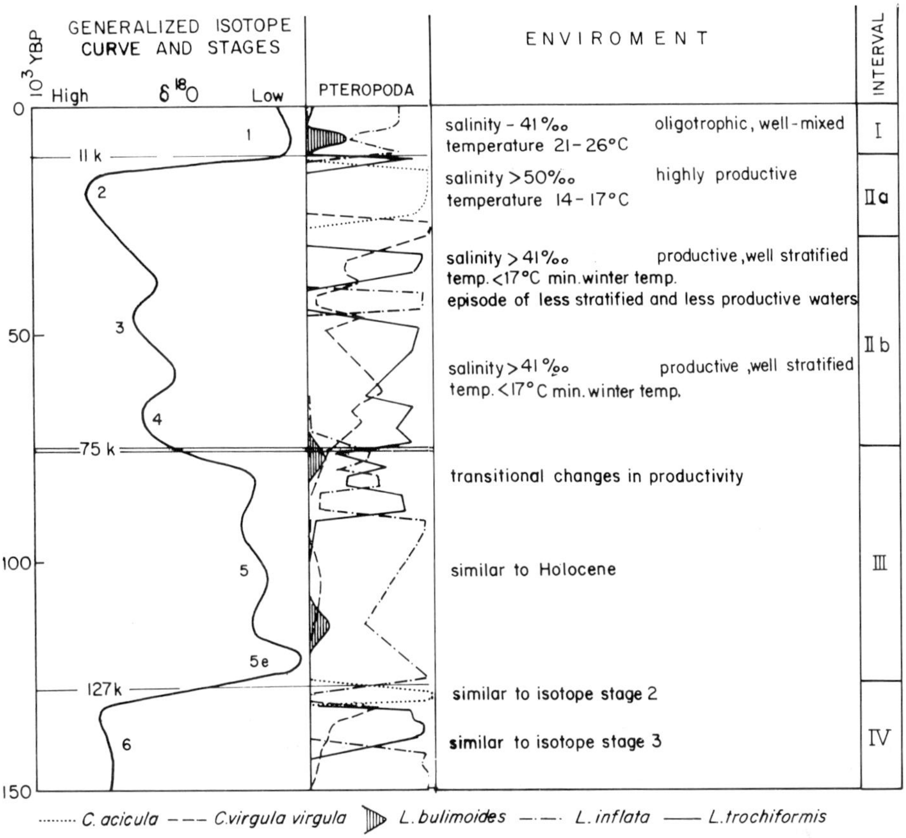

Fig. H.16. Interpretation of paleoceanographic changes in the Gulf of Aqaba during the Late Quaternary as indicated by Pteropoda and stable oxygen isotopes (Almogi-Labin 1982)

less well-stratified waters. The co-occurence of abundant *L. bulimoides* with very frequent *L. inflata* and *L. trochiformis* at the base of stage 4, preceding the strong increase of *L. trochiformis,* may suggest that *L. bulimoides* is intermediate between *L. inflata* and *L. trochiformis* as far as fertility of the sea is concerned. The fact that *L. bulimoides* is more abundant in cores 73 and 68, i.e., in the northermost parts of the Gulf of Aqaba and of the Red Sea respectively, may point to upwelling in those times in the northernmost part of the basins. The development of increasing aridity in the area (Gat and Margaritz 1980, Luz 1979) and decreasing fertility of the Red Sea–Gulf of Aqaba (Reiss et al. 1980) (as well as in the Eastern Mediterranean) may have been responsible for the near-disappearance of *L. bulimoides* from this region.

As pointed out before (Chap. E.3), the genus *Neogloboquadrina* is usually associated with fertile waters and is absent in the present-day Gulf of Aqaba, but frequent in the southern Red Sea and Gulf of Aden. *Neogloboquadrina pachyderma* (left-coiling) (whose ecological range is certainly wider than generally assumed) occurs – albeit rarely – in the *Atlantis* II cores exclusively in the upper part of stage 5 (Fig. H.8), together with a considerable increase in *Limacina trochiformis*, which indicates changes to somewhat higher productivity.

Diacria trispinosa, Clio inflexa and *Styliola subula* are characteristic of isotope stage 5 and they are all absent from the core record after the upper isotope stage 4 (Fig. H.11). All three pteropod species mentioned are absent from the present-day Arabian Sea. It has been suggested that the absence of *S. subula* may be due to the very low oxygen content of the waters below the thermocline. Possibly this barrier was established at the beginning of isotope stage 4.

The "nerithic" *Creseis acicula*, generally regarded as the most eurythermal and euryhaline pteropod species, is known to occur in seasonal dense swarms associated with high zooplankton productivity, as well as with nitrogen-fixing *Trichodesmium* blooms. The abundance of this species in the Glacial maxima of the *Atlantis* II core sequence and its association with rich diatom floras in stage 2 of core 68 (see above) indicate both more fertile and highly saline waters (and not necessarily very low sea level; Chen 1969).

Creseis virgula virgula is regarded in general as an oceanic, stenohaline (33–37‰) and stenothermal, epipelagic species. On the other hand, this species occurs at present in the hypersaline waters of the Gulf of Aqaba. The association of abundant *C. virgula virgula* with *Limacina trochiformis* on one hand and with *C. acicula* on the other in Glacial stages is strongly suggestive of fertility of the sea as the major determining factor. The same is true in the case of the ratios *L. inflata*/*L. trochiformis* and of *Creseis* spp./*L. inflata*, which show opposite trends (Figs. H.2 to H.4).

Clio convexa is generally characteristic of equatorial productive waters and lives at present in the Gulf of Aqaba down to 600 m depth, being present there during the short winter period when productivity is relatively high, especially in the southern basin of the Gulf (see Chap. E). *C. convexa* occurs in cores in the lower Interglacial stage 5 and in the middle of Postglacial stage 1, when conditions were similar, but apparently not identical to the present.

A summary of the paleoenvironmental interpretation of the *Atlantis* II core sequences in terms of pteropod ecology is given in Fig. H.16. This latter interpretation is in full agreement with those derived from stable oxygen isotopes, from coccolithophores and planktic Foraminifera (cf. Fig. H.11 and see also below with regard to benthic Foraminifera). It may be noteworthy that the *L. inflata* peak in Glacial stage 3, indicating low fertility of the sea, is bracketed between the peaks of *Globigerina bulloides-falconensis* indicative of high fertility; this peak is radiocarbon-dated in Red Sea cores at about 30,000 yr *BP*, but may be in fact older (~40,000 yr *BP*). It reflects an important episode within the last Glacial all over the Red Sea and Gulf of Aqaba.

The paleoenvironmentally significant *Atlantis* II core record of the **benthic Foraminifera** is summarized in Fig. H.11 (see also Figs. H.2 to H.4, and H.8, references in Halicz and Reiss 1981). In comparison with the Recent Gulf of Aqaba, the Holocene, and Interglacial stage 5, Glacial stages 2–4 and 6 are characterized by lower species diversity, by somewhat lower plankton/benthos ratios (with absence of planktic Foraminifera in Glacial maxima of stages 2 and upper 6), as well as by a dramatic decrease in numbers of specimens belonging to genera grouped in "rotaliformes" and nodosariacenas (see list above). This latter decrease in Glacial stages is compensated by a considerable increase in specimens of Buliminacea, as well as of Miliolidae (especially in stages 2–4).

Miliolaceans are known at present to be abundant among others in high-salinity areas (such as the Persian Gulf lagoons), in hypersaline mangroves and in sebkhas (Gulf of Aqaba). Thus, increase in salinity probably accounts for the high percentages of miliolaceans in the glacial Gulf of Aqaba–northern Red Sea (see also Reiss et al. 1980).

The trends of the **δ^{18}O curves derived from benthic Foraminifera** (Fig. H.17) parallel those of the planktic ones. This indicates that the benthic specimens are not reworked. Moreover, the benthic specimens in general and the miliolid ones in particular have heavier isotopic composition of the shell than the planktic ones (see below). This indicates that these specimens are autochtonous and not displaced from shallow waters, a conclusion supported by the scarcity or absence of taxa which are abundant in deeper parts of the photic zone.

Buliminacea (especially *Bulimina, Globobulimina, Bolivina*) occur in great abundance – regardless of depth – in marine environments with low oxygen, high primary production in the photic zone and high organic content of the sediments. Abundance of buliminaceans (including *Bulimina marginata*, the main buliminid in the Red Sea and its gulfs) is also observed in polluted shelf areas, with high nitrogen content of the muddy sediments (for references see Halicz and Reiss 1981).

B. marginata is frequent at present in the Gulf of Suez, which is rather poor in nutrients and highly oxygenated, in fine-grained sediments with high foraminiferal numbers and with the highest nitrogen content as compared to the Red Sea proper or the Gulf of Aqaba (Said 1950a). Buliminaceans are also abundant in the (generally well-oxygenated and nutrient-poor) Persian Gulf, in muddy sediments with high C_{org}/N ratios below 50 m depth. It is noteworthy that particularly abundant buliminacean assemblages (*Globobulimina*) occur in Quaternary deep-sea sediments of the Mediterranean, near and within sapropels, and inhabit at present a relatively deep basin in the northern Aegean Sea, where low-salinity surface waters from the Black Sea lead to stratification and low-oxygen bottom conditions.

As pointed out by Berger (1979) and by Douglas and Woodruff (1981), in many cases variable properties of water bodies like oxygen and nutrient content may be more important biologically than the conservative properties like temperature and salinity. Dissolved oxygen and calcium carbonate saturation levels are apparently the most important factors influencing the distribution of modern deep-sea benthic Foraminifera. On the other hand, these factors are interrelated and are a function of the fertility of the surface waters and the stirring of the ocean. Low oxygen levels are associated with high organic content of the sediments and for this reason the distribution of *Uvigerina, Bolivina, Fursenkoina* or *Globobulimina* found in low oxygen habitats may be due to food availability or trophic structure or some combination of minimum oxygen levels required for respiration and food. Thus, the organic content of the sediment and the oxygen level at the water–sediment interface or even within the uppermost layers of sediment, coupled with food and oxygen requirements of such species as *B. marginata* and other buliminids may be the major factors affecting their particular abundance in specimen-rich, low diversity assemblages.

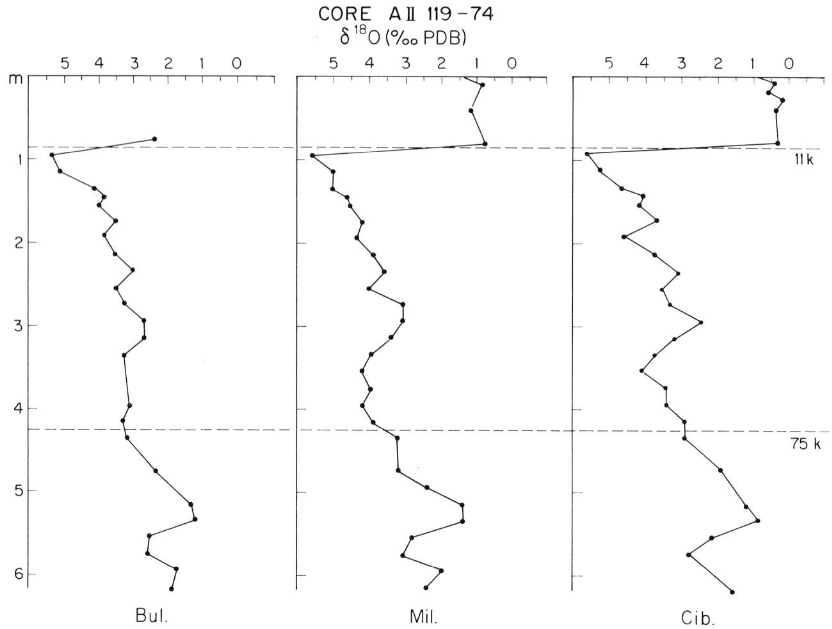

Fig. H.17. Stable oxygen isotope record of benthic Foraminifera from core 74. (*Bul.* Buliminacea, *Mil.* Miliolidae, *Cib.* Cibicididae) (Luz and Reiss 1983)

It would appear, therefore, that the benthic foraminiferal assemblages in the glacial intervals of the cores examined indicate lowered oxygen at the bottom. This is in agreement with the indications of higher fertility provided by the associated planktic foraminiferal, as well as by the pteropod and coccolithophorid assemblages.

Increased stratification of the water column in the buliminacean-rich layers is indicated by the difference in $\delta^{18}O$ between planktic and benthic Foraminifera; this difference is considerably larger in Glacial than in Interglacial or Holocene intervals (Fig. H.18). This stratification may be due either to temperature or to salinity or to both. Physical oceanographic considerations (Anati 1980) indicate that salinity between surface and deep waters, due to restricted water exchange through the straits during Glacial periods, is the major factor.

Figure H.17 shows the **stable oxygen isotope values** obtained from **benthic Foraminifera** in core 74. The samples examined are multispecific and grouped into three major groups, i.e., Miliolidae comprising several species of small, unornamented *Quinqueloculina, Triloculina* and *Spiroloculina*; Buliminacea comprising mainly varying proportions of *Brizalina* cf. *B. subargentea* Uchio, *B. amygdalaeformis* (Brady), *Bolivina hebes hebes* Macfadyen and *B. hebes suezensis* Said, as well as small numbers of *Bulimina marginata* d'Orbigny; and Cibicididae, comprising mainly *Cibicides mabahethi* Said and small numbers of *C. tabaensis* Perelis and Reiss.

It is important to note that the absolute $\delta^{18}O$ values of the three foraminiferal groups differ considerably and that only the miliolaceans seem to be close to iso-

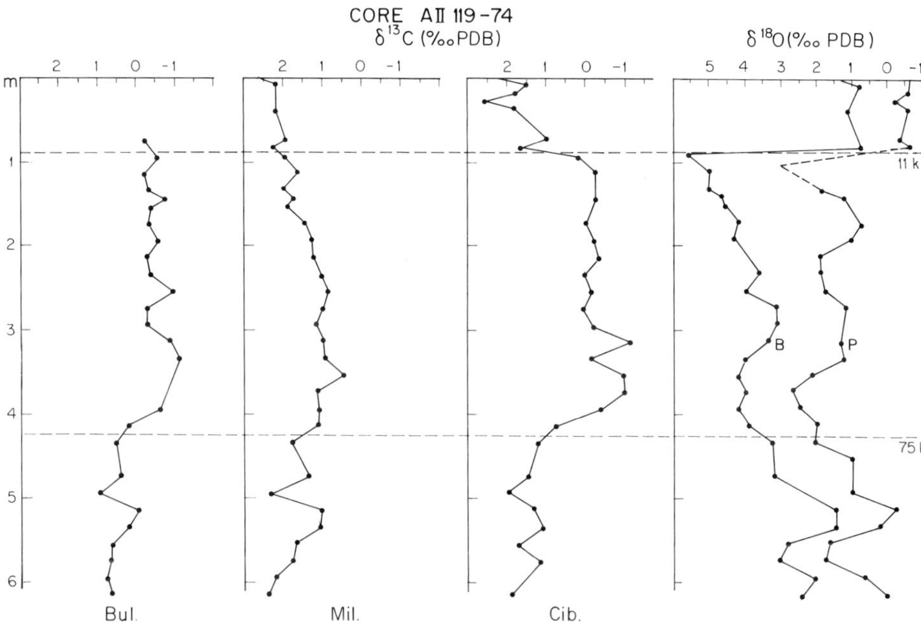

Fig. H.18. Stable carbon isotope record of benthic Foraminifera compared to the stable oxygen isotope record of planktic and benthic Foraminifera from core 74 (Luz and Reiss 1983)

topic equilibrium (based on post-Glacial data). Nevertheless, all three groups show very similar trends. Thus, regardless of the biological effect causing isotopic disequilibrium, the fluctuating isotopic ratios can be interpreted in terms of past environmental cnditions. The possibility is not exluded that the **fundamentally different mode of shell formation** in perforate-lamellar and porcelanous Foraminifera influences oxygen isotopic fractionation (Luz and Reiss 1983).

The **stable carbon isotopic composition of planktic foraminiferal shells** throughout the Glacial–Interglacial sequence does not show any significant changes as opposed to strong Glacial–Interglacial changes in $\delta^{13}C$ of **benthic Foraminifera**. The stable carbon isotopic composition of benthic foraminiferal shells obtained from the same samples as those shown in Fig. H.17 are given in Fig. H.18, alongside the averaged stable oxygen isotopic curves obtained from planktic and benthic Foraminifera. The Holocene specimens are all in disequilibrium if compared to present-day conditions in the deep water of the Gulf of Aqaba (Table H.3).

Departures from carbon isotopic equilibrium are common in perforate Foraminifera and might be related to internal cycling of organic carbon within the living Foraminifera. However, many porcelaneous species are known to be at or close to isotropic equilibrium (Luz and Reiss 1983, Luz et al. 1983, and references therein). As in the case of oxygen isotopes (see above) mode of shell formation may equally influence carbon isotope fractionation.

The apparent deviation from equilibrium (by 1.4‰) with the deep water of the miliolids may thus indicate their **infaunal habitat**. If this is true then they live in

Table H.3. Measured deep-water parameters and inferred ones for the Glacial interval (Luz and Reiss 1983)

	Present (and Post-Glacial)	Glacial (75,000–11,000 yr BP)
O_2 (mmol l^{-1})	0.2 [a]	0.03 [d]
ΣCO_2 (mmol l^{-1})	2.1 [b]	2.27 [d]
$\delta^{13}C$ of ΣCO_2 (‰ PDB)	1.5 [b]	−0.3
$\delta^{13}C$ of equilibrium calcite (‰ PDB)	3.4 [c]	1.6

[a] After Klinker et al. (1976). [b] After Shemesh et al. (1981)
[c] Calculated according to Emrich et al. (1970) for 20 °C. $\delta^{13}C$ equilibrium calcite $= \delta^{13}C$ dissolved $HCO_3^- + 1.15 + 0.035\,T$
[d] Calculated assuming $^{13}C_{org} = -23$‰

a microniche whose $\delta^{13}C$ of the total dissolved CO_2 ($\Sigma\ CO_2$) is 1.5–1.4 = 0.1‰ (Table H.3). Assuming that −23‰ organic carbon ($\delta^{13}C_{org}$) is oxidized in the sediment, then the resulting changes in both O_2 and CO_2 with respect to the deep water can be calculated from simple material balance considerations. Let D be the increase in $\Sigma\ CO_2$ or decrease of O_2 in mmol l^{-1} in the sediment and $\triangle$ the difference between $\delta^{13}C$ of miliolid calcitic shell and the calculated $\delta^{13}C$ of deep-water equilibrium-calcite (−1.4‰). At present, $\delta^{13}C$ in the CO_2 of the deep water is 1.5‰ and $\Sigma\ CO_2$ is 2.1 mmol l^{-1} (Table H.3).

$$\Delta + (\delta^{13}C_{CO_2})_{present} = \frac{D \times \delta^{13}C_{org} + (\delta^{13}C_{\Sigma CO_2} \times \Sigma CO_2)_{present}}{D + (\Sigma CO_2)_{present}}. \qquad (1)$$

By solving for D, a value of D = 0.13 mmol l^{-1} is obtained, indicating that the O_2 concentration in the miliolid microniche is at present 0.07 mmol l^{-1} or 1.6 ml l^{-1}. This estimate is not very sensitive to small changes in $\delta^{13}C_{org}$. For example for $\delta^{13}C_{org} = -26$‰, D is calculated at 0.11 mmol l^{-1} corresponding to 2.0 ml O_2 l^{-1} in the miliolid microniche.

Averaging the $\delta^{13}C$ values for Glacial and Interglacial intervals and comparing the Glacial–Interglacial changes in each group (Table H.4) it is of interest to note that the change in both miliolaceans and buliminaceans (despite the difference in absolute values) in the same (−1‰), while the strongest change occurs in the (largely monospecific) *Cibicides* assemblage (−1.8‰). This may support the suggestion that both **miliolaceans and buliminaceans share an infaunal habitat**, while the **epifaunal** *Cibidides* is affected by considerable changes in the deep water. It follows that equilibrium calcite with the Glacial deep-water had a $\delta^{13}C$ of 1.6‰ (3.4‰ present $+\triangle_{G\text{-}IG}$). Because temperature fractionation of carbon isotopes is relatively small, the same change ($\triangle_{G\text{-}IG}$) can be assigned to $\delta^{13}C$ of $\Sigma\ CO_2$ of Glacial and Interglacial times (Table H.3). One can now use Eq. (1) for calculating $\Sigma\ CO_2$ and O_2 concentration of the Glacial deep waters in the Gulf of Aqaba. Let D be the change in O_2 or CO_2 between the Glacial and the present and substitute $\triangle_{G\text{-}IG}$ for $\triangle$. Solving for D, one obtaines D = 0.17 mmol l^{-1} or 3.8 ml O_2 l^{-1}. Thus, the Glacial deep water in the Gulf of Aqaba had a **very low O_2-level, viz. about 0.7 ml l^{-1}**.

Table H.4. Average $\delta^{13}C$ (‰PDB) of benthic Foraminifera in Glacial and Interglacial intervals and the difference between these periods (Δ_{G-IG}), Atlantis II core 74 (Luz and Reiss 1983)

	Interglacial	Glacial	$\Delta G-IG$
Cibicididae	1.66	−0.15	−1.81
Buliminaceans	0.52	−0.49	−1.01
Miliolaceans	2.00	1.34	−0.66

Recalling that this estimate depends on the $\delta^{13}C$ of the oxidized organic carbon ($\delta^{13}C_{org}$) it is possible that at least part of the Glacial–Interglacial change results from a difference in $\delta^{13}C_{org}$. However, it is unlikely that this would greatly affect the estimated O_2 reduction. Calculation with Eq. (1) demonstrates that in order to decrease the change in O_2 level by one-half, $\delta^{13}C_{org}$ has to be about −45‰! In the same way the differences among the cibicidids, miliolaceans and buliminaceans in the change between Glacial and Interglacial $\delta^{13}C$ (Table H.4) perhaps may be related to changes in food sources with the latter's own (considerable) difference in $\delta^{13}C$. It is not known whether this latter consideration is realistic, and whether the spectrum of $\delta^{13}C$ of the available food sources for benthic Foraminifera is large enough. The fact that the lowering of $\delta^{13}C$ in the benthic Foraminifera corresponds to increase in fertility of the upper waters, points to **oxygen reduction, due to increased particulate flux to the deep water** as the most probable mechanism.

The increase in fertility of the photic zone during Glacials in the desert-enclosed Red Sea–Gulf of Aqaba is probably in the main the result of **heightened nutrient inflow from the Indian Ocean** due to increase of upwelling during Glacial times (see Vincent and Berger 1981). Some nutrient apport from land into the Gulf of Aqaba is also possible (see below).

Glacial–Interglacial changes in the chemistry of deep and interstitial waters, in addition to changes in stratification and salinity, are probably responsible for the observed changes in **carbonate mineralogy** in the Gulf of Aqaba and in the Red Sea. Thus, significant amounts of authigenic **Mg calcite** occur in sediments, especially in Interglacial and Holocene intervals. Precipitated **aragonite crusts on pteropod shells** occur in Glacial maxima, much more prominently so in the Red Sea proper. Generally, preservation of pteropods in Glacial intervals is excellent, while partial dissolution, as well as internal mold formation occur in Interglacial strata, especially of the Red Sea proper.

Extreme post-depositional **shape modifications of coccoliths by Mg calcite overgrowth**, leading to forms which strongly resemble discoasters, have been observed in Glacial and Interglacial intervals from the Gulf of Aqaba and especially from the Red Sea. These overgrown coccoliths occur together with well-preserved, dissolution-prone holococcoliths and pteropods. This is taken as an indication that the source of calcite necessary for secondary overgrowth in the Red Sea and Gulf of Aqaba is not previously dissolved calcareous organisms (as in the deep ocean, near the calcite lysocline), but the warm, highly saline waters, supersaturated with regard to $CaCO_3$.

The mineralogical study of the core samples reveals, as indicated above, differences between the Red Sea and the Gulf of Aqaba, especially during Glacial times. Other differences, mentioned already, are the absence of coccolithophorids and abundance of diatoms in the northernmost Red Sea during stage 2, when coccolithophorids are present and diatoms absent in the Gulf of Aqaba. Another prominent difference is the presence of widespread, mainly aragonitic **lithified layers** in the Red Sea and the absence of such layers in the Gulf of Aqaba.

Although the explanation of these differences requires further study, they may be connected with differences in salinity: the salinity in the Gulf of Aqaba may have been slightly lower than in the Red Sea, at least during the last Glacial maximum. If so, relatively higher freshwater contribution from land must be assumed. Indeed, a climate more humid than at present during the Glacial maximum for the Sinai Peninsula was advocated by various authors (see references in Reiss et al. 1980) and connected with the erosional pattern of the fringing reef assumed to have developed during the last Glacial low stand of sea level (Gvirtzman et al. 1977). Such increased run-off from land may have also contributed to higher fertility of the waters in the Gulf of Aqaba (see discussion in Winter, 1982a). Since beyond extremely high salinities the $\delta^{18}O$ varies only slightly with changes in salinity (Lloyd 1966), freshwater apport from Sinai into the Gulf of Aqaba during the Glacial maximum would not be reflected in the isotopic record of Pteropoda (planktic Foraminifera are absent at that time). At least in the Red Sea salinity reached such high values that aragonite precipitation could take place (Gevirtz and Friedman 1966).

The fact that increased fertility of the photic zone during Glacials is indicated both in the Gulf of Aqaba and in the Red Sea, suggests however, that the major source of increased nutrient supply was the Indian Ocean.

H.5 Paleoceanographic History

Micropaleontological and stable isotope analysis of deep-sea cores, in the light of knowledge derived from distribution and seasonality of living faunas and floras in relation to associated ecological parameters, facilitates the reconstruction of the **paleoceanographic history** of the Gulf of Aqaba and northernmost Red Sea over approximately the last 150,000 yr.

Combining all evidence available, one may conclude that during Glacial times **minimum winter temperature** of the upper waters was lower by not more than 4 °C (i.e. 17 °C), while **salinity** rose by more than 10‰ (to **more than 50‰**), as compared to the present, as a result of global cooling, sea-level changes and strait dynamics (both at Bab-el-Manbed and at Tiran). Reduced water exchange through the straits during Glacial times led to **stronger stratification** of the water column (due mainly to salinity) than at present, **longer residence time** of the deep water and higher input/output ratios of nutrients. The **increase in fertility of the photic zone** in the Glacial stratified Gulf of Aqaba is attributable mainly to higher fertility of the ocean in general and increased nutrient supply from the Red Sea, probably enhanced by upwelling at the latter's northern end, as well as possibly to some apport from Sinai (compare Issar and Bruins 1983).

Despite the increase in fertility of the upper waters during Glacial times, no *Globorotalia, Pulleniatina* or deep water pteropods penetrated into the northern Red Sea and into the Gulf of Aqaba; this may be due to a **south–north temperature gradient** stronger than at present and connected to generally cooler waters especially at greater depth, as well as to extremely **strong density gradients** (see Chap. E.3).

The increased fertility of the photic zone in the Gulf of Aqaba during Glacial times resulted in large amounts of organic material reaching the bottom, thus leading to **low oxygen conditions near the interface**. A short period of lowered productivity is indicated in middle isotope stage 3 (30,000–45,000 yr *BP*), both in the Gulf of Aqaba and in the Red Sea proper.

Although Interglacial isotope stage 5 is very similar to the Holocene, their characteristics are not identical. This is expressed both by differences in assemblage composition, as well as by stable oxygen isotope values (Fig. H.12). It would appear that stage 5 was characterized by less extreme conditions than the Holocene. From both micropaleontological and stable isotope data and based on the high-resolution multiple stratigraphic framework it seems that the **present-day conditions** of a "blue sea desert" were established in the Late Holocene, **about 4,000 yr** ***BP***.

The presence of various species – extant in the world oceans – of Coccolithophorida, planktic Foraminifera and Pteropoda, in core intervals which represent Glacial maxima with extreme salinity conditions for which no present-day analog exists, makes it possible to deduce **upper salinity tolerance limits** of these species from the paleoceanographic study of the Gulf of Aqaba.

References: Almogi-Labin (1982), Almogi-Labin and Reiss (1977), Anati (1980a), Assaf (1977), Berger (1979), Berggren (1969), Chen (1969), Degens and Ross (1969), Deuser et al. (1976), Douglas and Woodruff (1981), Emery (1963), Emrich et al. (1970), Gat and Magaritz (1980), Gevirtz and Friedman (1966), Gvirtzman et al. (1977), Halicz and Reiss (1981), Herman (1968), Klinker et al. (1976), Lloyd (1966), Luz (1979, 1982, 1984), Luz and Reiss (1983), Luz et al. (1983), Morcos (1970), Reiss (1977), Reiss et al. (1974, 1980, 1984), Rosenberg-Herman (1965), Said (1950a, b), Shemesh (1980), Shemesh et al. (1981), Vincent and Berger (1981), Winter (1982a,b), Winter et al. (1978, 1979, 1983).

References

Allan TD, Morelli C (1971) The Red Sea. In: Maxwell E (ed) The sea, vol. IV. Wiley, New York, pp 493–542

Almogi-Labin A (1981) Paleoceanographic significance of Pteropoda and planktonic Foraminiferida in the water column and sediments of the Gulf of Elat (Aqaba) and northern Red Sea. Ph D thesis, Hebrew Univ, Jerusalem, 158 pp (in Hebrew)

Almogi-Labin A (1982) Stratigraphic and paleoceanographic significance of Late Quaternary pteropods from deep-sea cores in the Gulf of Aqaba (Elat) and northernmost Red Sea. Mar Micropaleontol 7:53–72

Almogi-Labin A, Reiss Z (1977) Quaternary Pteropods from Israel. Rev Esp Micropaleontol 9:5–48

Amiran R (1969) Ancient pottery of the Holy Land from its beginnings in the neolithic period to the end of the iron age. Massada Press, Jerusalem

Anati DA (1974) Volume transports in the Gulf of Aqaba, L'Océanographie Physique de la Mer Rouge. CNEXO Publ No 2 Ser Act Colloq, IAPSO Symp, Paris, pp 165–173

Anati DA (1976) Balances and Transports in the Red Sea and the Gulf of Elat (Aqaba). Isr J Earth-Sci 24:104–110

Anati DA, (1980 a) A parametrization of the geometry of sea straits. Oceanol Acta 3:395–397

Anati DA (1980 b) The cross-axial thermohaline structure of the Gulf of Elat. Internal Rep Weizmann Inst Sci 1–10 (Rehovot)

Anati DA, Assaf G, Thompson RORY (1977) Laboratory models of sea straits. J Fluid Mech 81:341–351

Assaf G (1977) Sea straits and glacial periods in the Red Sea. Science 195:90

Assaf G, Anati DA (1974) Stress distribution in the Red Sea and in the Gulf of Aqaba. J Phys Oceanogr 4:663–668

Assaf G, Hecht A (1974) Sea straits: a dynamic model. Deep-Sea Res 21:947–958

Assaf G, Kessler Y (1976) Climate and energy exchange in the Gulf to Aqaba (Elat). Mon Weather Rev 104:383–385

Avraham S (1981) Population dynamics of planktic foraminifera in the Gulf of Elat. Unpublished report Mar Biol Lab, Elat 24 pp (in Hebrew)

Bé AWH (1977) An Ecological, Zoogeographic and Taxonomic Review of Recent Planktonic Foraminifera. In: Ramsay ATS (ed) Oceanic micropaleontology, vol I: Academic Press, London New York, pp 1–100

Bé AWH, Gilmer RW (1977) A Zoogeographic and Taxonomic Review of Euthecosomatous Pteropoda. In: Ramsay ATS (ed) Oceanic micropaleontology, vol I. Academic Press, London New York, pp 733–808

Bé AWH, Hutson WH (1977) Ecology of planktonic foraminifera and biogeographic patterns of life and fossil assemblages in the Indian Ocean. Micropaleontology 23:369–414

Bé AWH, Hemleben, Chr, Anderson OR, Spindler M, Hacunda J, Tuntivate-Choy S (1977) Laboratory and field observations of living planktonic foraminifera. Micropaleontology 23:155–179

Ben-Avraham Z, Garfunkel Z, Almagor G, Hall JK (1979 a) Continental Breakup by a Leaky Transform: The Gulf of Elat (Aqaba). Science 206:214–216

Ben-Avraham Z, Almagor G, Garfunkel Z (1979b) Sediments and structure of the Gulf of Elat (Aqaba)-northern Red Sea. Sediment Geol 23:239–267

Benayahu Y, Loya Y (1977) Space partitioning by stony corals, soft corals and benthic algae on the coral reefs of the northern Gulf of Eilat (Red Sea). Helgol Wiss Meeresunters 30:362–382

Berger WH (1979) Preservation of foraminifera. In: Foraminiferal ecology and paleoecology. SEPM Short Course No 6. Soc Econ Paleontol Mineral Houston, pp 105–155

Berger WH (1981) Paleoceanography: the deep-sea record. In Emiliani C (ed) The oceanic lithosphere. The sea, vol VII. Wiley, New York, pp 1437–1519

Berger HW, Bé AWH, Vincent E (eds) (1981) Oxygen and carbon isotopes in Foraminifera. Paleogeogr Paleoclimatol Paleoecol 33:1–277

Berggren WA (1969) Micropaleontological investigations of the Red Sea cores – summation and synthesis of results. In: Degens ET, Ross DA (eds), Hot brines and recent heavy metal deposits in the Red Sea. Springer, Berlin Heidelberg New York, pp 282–298

Berthold W-U (1976) Biomineralisation bei milioliden Foraminiferen und die Matrizen-Hypothese. Naturwissenschaften 63:196

Billman H, Hottinger L, Oesterle H (1980) Neogene to Recent rotaliid foraminifera from the Indopacific Ocean: their canal system, their classification and their stratigraphic use. Schweiz Palaeontol Abh 101:71–113

Buchardt B, Hansen H-J (1977) Oxygen isotope fractionation and algal symbiosis in benthic foraminifera from the Gulf of Elat, Israel. Bull Geol Soc Den 26:185–194

Chen C (1969) Pteropods in the hot brines sediments of the Red Sea. In: Degens ET, Ross DA (eds) Hot brines and recent heavy metal deposits in the Red Sea. Springer, Berlin Heidelberg New York, pp 313–316

Cifelli R, Smith RK (1974) Distributional patterns of planktonic Foraminifera in the Western North Atlantic. J Foram Res 4:112–125

Cohen S (1975) Red Sea divers guide. Taracoda, Int Guides Albums, Tel Aviv

Collins AC (1958) Foraminifera. Scient Repts Great Barrier Reef Exped 1928–1929, Brit Museum Nat Hist London VI:335–437

Dan J, Yaalon DH (1982) Automorphic saline soils in Israel. In: Yaalon DH (ed) Aridic soils and geomorphic processes. Catena Suppl 1:103–115

Deacon, GER (1952) Preliminary hydrological report, The "Manihine Expedition to the Gulf of Aqaba." Bull Br Mus Nat Hist Zool 1:159–162

Degens ET, Ross DA (eds) (1969) Hot brines and recent heavy metal deposits in the Red Sea. Springer, Berlin Heidelberg New York

Deuser WG, Ross EH, Waterman LS (1976) Glacial and pluvial periods: Their relationship revealed by Pleistocene sediments of the Red Sea and Gulf of Aden. Science 191:1168–1170

Douglas R, Woodruff F (1981) Deep-Sea Benthic Foraminifera. In: Emiliani C (ed) The oceanic lithosphere. The sea, vol VII. Wiley, New York, pp 1233–1327

Eggink JW, Baumfalk YA (1983) The exceptional reproduction and embryonic morphology of *Orbitoides gensacicus* (Late Cretaceous, S. France) J Foram Res 13 (3):179–190

Ehrlich A (1983) The distribution of living diatom assemblages in some saline water bodies of Sinai-Paeloecological implications for the diatomiferous layers of the Pleistocene Lisan marls. Isr Geol Surv Bull 72:19

Emery KO (1963) Sediments of the Gulf of Aqaba (Elat). In: Miller RL (ed) Papers in marine geology. Macmillan, New York, pp 257–273

Emrich K, Ehhalt DH, Vogel JC (1970) Carbon isotope fractionation during the precipitation of calcium carbonate. Earth Planet Sci Lett 8:363–371

Epstein S, Buchsbaum HA, Lowenstam HA, Urey HC (1953) Revised carbonate-water isotopic temperature scale. Bull Geol Soc Amer 64:1315–1325

Erez J (1972) Multivariate analysis of biogenic constituents in sediments of Ras Burka, Gulf of Elat. M Sc thesis, Dep Geol, Hebrew Univ, Jerusalem (in Hebrew)

Erez J (1978) Vital effect on stable isotope-composition seen in foraminifera and coral skeletons. Nature (London) 273:199–202

Erez J (1983) Calcification rates, photosynthesis and light in planktonic Foraminifera. In Westbroek P, deJong WE (eds) Biomineralization and biological metal accumulations, Reidel, Dordrecht, pp 307–312

Erez J, Gill D (1977) Multivariate Analysis of Biogenic Constituents in Recent Sediments off Ras Burka, Gulf of Elat, Red Sea. Math Geol 9:77–98

Erez J, Luz B (1982) Temperature control of oxygen-isotope fractionation of cultured planktonic foraminifera. Nature (London) 297:220–222

Erez J, Luz B (1983) Experimental paleotemperature equation for planktonic foraminifera. Geochim Cosmochim Acta 47:1025–1031

Erez J, Peguine C (1982) Project light. Unpublished report Mar Biol Lab, Elat

Erez J, Reches Z, Garfunkel Z (1983) Tectonic-sedimentologic underwater survey of the shallow margins of the Gulf of Elat. Abstr Isr Geol Soc Annu Meet 1983, Nazaret, pp 20–21

Eyal M, Bartov Y, Shimron AE and Bentor YK (1980) Sinai-Geol Map 1 : 500,000. Geol Survey Israel, Jerusalem

Ferber I (1972) Intertidal occurrence of the heart-urchin *Lovenia elongata* (Echinodermata, Echinoidea). Sci Newslett Mar Biol Lab Elat 2:10

Fermont WJJ (1977a) Biometrical investigation of the genus *Operculina* in recent sediments of the Gulf of Elat. Utrecht Micropaleontol Bull 15:171–204

Fermont WJJ (1977b) Depth gradient in interval parameters of *Heterostegina* in the Gulf of Elat. Utrecht Micropaleontol Bull 15:149–163

Fermont, WJJ, Kreulen R, Zwaan GJ van der (1983) Morphology and stable isotopes as indicators of productivity and feeding patterns in Recent *Operculina ammonoides* (Gronovius). J Foram Res 13:122–128

Frenkel H (1974) Observations on some Shallow Benthonic Foraminifera in the Gulf of Elat, Israel. Isr J Earth-Sci 23:63–67

Fricke H (1982a) Vorstoß an die Grenze des Lichts. Geo 2:8–26

Fricke H (1982b) Lawinen am Meeresgrund. Geo 11:62–76

Fricke H, Hottinger L (1983) Coral bioherms below the euphotic zone in the Red Sea. Mar Ecol Prog Ser 11:113–117

Fricke H, Schuhmacher H (1984) The depth limits of Red Sea stony corals: an ecophysiological problem (A deep diving survey by submersible). Mar Ecol in press

Friedman GM (1968) Geology and geochemistry of reefs, carbonate sediments and waters, Gulf of Aqaba (Elat), Red Sea. J Sediment Petrol 38:895–919

Gabrie C (1982) Sédimentologie de quelques recifs coralliens frangeants de la Mer Rouge et de l'Océan Indien Occidental. Traitement mathematique des données Thèse, Univ Aix-Marseille II, Fac Sci

Garfunkel Z (1981) Internal structure of the Dead Sea leaky transform (rift) in relation to plate kinematics. In: Freund R, Garfunkel Z (eds) The Dead Sea rift. Tectonophysics 80:81–108

Garfunkel Z, Zak I, Freund R (1981) Active faulting in the Dead Sea rift. In: Freund R, Garfunkel Z (eds) The Dead Sea rift. Tectonophysics 80:1–26

Gartner S (1981) Calcareous nannofossils in marine sediments. In: Emiliani C (ed) The oceanic lithosphere. The sea, vol VII. Wiley, New York, pp 1145–1177

Gat JR, Magaritz M (1980) Climatic Variations in the Eastern Mediterranean Sea Area. Naturwissenschaften 67:80–87

Gavish E, Krumbein W, Tamir N (1978) Recent clastic (carbonate) sediments and sabkhas marginal to the Gulfs of Elat and Suez. Abstr 10th Int Congr Sedimentol, Jerusalem, pp 309–329

Geitzenauer KR, Roche MB, McIntyre A (1976) Modern Pacific coccolith assemblages: derivation and application to Late Pleistocene paleotemperature analysis. Geol Soc Am Mem 145:423–448

Gevirtz JL, Friedman GM (1966) Deep Sea carbonate sediments of the Red Sea and their implications on marine lithification. J Sedimentol Petrol 36:143–151

Gorsky G (1975) Distribution of Euphausiacea (Crustacea) in the Gulf of Elat, Red Sea. M Sc thesis, Hebrew Univ, Jerusalem (in Hebrew)

Gvirtzman G, Buchbinder B (1978) Recent and Pleistocene Coral Reefs and Coastal Sediments of the Gulf of Elat. 10th Int Congr Sedimentol, vol II, Jerusalem, Guideb, pp 162–191

Gvirtzman G, Buchbinder B, Sneh A, Nir Y., Friedman GM (1977) Morphology of the Red Sea fringing reefs: a result of the erosional pattern of the last-glacial low-stand sea level and the following Holocene recolonization. Mem BRGM 89:480–491

Halicz E, Reiss Z (1979) Recent Textulariidae from the Gulf of Elat (Aqaba), Red Sea. Rev Esp Micropaleontol XL (2):295–320

Halicz E, Reiss Z (1981) Paleoecological relations of foraminifera in a desert-enclosed sea – the Gulf of Aqaba (Elat), Red Sea. Mar Ecol 2:15–34

Halicz E, Noy N, Reiss Z (1984) Foraminifera from Shura Arwashie Mangrove. In: Por FD, Dor I (eds) Aquatic biology of the mangal ecosystem. Developments in Hydrobiology. Junk, Den Haag: 145–150

Hall J (1975) Bathymetric chart of the Straits of Tiran. Isr J Earth-Sci 24:69–72
Hall J, Ben-Avraham Z (1978) New bathymetric map of the Gulf of Elat (Aqaba). Abstr 10th Int Congr Sedimentol, vol I, Jerusalem, p 285 and map 1:250,000
Hallock P (1979) Trends in test shape with depth in large, symbiont-bearing foraminifera. J Foram Res 9 (1):61–69
Hallock P (1981) Light dependence in Amphistegina. J Foram Res 11 (1):40–46
Hallock P, Hansen HJ (1979) Depth adaptation in Amphistegina: change in lamellar thickness. Bull Geol Soc Den 27:99–104
Hallock P, Larsen, AR (1979) Coiling direction in Amphistegina Mar Micropaleontol 4:33–44
Hansen HJ (1975) On feeding and supposed buoyancy mechanisms in four Recent globigerinid Foraminifera from the Gulf of Elat, Israel. Rev Esp Micropaleontol 7:325–339
Hansen HJ, Buchardt B (1977) Depth-distribution of *Amphistegina* in the Gulf of Elat, Israel. Utrecht Micropaleontol Bull 15:205–224
Hansen HJ, Dalberg P (1979) Symbiotic algae in milioline foraminifera: CO_2 uptake and shell adaptations. Bull Geol Soc Den 28:47–55
Hansen HJ, Lykke-Andersen AL (1976) Wall structure and classification of fossil and recent elphidiid and nonionid Foraminifera. Fossils Strata 10:1–37
Hansen HJ, Reiss Z (1971) Electron microscopy of Rotaliacean wall structures. Bull Geol Soc Den 10:329–346
Hecht A, Anati DA (1983) A description of the Straits of Tiran in winter 1978. Isr J Earth Sci 32:149–164
Herman Y (1968) Evidence of climatic changes in Red Sea cores. In: Means of correlations of quaternary successions. Congr Int Assoc Quat Res 8:325–348
Honjo S, Okada H (1974) Community structure of coccolithophores in the photic layer of the mid-Pacific. Micropaleontology 20:209–230
Hottinger L (1972) *Pseudochama cornucopiae* (Linné) in Dahab, a model of rudist "reefs"? Sci Newslett Mar Biol Lab Elat 2:6–7
Hottinger L (1977a) Distribution of larger Peneroplidae, *Borelis* and Nummulitidae in the Gulf of Elat, Red Sea. Utrecht Micropaleontol Bull 15:35–110
Hottinger L (1977b) Foraminifères operculiniformes. Mem Mus Natl Hist Nat C 40:1–159
Hottinger L (1978a) Comparative anatomy of elementary shell structures in selected larger foraminifera. In: Hedley RH, Adams CG (eds) Foraminifera, vol 3, Academic Press, London New York, pp 203–266
Hottinger L (1978b) Early diagenesis as a main ecological factor of shallow benthic life in the Gulf of Elat, Red Sea. Abstr 10th Int Congr Sedimentol,vol I, Jerusalem, pp 313–314
Hottinger L (1980) Repartition comparée des grands foraminifères de la Mer Rouge et de l'Océan Indien. Ann Univ Ferrara (N S Sez IX) Suppl VI: 1–13
Hottinger L (1981a) Fonctions de la disposition alternante des loges chez les foraminifères et la structure d'*Omphalocyclus*. Cah Micropaleontol 1981/4:45–55
Hottinger L (1981b) Funktion und Funktionswandel der Großforaminiferen-Schalen. In: Reif W (ed) Funktionsmorphologie. Paläontol Kursb 1. Paläontol Ges, München, pp 159–172
Hottinger L (1981c) The resolution power of the biostratigraphic clock based on evolution and its limits. In: Martinell J (ed) Internatl symp concepts meth Paleont. Barcelona, contr papers. Paleontol Dept Barcelona Univ Barcelona, pp 233–242
Hottinger L (1982a) Larger Foraminifera, Giant Cells with a Historical Background. Naturwissenschaften 69:361–371
Hottinger L (1982b) Dem Meer Geheimnisse entlocken. Mag Ciba-Geigy 2/82:29–33
Hottinger L (1983a) Neritic macroid genesis, an ecological approach. In: Peryt TM (ed) Coated grains. Springer, Berlin Heidelberg New York, pp 38–55
Hottinger L (1983b) Processes determining the distribution of larger foraminifera in space and time. In: Meulenkamp JE (ed) Reconstruction of marine paleoenvironments. Utrecht Micropaleontol Bull 30:239–253
Hottinger L (1984a) Les organismes constructeurs sur la plateforme du Golfe d'Aqaba (Mer Rouge) et les méchanismes régissant leur répartition. 1 Congr Int Paleoécol Lyon 1983, Geobios mem spec 8:241–249
Hottinger L (1984b) Role des organismes dans la génèse des sediments et des roches; discussion sur les peuplements des rudistes. 1 Congr Int Paleoecol Lyon 1983, Geobios mem spec 8:271–272

Hottinger L (1984c) Foraminifères de grande taille: signification des structures complexes de la coquille. 2nd Int Symp Benthic Foram Pau 1983, Pau Bordeaux, pp 309–315
Hottinger L, Dreher D (1974) Differentiation of protoplasm in Nummulitidae (Foraminifera) from Elat, Red Sea. Mar Biol 25:41–61
Hottinger L, Drobne K (1980) Early Tertiary conical imperforate Foraminifera. Razprave Ljubljana IV 22/3:189–276
Hottinger L, Leutenegger S (1980) The structure of calcarinid foraminifera. Schweiz Palaeontol Abh 101:115–154
Hottinger L, Levinson G (1978) Cementation of reefs in the Gulf of Elat (Red Sea) by Acervulinid Foraminifera. 10th Int Congr Sedimentol, Jerusalem, Abstr 315
Issar AS, Bruins HJ (1983) Special climatological conditions in the deserts of Sinai and the Negev during the latest Pleistocene. Paleogeogr Paleoclimatol Paleoecol 43:63–72
Jerlov NG (1968) Optical oceanography. Elsevier, Amsterdam
Kimor B (1973) Plankton relations of the Red Sea, Persian Gulf and Arabian Sea. In: Zeitschel B (ed) Ecological studies. Analysis and synthesis vol III. Springer, Berlin Heidelberg New York, pp 221–232
Kimor B, Golandsky B (1977) Microplankton of the Gulf of Elat: aspects of seasonal and bathymetric distribution. Mar Biol 42:55–67
Kimor B, Golandsky-Baras B (1981) Distribution and ecology of the tintinnids in the Gulf of Elat (Aqaba), Red Sea. J Plankton Res 3:445–459
Kimor B, Wood EJF (1975) A plankton study in the Eastern Mediterranean Sea. Mar Biol 29:321–333
Kitazato H (1981) Observation of behavior and mode of life of benthic foraminifers in the laboratory. Geosci Rep Shizuoka Univ 6:61–71
Klinker J, Kropach C, Harpaz H, Levanon I, Reiss Z (1975) DCPE-Data Collecting Program in the Gulf of Elat, Rep No 1. In: Reiss Z, Paperna I (eds) 4th Rep. H. Steinitz Mar Biol Lab, Elat, pp 46–141
Klinker J, Reiss Z, Kropach C, Levanon I, Harpaz H, Halicz E, Assaf G (1976) Observations on the circulation pattern in the Gulf of Elat (Aqaba), Red Sea. Isr J Earth-Sci 25:85–103
Klinker J, Reiss Z, Kropach C, Levanon I, Harpaz H, Shapiro Y (1978) Nutrients and biomass distribution in the Gulf of Aqaba (Elat), Red Sea. Mar Biol 45:53–64
Kremer BP, Schmaljohann RE, Röttger R (1980) Features and nutritional significance of photosynthates produced by unicellular algae symbiontic with larger Foraminifera. Mar Ecol Prog Ser 2:225–228
Krey J, Babenerd B (1976) Atlas of the International Indian Ocean Expedition. Phytoplankton Production. IOC-UNESCO Cartogr Off, Schleswig-Holstein, FRG
Krumbein WE (1977) Primary production, mat formation and lithification: Contribution of oxygenic and facultative anoxygenic cyanobacteria. In: Flügel E (ed) Fossil algae. Springer, Berlin Heidelberg New York, pp 37–56
Krumbein WE, Cohen Y (1974) Biogene, klastische und evaporitische Sedimentation in einem mesothermen monomiktischen ufernahen See (Golf von Aqaba). Geol Rundsch 63 (3):1035–1065
Krumbein WE, Cohen Y, Shilo M (1977) Solar Lake (Sinai). 4. Stromatolitic cyanobacterial mats. Limnol Oceanogr 22 (4):635–656
ter Kuile B, Erez J (1984) In-situ growth rate experiments on the symbiont-bearing Foraminifera *Amphistegina lobifera* and *Amphisorus hemprichii.* M Sc thesis (ter Kuile), Univ Groningen
Lacroix E (1941) Les Orbitolites du golfe d'Akaba. Bull Inst Oceanogr Monaco 794:792–799
Larsen AR (1976) Studies of recent Amphistegina. Taxonomy and some ecological aspects. Isr J Earth-Sci 25:1–26
Larsen AR (1978) Phylogenetic and paleobiogeographical trends in the foraminiferal genus Amphistegina. Rev Esp Micropaleontol X (2):217–243
Larsen AR, Drooger CW (1977) Relative thickness of the test in the Amphistegina species of the Gulf of Elat. Utrecht Micropaleontol Bull 15:225–239
Lee JJ (1980) Nutrition and physiology of the foraminifera. In: Levandowski M, Hutner SH (eds) Biochemistry and physiology of protozoa, vol III. Academic Press, London New York, pp 43–66
Lee JJ (1983) Perspective on algal edosymbionts in larger foraminifera. Int Rev Cytol Suppl 14:49–77
Lee JJ, Bock WD (1976) The importance of feeding in two species of Soritid Foraminifera with algal symbionts. Bull Mar Sci 26 (4):530–537

Lee JJ, McEnery ME, Kahn EG (1979a) Symbiosis and the evolution of larger foraminifera. Micropaleontology 25 (2):118–140

Lee JJ, McEnery ME, Shilo M, Reiss Z (1979b) Isolation and cultivation of diatom symbionts from larger Foraminifera (Protozoa) Nature (London) 280:NO 5717, 57–58

Lee JJ, McEnery ME, Lee JM, Reidy JJ, Garrison JR, Röttger R (1980a) Algal symbionts in larger foraminifera. In: Schwemmler W, Schenk HEA (eds) Endosymbiosis and cell biology, vol I. Walter de Gryter, Berlin New York, pp 113–124

Lee JJ, McEnery ME, Röttger R, Reimer ChW (1980b) The isolation, culture and identification of endosymbiotic diatoms from *Heterostegina depressa* D'Orbigny and *Amphistegina lessonii* D'Orbigny (Larger Foraminifera) from Hawaii. Bot Mar XXIII:297–302

Lee JJ, Reimer ChW, Mc Enery ME (1980c) The identification of diatoms isolated as endosymbionts from larger Foraminifera from the Gulf of Eilat (Red Sea) and the description of two new species, *Fragilaria shiloi* sp. nov. and *Navicula reissii* sp. nov. Bot Mar XXIII:41–48

Lee JJ, McEnery M, Garrison J (1980d) Experimental studies of larger foraminifera and their symbionts from the Gulf of Elat. J Foram Res 10:31–47

Lehmann R (1961) Strukturanalyse einiger Gattungen der Subfamilie Orbitolitinae. Eclogae geol Helv 54 (2):597–667

Leutenegger S (1977a) Ultrastructure de foraminifères perforés et imperforés ainsi que de leurs symbiotes. Cah Micropaleontol 1977/3:1–52

Leutenegger S (1977b) Reproduction cycles of larger foraminifera and depth distribution of generations. Utrecht Micropaleontol Bull 15:26–34

Leutenegger S (1977c) Symbiosis between larger foraminifera and unicellular algae in the Gulf of Elat, Red Sea. Utrecht Micropaleontol Bull 15:241–244

Leutenegger S (1977d) Ultrastructure and motility of dinophyceans symbiotic with larger, imperforated Foraminifera. Mar Biol 44:157–164

Leutenegger S (1983) Specific host-symbiont relationship in larger foraminifera. Micropaleontology 29 (2):111–125

Leutenegger S (1984) Symbiosis in benthic Foraminifera – specificity and host adaptations. J Foram Res 14/1:16–35

Leutenegger S, Hansen HJ (1979) Ultrastructural and radiotracer studies on pore function in foraminifera. Mar Biol 54:11–16

Levanon-Spanier I, Padan E, Reiss Z (1979) Primary production in a desert-enclosed sea – the Gulf of Elat (Aqaba), Red Sea. Deep-Sea Res 26:673–685

Levinson GE (1976) A Study of Acervulina and acervuline related Foraminifera. Dipl thesis, Univ Basel

Lévy A (1977) Revision of recent Soritidae from the Bahamas. A new genus: *Androsina*. Bull Cent Rech Explor Prod Elf-Aquit 1 (2):393–449

Liebau A (1978) Abrieb-geschützte Foraminiferen. Neu Jahrb Geol Paläontol Abh 157:119–122

Lloyd RM (1966) Oxygen isotope enrichment of sea water by evaporation. Geochim. Cosmochim Acta 30:801–814

Loya Y, Slobodkin BL (1971) The coral reefs of Eilat (Gulf of Eilat, Red Sea). Symp Zool Soc London 28:117–139

Luksch J (1901) Expedition S.M. Pola in das Rothe Meer. Denkschr Akad. Wiss XVIII Phys Unters LXIX:337–398

Luz B (1979) Paleo-oceanography of the post glacial eastern Mediterranean. Nature London 278:847–848

Luz B (1982) Paleoclimatic interpretation of the last 20,000 yr record of deep-sea cores around the Middle East. In: Bintliff JL, Zeist W van (eds), Paleoclimates, palaeoenvironments and human communities in the Eastern Mediterranean Region in later prehistory. Brit Archaeol Repts Int Ser, Oxford 133:41–61

Luz B (1984) Isotopic and mineralogical records from Late Pleistocene sediments of the N. Red Sea. Paleocol Afr 16:53

Luz B, Reiss Z (1983) Stable carbon isotopes in Quaternary Foraminifera from the Gulf of Aqaba (Elat), Red Sea. Utrecht Micropaleontol Bull 30:129–140

Luz B, Reiss Z, Zohary-Segev T (1983) Seasonal variation of stable isotopes in *Amphisorus* from the Gulf of Aqaba (Elat), Red Sea. Utrecht Micropaleontol Bull 30:229–238

McEnery ME, Lee JJ (1981) Cytological and fine structural studies of three species of symbiont-bearing larger foraminifera from the Red Sea. Micropaleontology 27:71–83
McIntyre A (1969) The coccolithophorida in Red Sea sediments. In: Degens ET, Ross DA (eds). Hot brines and recent heavy metal deposits in the Red Sea. Springer, Berlin Heidelberg New York, pp 239–305
McIntyre A, Bé AWH (1967) Modern Coccolithophoridae of the Atlantic Ocean – I. Placoliths and Cyrtoliths. Deep-Sea Res 14:561–597
Mergner H (1971) Structure, ecology and zonation of Red Sea reefs (in comparison with South Indian and Jamaica Reefs). Symp Zool Soc, London 28:141–161
Mergner H (1979) Quantitative ökologische Analyse eines Rifflagunenareals bei Aqaba (Golf von Aqaba, Rotes Meer). Helgol Wiss Meeresunters 32:476–507
Mergner H (1980) Ecology of a fore reef area near Aqaba (Red Sea). Proc Symp Coast Mar Environ Red Sea, Gulf Aden Trop West Indian Ocean. Univ Khartoum, Sudan, pp 77–86
Mergner H, Mastaller M (1980) Ecology of a reef lagoon area near Aqaba (Red Sea). Proc Symp Coast Mar Environ Red Sea, Gulf Aden Trop West Indian Ocean, Univ Khartoum, Sudan, pp 39–76
Mergner H, Schuhmacher H (1974) Morphologie, Ökologie und Zonierung von Korallenriffen bei Aqaba (Golf von Aqaba, Rotes Meer). Helgol Wiss Meeresunters 26:238–358
Mergner H, Svoboda A (1977) Productivity and seasonal changes in selected reef areas in the Gulf of Aqaba (Red Sea). Helgol Wiss Meeresunters 30:383–399
Meulenkamp JE (1977) Biometrical data on *Heterocyclina.* Utrecht Micropaleontol Bull 15:165–170
Mikkelsen N (1973) Calcareous nannoplankton in the Gulf of Elat. Isr J Earth-Sci 22:269–271
Mohamed AF (1940) The Egyptian exploration of the Red Sea. Proc R Soc London Ser B 128:306–316
Morcos SA (1970) Physical and chemical oceanography of the Red Sea. Oceanogr Mar Biol Rev 8:73–202
Müller C (1976) Nannoplankton-Gemeinschaften aus dem Jungquartär des Golfes von Aden und des Roten Meeres. Geol Jb Hannover D 17:33–77
Müller-Merz E (1980) Strukturanalyse ausgewählter rotaloider Foraminiferen. Schweiz palaeont Abh 101:5–39
Murray SP, Hecht A, Babcock A (1984) On the mean flow in the Tiran Strait in Winter. J Mar Res in press
Muscatine L, Porter JW (1977) Reef corals: Mutualistic symbioses. Adapted to nutrient-poor environments. Bio Science 27/7:454–460
Natour RM, Nienhuis H (1980) Some phytoplanktonic studies in Aqaba, gulf of Jordan. Proc Symp Coast Mar Environ Red Sea, Gulf Aden Trop West Indian Ocean: Univ Khartoum, Sudan, pp 489–500
Neumann AC, McGill, DA (1962) Circulation of the Red Sea in early summer. Deep-Sea Res 8:223–235
Noy N (1982) Environment influence on population composition and stable isotopes of carbon and oxygen in shells of benthic foraminifera. M Sc thesis, Hebrew Univ, Jerusalem (in Hebrew)
Ogdon IC (1980) Faunal relationships in Caribbean seagrass beds. In: Philipps RC, McRoy CP (eds) Handbook of seagrass biology. Garland, New York London, pp 173–198
O'Herne L (1974) A reconsideration of *Amphistegina lessonii* d'Orbigny, 1826, sensu Brady, 1884 (Foraminifera). Scr Geol 26:1–53
Okada H, Honjo S (1973) The distribution of oceanic coccolithophorids in the Pacific. Deep-Sea Res 20:355–374
Okada H, Honjo S (1975) The distribution of coccolithophores in marginal seas along the Western Pacific Ocean and the Red Sea. Mar Biol 31:271–285
Okada H, McIntyre A (1977) Modern coccolithophores of the Pacific and North Atlantic oceans. Micropaleont. 23:1–55
Okada H, McIntyre A (1979) Seasonal distribution of modern Coccolithophores in the Western Atlantic Ocean. Mar Biol 54:319–328
Oren OH (1962) A note on the hydrography of the Gulf of Eylath. Sea Fish Stn Haifa Bull 30:3–14
Paldor N, Anati DA (1979) Seasonal changes of temperature and salinity in the Gulf of Elat (Aqaba). Deep-Sea Res 26:661–672
Patzert WC (1972) Seasonal variations in structure and circulation in the Red Sea. Thesis, Hawaii Inst Geophys, Univ Hawaii

Perath I (1982) Tectonic shorelines, southern Gulf of Eilat: Recent crescentic slumping? Abstr Isr Geol Soc Annu Meet, Jerusalem, pp 50–53
Perelis L (1974) Foraminiferal thanatocoenoses in the Wadi Taba area, Gulf of Elat, Israel. Thesis Dep Geol, Hebrew Univ, Jerusalem
Perelis L, Reiss Z (1975) Cibicididae in recent sediments from the Gulf of Elat. Isr J Earth Sci 24:73–96
Peter KJ, Paulinose VT (1978) Swarming of *Creseis acicula* Rang (Pteropoda) in the Bay of Bengal. Indian J Mar Sci 7:126–127
Platt T, Subba Rao DV, Irwin B (1983) Photosynthesis of picoplankton in the oligotrophic ocean. Nature (London) 301:702–704
Por FD, Lerner-Seggev R (1966) Preliminary data about benthic fauna of the Gulf of Elat (Aqaba), Red Sea. Isr J Zool 15:38–50
Por FD, Dor I, Amir A (1977) The mangal of Sinai: limits of an ecosystem. Helgoländer Wiss Meeresunters 30:295–314
Prado-Por MS (1983) Diversity and dynamics of Copepoda Calanoida in the northern Gulf of Elat (Aqaba), Red Sea. Oceanol Acta 6:139–146
Reiss Z (1959) Notes on Foraminifera from the Gulf of Elat, Ras Muhamed and Tiran. Sea Fish Stn Haifa Bull 20:1–3
Reiss Z (1963) Reclassification of perforate Foraminifera. Bull Israel Geol Surv 35:1–111
Reiss Z (1974) Steinitz Mar Biol Lab, Elat. Rep Hebrew Univ, Jerusalem
Reiss Z (1977) Foraminiferal research in the Gulf of Elat-Aqaba. Utrecht Micropaleontol Bull 15:7–26
Reiss Z (1979) DCPE – Data collecting program in the Gulf of Elat. Rep No 4. In: Shilo M, Cohen Y (eds) 7th Rep H Steinitz Mar Biol Lab, Elat, pp 22–36
Reiss Z, Gvirtzman G (1966) Borelis from Israel. Eclogae geol Helv 59 (1):437–447
Reiss Z, Halicz E (1976) Phenotypy in planktonic Foraminiferida from the Gulf of Elat. Isr J Earth-Sci 25:27–39
Reiss Z, Klug K, Merling P (1961) Recent Foraminifera from the Mediterranean and Red Sea coasts of Israel. Isr Geol Surv Bull 32:27–28
Reiss Z, Halicz E, Perelis L (1974) Planktonic Foraminiferida from Recent sediments in the Gulf of Elat. Isr J Earth-Sci 23:69–105
Reiss Z, Kropach C, Levanon I, Klinker J, Harpaz H, Shapiro Y, Ben-Avraham Z (1976) DCPE-Data collecting program in the Gulf of Elat, Rep No 2. In: Reiss Z, Paperna I (eds) 5th Rep H Steinitz Mar Biol Lab, Elat, pp 22–67
Reiss Z, Levanon I, Raz G, Harpaz H, Ben-Avraham Z, Padan E (1977 a) DCPE-Data collecting program in the Gulf of Elat, Rep No 3. In: Shilo M, Paperna I (eds) 6th Rep H Steinitz Mar Biol Lab, Elat, pp 31–54
Reiss Z, Leutenegger S, Hottinger L, Fermont WJJ, Meulenkamp JE, Thomas E, Hansen HJ, Buchardt B, Larsen AR, Drooger CW (1977 b) Depth-relations of Recent larger Foraminifera in the Gulf of Aqaba-Elat. Utrecht Micropaleontol Bull 15:1–244
Reiss Z, Hottinger L, Hansen HJ, Drooger CW, Leutenegger S, Larsen AR, Buchardt B, Fermont WJJ, Thomas E, Meulenkamp JE, Luz B, Halicz E, Erez J, Perelis L, Zweig-Strykowski M (1978) Sedimentological significance of foraminiferal ecology in a desert-enclosed sea: the Gulf of Aqaba (Elat) Abstr 10th Int Congr Sedimentol, vol II, Jerusalem, pp 541–542
Reiss Z, Luz B, Almogi-Labin A, Halicz E, Winter A, Wolf M (1980) Late Quaternary paleoceanography of the Gulf of Aqaba (Elat), Red Sea. Quat Res 14:294–308
Reiss Z, Luz B, Almogi-Labin A, Halicz E, Winter A, Erez J (1984) Paleoceanography of the Gulf of Aqaba during the last 150,000 years. Paleoecol Afr Balkema, Rotterdam 16:55–64
Riedl RJ, Huang N, Machan R (1972) The subtidal pump: a mechanism of interstitial water exchange by wave action. Mar Biol 13 (3):210–221
Rosen RB (1971) The distribution of reef coral genera in the Indian Ocean. Symp Zool Soc London 28:263–299
Rosenberg-Herman Y (1965) Études des sédiments quaternaires de la Mer Rouge. Thèses, Univ Paris, A (1123):339–430
Roth P, Berger W (1975) Distribution and dissolution of coccoliths in the South and Central Pacific. Cushman Found Foram Res Spec Publ 13:87–113

Röttger R (1972) Die Bedeutung der Symbiose von *Heterostegina depressa* (Foraminifera, Nummulitidae) für Siedlungsdichte und Karbonatproduktion. Verh Ber Jahresvers Dtsch Zool Ges 65:42–47

Röttger R, Berger WH (1972) Benthic Foraminifera: morphology and growth in clone cultures of *Heterostegina depressa*. Mar Biol 15:89–94

Röttger R, Irwan A, Schmaljohann R, Franzisket L (1980) Growth of the symbiont-bearing foraminifera *Amphistegina lessonii* d'Orbigny and *Heterostegina depressa* d'Orbigny (Protozoa). In: Schwemmler W, Schenk HEA (eds) Endocytobiology, endosymbiosis and cell biology, vol I. Walter de Gruyter, Berlin New York, pp 125–132

Rückert-Hilbig A (1983) Megalospheric Gamonts of *Rosalina globularis* d'Orbigny, *Cymbaloporetta bulloides* (d'Orb.) and *C. millettii* (H. and E.) (Foraminifera) with differently constructed swimming apparatus. Tübinger Mikropaleontol Mitt 1:1–69

Said R (1949) Foraminifera of the northern Red Sea. Cushman Lab Foram Res Spec Publ 26:1–44

Said R (1950a) Additional Foraminifera from the Northern Red Sea. Contrib Cushman Found Foram Res I (1):4–9

Said R (1950b) The distribution of foraminifera in the northern Red Sea. Contrib Cushman Found Foram Res 1:9–29

Sakthivel M (1973) Studies on *Limacina inflata* d'Orbigny (Thecosomata, Gastropoda) in the Indian Ocean. In: Zeitschel B, Gerlach SA (eds) The biology of the Indian Ocean. Springer, Berlin Heidelberg New York, pp 383–397

Saks NM (1981) Growth, productivity and excretion of *Chlorella* spp. Endosymbionts from the Red Sea: Implications for Host Foraminifera. Bot Mar XXIV:445–449

Savin SM, Yeh HW (1981) Stable isotopes in ocean sediments. In: Emiliani C (ed). The oceanic lithosphere, vol VII. The Sea. Wiley, New York, pp 1521–1554

Scheer G, Pillai CSG (1983) Report on the stony corals from the Red Sea. Zoologica 45: H 133:1–198

Scheidegger G (1972) *Heterocyathus*, an ahermatypic Madreporarian in the Halophila environment of the Gulf of Elat, Red Sea. Sci Newslett Mar Biol Lab Elat, 2:8–9

Schmaljohann R, Röttger R (1978) The ultrastructure and taxonomic identity of the symbiontic algae of *Heterostegina depressa* (Foraminifera: Nummulitidae). J Mar Biol Assoc UK 58:227–237

Shemesh A (1980) Carbonate system in selected regions in the north of the Gulf of Elat. M Sc thesis Dept. Hebrew Univ. Jerusalem pp 1–64 (in Hebrew)

Shemesh A, Luz B, Erez J (1981) $\delta^{13}C$, O_2 and the carbonate system in the Gulf of Elat. Abstr E.C.O.G. VII. 1 p

Shemesh A, Luz B, Erez J (in press) Carbon Isotopes, dissolved oxygen and the carbonate system in the Gulf of Aqaba (Elat). Limnol Oceanogr

Sournia A (1977) Notes on primary productivity of coastal waters in the Gulf of Elat (Red Sea). Int Rev Gesamten Hydrobiol 62:813–819

Spindler M, Röttger R (1973) Der Kammerbauvorgang der Großforaminifere *Heterostegina depressa* (Nummulitidae). Mar Biol 18:146–159

Spindler M, Hemleben C, Bayer U, Bé AWH, Anderson OR (1979) Lunar periodicity of reproduction in the planktonic Foraminifer *Hastigerina pelagica*. Mar Ecol Prog Ser 1:61–64

Spoel S van der (1967) Euthecosomata, a group with remarkable developmental stages. (Gastropoda, Pteropoda). J Noorduijn Zoon NV Gorinchem 1–375

Steiner P (1982) Zonierung von Bryozen im Roten Meer bei Elat. Dipl thesis, Univ Basel

Thomas E (1977) Biometric study of recent *Planorbulinella* from the Gulf of Aqaba (Elat). Utrecht Micropaleontol Bull 15:111–163

Van Straaten LM (1960) Marine mollusc shell assemblages of the Rhône Delta. Geol Mijnbouw 39:105–109

Vincent E, Berger WH (1981) Planktonic Foraminifera and their use in paleoceanography. In: Emiliani C (ed) The oceanic lithosphere. The Sea, vol VII, Wiley, New York, pp 1025–1119

Wefer C, Berger WH (1980) Stable isotopes in benthic foraminifera and seasonal variation in large tropical species. Science 209:803–805

Weikert H (1980) On the plankton of the central Red Sea. A first synopsis of results obtained from the cruises Meseda I and Meseda II. Proc Symp Coast Environ Red Sea, Gulf Aden Trop West Indian Ocean. Univ Khartoum, Sudan, pp 135–167

Winter A (1982a) Paleoenvironmental interpretation of Quaternary coccolith assemblages from the Gulf of Aqaba (Elat), Red Sea. Rev Esp Micropaleontol 14:291–314
Winter A (1982b) Extreme post-depositional shape modification in Red Sea coccoliths. Micropaleontology 28:319–323
Winter A, Reiss Z, Luz B (1978) Living *Gephyrocapsa protohuxleyi* McIntyre in the Gulf of Elat (Aqaba). Mar Micropaleontol 3:295–298
Winter A, Reiss Z, Luz B (1979) Distribution of living coccolithophore assemblages in the Gulf of Elat (Aqaba) Mar Micropaleontol 4:197–223
Winter A, Almogi-Labin A, Erez J, Halicz E, Luz B, Reiss Z (1983) Salinity tolerance of marine organisms deduced from Red Sea Quaternary record. Mar Geol 53:17–22
Wonders AAH (1980) Middle and Late Cretaceous planktonic Foraminifera of the western Mediterranean area. Utrecht Micropaleontol Bull 24:1–157
Wyrtki K, Bennett EB, Rochford DJ (1971) Oceanographic atlas of the international Indian Ocean expedition. Nat Sci Found, Washington
Zieman J, Wetzel RG (1980) Productivity in seagrasses: Methods and rates. In: Philipps RC, McRoy CP (eds) Handbook of seagrass biology. Garland, New York London, pp 87–116
Zmiri A, Kahan D, Reiss Z (1972) Aspects of thermotactic response in Foraminifera. J Protozool 19:56
Zmiri A, Kahan D, Hochstein S, Reiss Z (1974) Phototaxis and thermotaxis in some species of *Amphistegina* (Foraminifera). J Protozool 21:133–138
Zohary T, Reiss Z, Hottinger L (1980) Population dynamics of *Amphisorus hemprichii* (Foraminifera) in the Gulf of Elat (Aqaba), Red Sea. Eclogae geol Helv 73:1071–1094
Zweig-Strykowski M (1973) Significance of foraminiferal thanatocoenoses in the Coral Island Region, Gulf of Elat. Thesis, Hebrew Univ Jerusalem
Zweig-Strykowski M, Reiss Z (1975) Bolivinitidae from the Gulf of Elat. Isr J Earth-Sci 24:97–111

Taxonomic Index

Page numbers given in **boldface** refer to illustrations of relevance for the identification of the taxon.

Subject Index

The main ecological factors, groups of organisms and some other important subjects are given in this index in **boldface**. The keywords listed with each subject survey the data and processes discussed in connection with each main item. The page numbers listed refer to the text as well as to the content of figures and tables. Geographical names and geologic time units are distinguished from other subjects by capital letters. The names of organisms are listed in the taxonomic index below.